Hrachya Nersisyan · Christian Toepffer · Günter Zwicknagel

Interactions Between Charged Particles in a Magnetic Field

Hrachya Nersisyan · Christian Toepffer
Günter Zwicknagel

Interactions Between Charged Particles in a Magnetic Field

A Theoretical Approach to Ion Stopping in Magnetized Plasmas

With 36 Figures

Springer

Dr. Hrachya Nersisyan
National Academy of Sciences of Armenia
Theoretical Physics Division
Institute of Radiophysics and Electronics
Alikhanian Brothers Str. 1
378410 Ashtarak, Armenia
E-mail: *hrachya@irphe.am*

Professor Dr. Christian Toepffer
Universität Erlangen
Institut für Theoretische Physik II
Staudtstr. 7
91058 Erlangen, Germany
E-mail: *toepffer@theorie2.physik.uni-erlangen.de*

Dr. Günter Zwicknagel
Universität Erlangen
Institut für Theoretische Physik II
Staudtstr. 7
91058 Erlangen, Germany
E-mail: *zwicknagel@theorie2.physik.uni-erlangen.de*

Library of Congress Control Number: 2007920192

ISBN 978-3-540-69853-1 Springer Berlin Heidelberg New York

Springer is a part of Springer Science+Business Media

springer.com

Typesetting: Camera ready by author
Production: LE-TEX Jelonek, Schmidt & Vöckler GbR, Leipzig
Cover: WMX Design GmbH, Heidelberg

SPIN 11401186 57/3100/YL - 5 4 3 2 1 0 Printed on acid-free paper

To: Eranuhi and Bardassar Nersisyan and Christel

Preface

The present theory for the penetration of charged particles through matter rests on the fundamental work by Bohr, Bethe and Bloch. Due to the progress in the acceleration and detection of particles since the early years of the last century an increasing range of applications has emerged, both in fundamental physics but also in fields like medical radiology, materials science, nuclear fission and fusion technology and many others. These developments require a detailed and quantitative knowledge about the processes occuring during the passage of charged particles through matter. A comprehensive overview on the present status of the theory of stopping of heavy ions in matter has, e.g., been presented in recent monographs by Sigmund.

However, in recent decades considerable interest has been directed towards applications in which strong magnetic fields play a dominant role. Then the cyclotron radius of the target electrons is the smallest relevant length scale and their cyclotron period the smallest time scale of the problem. While the fields occuring in magnetic confinement fusion devices are still marginal in this sense, stronger fields are employed to guide the electrons in the cooling sections of particle storage rings and even more so for the deceleration of heavy ions and antiprotons in traps. In the rest frame of the beams the cooling/deceleration process may be viewed as the stopping of ions in an electron plasma. Except for recombination, changes in the internal structure are not so important in these applications. The emphasis of this book lies therefore on the interaction of ions with a magnetized electron plasma.

The original results presented in this book could not have been obtained without the generous support by:

- Alexander von Humboldt Stiftung
- Bundesministerium für Bildung und Forschung (BMBF)
- CERN
- Gesellschaft für Schwerionenforschung (GSI)
- Deutscher Akademischer Austauschdienst (DAAD)
- Centre National de la Recherche Scientifique (CNRS), Université Paris-Sud
- Armenian National Science and Education Foundation (ANSEF).

The authors are also grateful for the advice offered in many enlightening discussions with our colleagues Jacques Bosser, Christian Carli, Claude Deutsch, Frank Herfurth, Heinz-Jürgen Kluge, Gilles Maynard, Dieter Möhl, Wolfgang Quint, Paul-Gerhard Reinhard, Markus Steck, Andreas Wolf and our students Bernd Möllers and

Markus Walter. Large parts of the manuscripts have been meticulously prepared by Claudia Schlechte.

Erlangen,
November 2006

Hrachya Nersisyan
Christian Toepffer
Günter Zwicknagel

Contents

1 Introduction

The interaction of charged particles with matter has been an issue of extensive investigations throughout the whole last century. Its theoretical treatment starts with the classical description of the energy loss of fast projectiles considered by Bohr [24]. Later a quantum mechanical treatment of the energy transfer to bound electrons was established by Bethe [18] and refined by Bloch [23]. Further considerable improvements of the theoretical description have been achieved by Fermi and Teller [41] and finally by Lindhard [79]. The present status of the theory has, e.g. been reviewed in the monographs by Sigmund [118, 119]. Till nowadays an enormous number of publications are dedicated to specific questions on the energy loss for a variety of possible projectile and target conditions. Recent applications are the energy transfer to pellets for inertial confinement fusion, electron cooling of heavy ion beams as well as the deceleration of particle beams in traps.

The interaction of a test particle with a medium of charged particles like the stopping of an ion in an electron plasma or its deceleration in a trap can be described in two approaches which are complementary to each other. The dielectric theory (DT) is a continuum theory in which the response of charge and current densities to external perturbations is calculated. While this requires cut–offs at small distances (or large wavenumbers in Fourier space) to exclude hard collisions of close particles, the collectivity of the excitation can be taken into account. In the binary collision approximation (BC), on the other hand, the motion of the ion is described as the aggregate of subsequent pairwise interactions with the target electrons. This requires cut–off parameters at large distances (corresponding to small wave numbers in Fourier space) to account for screening.

For a physically meaningful comparison between DT and BC we adhere in this book to the following terminology: The basic, but generally unobserved quantity in BC is the energy or velocity transfer ΔE_i or Δv_i, respectively, to the test particle in a collision with specific initial data. Averaging with respect to quantities like the phase angle φ of the cyclotron motion and integration with respect to the impact parameter s yields the energy loss $\mathrm{d}E_i/\mathrm{d}l$ of the test particle with monochromatic electrons. Here $\mathrm{d}\boldsymbol{l} = \boldsymbol{v}_i \mathrm{d}t$ is the path element of the test particle moving with velocity $\boldsymbol{v}_i$ in a time interval $\mathrm{d}t$. Then averaging with respect to the electron velocity distribution $f(\boldsymbol{v}_e)$ yields the stopping power S and the stopping force or drag force $\boldsymbol{\mathcal{F}}$,

$$S = -\boldsymbol{\mathcal{F}} \cdot \hat{\boldsymbol{v}}_i = -\frac{\mathrm{d}\mathcal{E}_i}{\mathrm{d}l} = -\left\langle \frac{\mathrm{d}E_i}{\mathrm{d}l} \right\rangle . \tag{1.1}$$

In Chap. 2 we review the previous work on this subject, present the methods which were employed and their results and point out the inherent problems in approximation methods like perturbation expansion and linearization. These are associated with the infinite range of the Coulomb interaction and the role of the collective excitations in the target, large velocity transfers in hard collisions and the transition to a quasi-one-dimensional electron motion for strong magnetic fields. In contrast to the field-free case the BC depends then on the sign of the interaction between the test particle and the electrons. Consider, e.g., a test particle moving parallel to the magnetic field, while the electrons move like beads on a wire along the field lines. In the attractive case no velocity transfer takes place at all, while it is maximal in the repulsive case. This indicates a failure of the perturbation expansion. Even in the attractive case caution is indicated. For small ion velocities the stopping power has the logarithmic behavior typical for one-dimensional problems unless one accounts for the velocity dependence of the lower cut–offs. In any case the validity of the more analytical approaches, perturbation expansion in BC and linearization in DT must be checked by numerical simulations. To this end we employ classical trajectory Monte-Carlo (CTMC) and particle-in-cell (PIC) simulations.

It turns out to be advantageous to include the cut–offs, which are physically motivated by (dynamic) screening and quantum diffraction already on the level of the interaction potentials. For this purpose we have developed analytical methods in which the exact form of the interaction potential must be specified only at the end of the calculation of the stopping power.

This program is carried out in Chap. 3 for the BC. Closed expressions for the averaged energy transfer in second-order perturbation theory are obtained for the limiting cases of weak and strong magnetic fields and for parallel ion motion in arbitrary magnetic fields. The validity of the perturbation expansion and the appearance of chaotic regimes are studied by comparison with CTMC calculations. Knowing the velocity transfers in the binary collisions one can also calculate the velocity diffusion of charged particles in a magnetic field, i.e. the straggling.

In Chap. 4 we turn to the DT, which is formulated in linear response (LR) by calculating first the dielectric function, which involves an integration in velocity space. The zeroes of this function describe the excitation modes of the target. Then the imaginary part of the inverse of the dielectric function is integrated in Fourier space for the stopping power. Closed expressions are obtained in the limits of small and large projectile velocities in a weak as well as in a strong magnetic field. For intermediate fields the stopping power can only be evaluated in closed form under the assumption of weakly interacting electrons with a vanishing plasma frequency. Then the stopping power does not receive any contribution from dynamic collective plasma modes, but the collectivity can be reintroduced by replacing the Coulomb interaction between the ion and the target electrons by a screened interaction. From its very concept this reduced linear response (RLR) should be equivalent to the BC with a screened electron-ion interaction, and such a conformity is verified explicitly. As in the BC there emerges a logarithmic anomalous behavior of the stopping power at low ion velocities and strong magnetic fields both in the LR and the RLR versions of the DT. As the averaging with respect to the electron velocity distribution is done first when calculating the dielectric function this cannot be avoided by employing velocity dependent cut–offs in the later spatial (Fourier) integrations. Insofar the LR is less flexible than the BC.

The velocity dependence of the cut–off at small distances is suggested by quantum diffraction. An ab initio quantum treatment of the stopping power is presented in Chap. 5. This involves the equation of state and the dielectric function of a magnetized quantum plasma and their semiclassical limits. In the framework of the quantum BC the conformity between the RLR and the transition to the classical case are shown.

In Chap. 6 some applications are discussed. Electron cooling is a powerful technique to improve the phase space structure of ion beams in storage rings. In the cooling section the ion beam is superimposed by a comoving electron beam. Due to the acceleration in the electron gun the velocity distribution of the electrons in the rest frame of the beams is highly anisotropic, the temperature parallel to the beam and its magnetic guiding field is lower by some orders of magnitude than the transverse temperature. As the transverse motion of the electrons is quenched by the magnetic field, it is rather small longitudinal velocity of the electrons which sets the scale for the velocity dependence of the stopping power.

Another recent application is the deceleration of heavy ions or antiprotons by electrons in traps for the purpose of precision experiments on QED and symmetries, respectively. Here the particles move under the influence of the external electric and magnetic fields, the mean fields produced by the particles themselves and the drag force $\mathcal{F}$. The stopping of the heavy particles is accompanied by a heating of the electrons, which in turn loose energy by radiation. The solution of the coupled equations yields the time in which the particles come to rest for the precision experiments. For ions the stopping process must be faster than the recombination. This can be achieved under realistic experimental conditions.

To conclude we propose in Chap. 7 a pragmatic approach for a calculation of the stopping power. In the presence of a magnetic field there exist no closed solutions. Even approximate treatments like second order BC or linearized DT can only be evaluated in closed form in certain limiting cases like parallel ion motion, an infinitely strong magnetic field etc. The validity of these approximations is critically examined as there exists no universal parameter of smallness. Unphysical divergences can be suppressed by using physically well motivated velocity dependent cut–off parameters. An explicit comparison with numerical simulation validates the linearization of the DT and the perturbative BC except for a very slow motion of the ions transverse to the magnetic field. Fortunately this region of parameter space is unimportant for the present experiments on ion beam cooling and trapping.

2 Previous Work, Status and Overview

2.1 Energy Loss in an Unmagnetized One–Component–Plasma (OCP)

There are basically two complementary approaches to describe the energy loss of charged particles in matter. In the dielectric treatment (DT) the response of the target's charge and current densities to the perturbation caused by the passing projectile is calculated. Such a continuum theory requires cut–offs to exclude hard collisions of close particles, but the collectivity of the excitation can be taken into account. In the binary collision approximation (BC), on the other hand, the energy loss of the projectile is the aggregate of subsequent pairwise interactions with the target particles. This requires cut–offs at large distance to account for screening. We will first treat the case of an unmagnetized target plasma. This will illustrate the basic ideas of both approaches and serves to highlight the difficulties introduced by the presence of an external magnetic field.

We consider the stopping power S, which is the energy loss of an ion moving with velocity $\boldsymbol{v}_i$ along a path element $\mathrm{d}l = v_i \mathrm{d}t$

$$S = -\frac{\mathrm{d}\mathcal{E}_i}{\mathrm{d}l} = -\frac{1}{v_i}\frac{\mathrm{d}\mathcal{E}_i}{\mathrm{d}t} = -\boldsymbol{\mathcal{F}} \cdot \hat{\boldsymbol{v}}_i \,. \tag{2.1}$$

Here $\boldsymbol{\mathcal{F}}$ is the stopping force, i.e. the drag exerted by the target on the ion. Its dependence on v_i is easily obtained in the dielectric picture shown in Fig. 2.1.

The passing ion creates in its wake a polarization cloud with the typical radius of the Debye screening length $\lambda_\mathrm{D} = v_\mathrm{th}/\omega_p$, where v_th is the thermal velocity of the target particles and the time scale is set by the inverse plasma frequency ω_p^{-1}. Near the center of the cloud the electric field increases linearly with the distance, outside the cloud it decreases with an inverse square law. For small velocities $v_i \lesssim v_\mathrm{th}$ the ion will still be within its polarization cloud after the typical time ω_p^{-1}, so the drag force will increase linearly with v_i. For large velocities $v_i \gtrsim v_\mathrm{th}$ the ion has left its polarization cloud before its formation has been completed. The drag force will then be inverse proportional to the square of the ion velocity. This gross, schematic behavior is therefore described by

$$|\boldsymbol{\mathcal{F}}| = -\frac{\mathrm{d}\mathcal{E}_i}{\mathrm{d}l} \propto \begin{cases} v_i\,, & v_i < v_\mathrm{th} \\ v_i^{-2}\,, & v_i > v_\mathrm{th} \end{cases} \,. \tag{2.2}$$

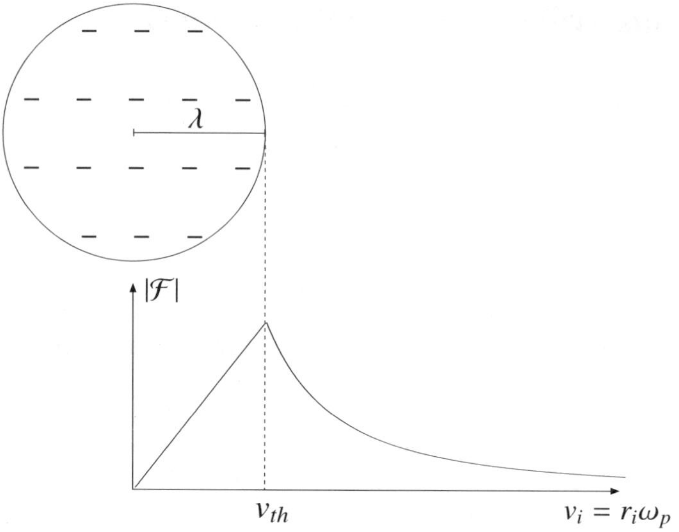

Fig. 2.1. Schematic dielectric model for the stopping force.

It follows immediately that a swift projectile is stopped most effectively by the fastest target particles, i.e. the electrons and that nuclear stopping becomes only important at very low projectile velocities [118]. We will therefore concentrate on electron stopping in the following.

In order to gain more insight we rederive (2.2) in the complementary binary collision model. Let $\boldsymbol{F}(\boldsymbol{r}(t)) = F(r(t))\hat{\boldsymbol{r}}(t)$ be a central force exerted by the ion on the electron, $\boldsymbol{n}$ a unit vector in the direction of closest approach and θ the polar angle between relative distance $\boldsymbol{r}(t)$ and $\boldsymbol{n}$. For reasons of symmetry, see Fig. 2.2, only the component $\boldsymbol{n}(\boldsymbol{n}\cdot\boldsymbol{F}) = \boldsymbol{n}F(r)\cos\theta$ contributes to the energy loss. Because of angular momentum conservation the energy transfer (3.21) can be written as

$$\Delta E_i = -\int_{-\infty}^{\infty} \mathrm{d}t\, \boldsymbol{v}_i \cdot \boldsymbol{F}(\boldsymbol{r}(t)) = -2\,\frac{m}{l}\,\boldsymbol{v}_i \cdot \boldsymbol{n} \int_0^{\theta_{\max}} r^2 F(r)\cos\theta\,\mathrm{d}\theta\,. \tag{2.3}$$

Here $-\boldsymbol{F}$ is the force acting on the ion and $l = mr^2 d\theta/dt$ is the angular momentum and the reduced mass has been approximated by the electron mass m. For the Coulomb interaction (3.1) with $F(r) = -Z\not{e}^2/r^2$ (where $\not{e}^2 = e^2/(4\pi\varepsilon_0)$ with the permittivity of the vacuum ε_0) the explicit dependence on the radial distance cancels, so that

$$\Delta E_i = \frac{2Z\not{e}^2 m}{l}\,\boldsymbol{v}_i \cdot \boldsymbol{n} \int_0^{\theta_{\max}} \mathrm{d}\theta \cos\theta = \frac{2Z\not{e}^2 m}{l}\,(\boldsymbol{n}\cdot\boldsymbol{v}_i)\sin\theta_{\max}\,. \tag{2.4}$$

The direction $\theta_{\max}$ of the asymptotic motion and the scattering angle $\Theta = \pi - 2\theta_{\max}$ are related to the excentricity ε of the hyperbolic trajectory by [48, Chap. 3]

$$\cos\theta_{\max} = \sin\frac{\Theta}{2} = \frac{1}{\varepsilon} = \frac{1}{1+(s/s_{90})^2}\,. \tag{2.5}$$

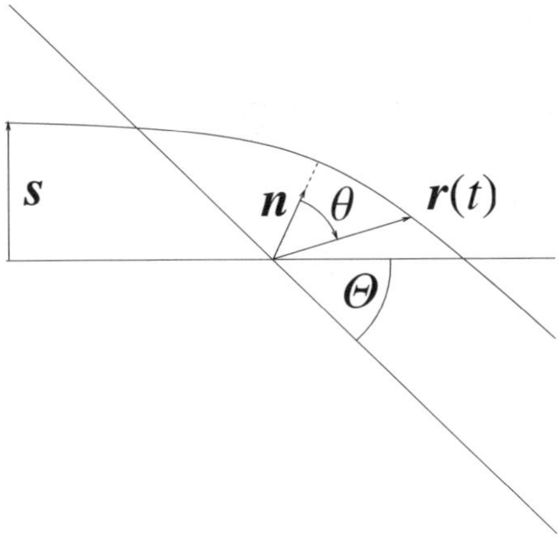

Fig. 2.2. Trajectory of relative motion in an electron-ion collision.

Here s is the impact parameter which is related to the scattering angle Θ and the relative velocity v_r by

$$s = \frac{|Z|e^2}{mv_r^2} \cot \frac{\Theta}{2} \; . \tag{2.6}$$

The impact parameter $s_{90} = |Z|e^2/mv_r^2$ separates the regimes of soft ($s > s_{90}$) and hard ($s < s_{90}$) collisions. With these relations the energy transfer (2.4) can be expressed as

$$\Delta E_i = \frac{2Ze^2}{v_r s_{90}} \, (\boldsymbol{n} \cdot \boldsymbol{v}_i) \sin \frac{\Theta}{2} \; . \tag{2.7}$$

Averaging $\boldsymbol{n}$ with respect to the directions of $\hat{\boldsymbol{s}}$ yields a factor $(\boldsymbol{n} \cdot \boldsymbol{v}_i)\boldsymbol{v}_r = \boldsymbol{v}_r \sin \Theta/2$. Thus

$$\langle \Delta E_i \rangle_{\hat{s}} = \frac{2Ze^2}{v_r^2} \, \frac{s_{90}}{s_{90}^2 + s^2} \, (\boldsymbol{v}_r \cdot \boldsymbol{v}_i) = \frac{2Z^2 e^4}{mv_r^4(s_{90}^2 + s^2)} \, (\boldsymbol{v}_r \cdot \boldsymbol{v}_i) \; . \tag{2.8}$$

The energy transfer (2.8) refers to a binary collision with an initial relative velocity $\boldsymbol{v}_r = \boldsymbol{v}_e(t \to -\infty) - \boldsymbol{v}_i = \boldsymbol{v}_{e0} - \boldsymbol{v}_i$ and an impact parameter $\boldsymbol{s} \perp \boldsymbol{v}_{e0}$. Assuming monochromatic electrons with a velocity $\boldsymbol{v}_{e0}$ and a number density n_e an ion collides during a time interval Δt with $2\pi n_e v_r s \mathrm{d}s \Delta t$ electrons with impact parameters between s and $s + \mathrm{d}s$. The energy loss of an ion on a path length $\mathrm{d}l = v_i \mathrm{d}t$ is

$$\frac{\mathrm{d}E_i}{\mathrm{d}l} = \frac{1}{v_i} \frac{\mathrm{d}E_i}{\mathrm{d}t} = \frac{2\pi n_e v_r}{v_i} \int_0^{s_{\max}} \mathrm{d}s s \langle \Delta E_i \rangle_{\hat{s}} \; , \tag{2.9}$$

where $\langle \ldots \rangle_{\hat{s}}$ indicates averaging with respect to the azimuthal angle of $\boldsymbol{s}$. Insertion of (2.8) yields

$$\frac{\mathrm{d}E_i}{\mathrm{d}l} = \frac{4\pi Z^2 n_e e^4}{mv_r^3} \, (\boldsymbol{v}_r \cdot \hat{\boldsymbol{v}}_i) \int_0^{s_{\max}} \frac{s\mathrm{d}s}{s_{90}^2 + s^2} = \frac{4\pi Z^2 n_e e^4}{mv_r^3} \, (\boldsymbol{v}_r \cdot \hat{\boldsymbol{v}}_i) \, \Lambda(s_{\max}, s_{90}) \tag{2.10}$$

with the modified Coulomb logarithm [124]

$$\Lambda(s_{\max}, s_{\min}) = \frac{1}{2} \ln\left(1 + \frac{s_{\max}^2}{s_{\min}^2}\right) . \tag{2.11}$$

Here the cut–off $s_{\max}$ is required by the infinite range of the Coulomb interaction. A physical model is provided by screening, i.e. $s_{\max}$ is set equal to the Debye length λ_D or suitable modifications thereof, see Sect. 3.2. On the other hand no lower cut–off is required in the s-integral (2.10) as hard collisions with $s < s_{90}$ are treated exactly. We note that the modified Coulomb logarithm $\Lambda(s_{\max}, s_{90})$ is strongly self–cutting as it behaves like $O(v_r^4)$ in the limit $v_r \to 0$.

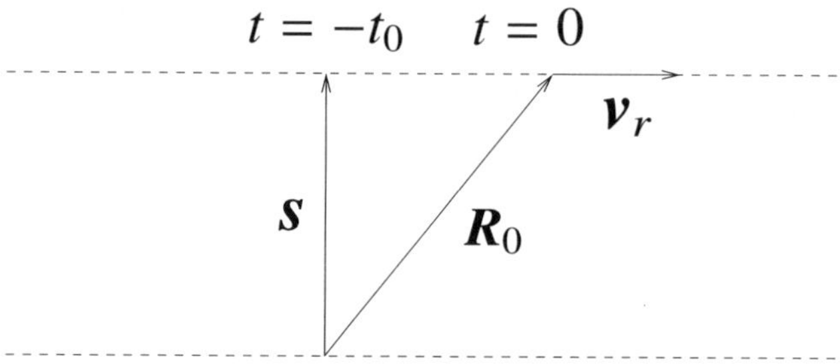

Fig. 2.3. Zero order trajectory of relative motion (2.12).

For more general forces than inverse square and in the presence of a magnetic field the exact closed solution (2.4) is not possible. We must seek for approximate treatments. A systematic perturbation expansion for some central interaction potentials without magnetic field has been formulated by Lehmann and Leibfried [78], see also [119]. We start in zero order from uniform motion

$$\boldsymbol{v}_0 = \boldsymbol{v}_r \, , \quad \boldsymbol{r}_0(t) = \boldsymbol{R}_0 + \boldsymbol{r}(t) = \boldsymbol{s} + \boldsymbol{v}_r(t + t_0) \, . \tag{2.12}$$

It should be noted that the impact vector $\boldsymbol{s}$ is here equal to the vector of closest approach, see Fig. 2.3. The first order correction to the relative velocity is

$$\boldsymbol{v}_1(t) = \frac{1}{m} \int_{-\infty}^{t} \boldsymbol{F}_0(\boldsymbol{r}_0(t'))\mathrm{d}t' \tag{2.13}$$

with

$$\boldsymbol{F}_0(\boldsymbol{r}_0(t)) = -Z\not{e}^2 \frac{\boldsymbol{r}_0(t)}{r_0^3(t)} = \frac{\boldsymbol{r}_0(t)}{r_0(t)} F(r_0(t)) \tag{2.14}$$

and $F_0(r_0(t)) = -Z\not{e}^2/r_0^2(t)$. For symmetry reasons $\boldsymbol{v}_1(t \to \infty)$ is parallel to $\boldsymbol{s}$, thus

$$\left\langle \Delta E_i^{(1)} \right\rangle_{\hat{s}} = \left\langle - \int_{-\infty}^{\infty} \mathrm{d}t' \boldsymbol{v}_i \cdot \boldsymbol{F}_0(\boldsymbol{r}_0(t')) \right\rangle_{\hat{s}} = 0 \, . \tag{2.15}$$

The first–order force can be expanded as

$$\boldsymbol{F}_1(\boldsymbol{r}_0(t), \boldsymbol{r}_1(t)) = \boldsymbol{r}_1 \frac{F(r_0)}{r_0} + \frac{\boldsymbol{r}_0(\boldsymbol{r}_0 \cdot \boldsymbol{r}_1)}{r_0} \frac{\partial}{\partial r_0}\left(\frac{F(r_0)}{r_0}\right) , \tag{2.16}$$

where the time arguments have been dropped for simplicity. Here

$$\boldsymbol{r}_1(t) = \frac{1}{m} \int_{-\infty}^{t} \boldsymbol{F}_0(\boldsymbol{r}_0(t'))(t - t')\mathrm{d}t' \tag{2.17}$$

is the first–order trajectory correction. The second–order energy transfer is then

$$\left\langle \Delta E_i^{(2)} \right\rangle_{\hat{s}} = \left\langle - \int_{-\infty}^{\infty} \mathrm{d}t' \boldsymbol{v}_i \cdot \boldsymbol{F}_1(\boldsymbol{r}_0(t'), \boldsymbol{r}_1(t')) \right\rangle_{\hat{s}} , \tag{2.18}$$

and inspection shows that only the terms

$$\begin{aligned} \langle \boldsymbol{v}_i \cdot \boldsymbol{r}_0(t) \rangle_{\hat{s}} &= (\boldsymbol{v}_i \cdot \boldsymbol{v})\,(t + t_0) , \\ \langle (\boldsymbol{v}_i \cdot \boldsymbol{r}_0(t))(\boldsymbol{r}_0(t) \cdot \boldsymbol{r}_0(t')) \rangle_{\hat{s}} &= s^2\,(\boldsymbol{v}_i \cdot \boldsymbol{v})\,(t + t_0) + v^2(t + t_0)^2(t' + t_0)\,(\boldsymbol{v}_i \cdot \boldsymbol{v}) \end{aligned} \tag{2.19}$$

survive the $\hat{s}$–averaging. The remaining nested integrals are of the type

$$\int_{-\infty}^{\infty} \mathrm{d}t f(t) \int_{-\infty}^{t} \mathrm{d}t' g(t') = \frac{1}{2} \left(\int_{-\infty}^{\infty} \mathrm{d}t f(t)\right)\left(\int_{-\infty}^{\infty} \mathrm{d}t' g(t')\right) \tag{2.20}$$

as the integrands are either both even or both odd. A straightforward calculation yields then for the Coulomb case

$$\left\langle \Delta E_i^{(2)} \right\rangle_{\hat{s}} = \frac{2Z^2 e^4}{m v_r^4 s^2}\,(\boldsymbol{v}_r \cdot \boldsymbol{v}_i) , \tag{2.21}$$

which should be compared to the exact result (2.8). Here the second–order pole for $s \to 0$ indicates the inability to handle hard collisions in perturbation theory. The integration with respect to the impact parameter must now exclude hard collisions by introducing a lower cut–off $s_{\min}$

$$\frac{\mathrm{d}E_i}{\mathrm{d}l} = \frac{4\pi Z^2 n_e e^4}{m v_r^3}\,(\boldsymbol{v}_r \cdot \hat{\boldsymbol{v}}_i)\; L\,(s_{\max}, s_{\min}) \tag{2.22}$$

with the conventional Coulomb logarithm [72]

$$L\,(s_{\max}, s_{\min}) = \ln \frac{s_{\max}}{s_{\min}} . \tag{2.23}$$

With the apparently reasonable choice $s_{\min} = s_{90} \propto v_r^{-2}$ this Coulomb logarithm becomes negative at small relative velocities. This can cause artifacts when integrating over the velocity distribution of the electrons, see Sect. 3.10. In contrast, the modified Coulomb logarithm (2.11) is self–cutting for $v_r \to 0$. Conversely, the regularization $1/s \to s/(s_{90}^2 + s^2)$ may be used as a *recipe* to account for hard collisions beyond the second–order regime for problems more general than Coulombic.

In most applications the electrons are not monochromatic, so $\mathrm{d}E_i/\mathrm{d}l$ must be averaged with respect to the electron velocity distribution $f_0(\boldsymbol{v}_e)$. (Of course the argument is here the asymptotic velocity of the electrons $\boldsymbol{v}_{e0}$, we drop the subscript zero to simplify the notation). The stopping power is then

$$\frac{\mathrm{d}\mathcal{E}_i}{\mathrm{d}l} = \left\langle \frac{\mathrm{d}E_i}{\mathrm{d}l} \right\rangle = \frac{4\pi Z^2 n_e e^4}{m} \hat{\boldsymbol{v}}_i \cdot \int \mathrm{d}\boldsymbol{v}_e \frac{\boldsymbol{v}_e - \boldsymbol{v}_i}{|\boldsymbol{v}_e - \boldsymbol{v}_i|^3} f_0(\boldsymbol{v}_e) \Lambda\left(s_{\max}, s_{\min}\right) . \tag{2.24}$$

If the Coulomb logarithm varies sufficiently slowly with the electron velocity, its average value Λ_{th} can be taken out of the integral and the remaining problem is analoguous to the evaluation of the field associated with the pseudopotential $H(\boldsymbol{v}_i)$ [85, 112] of the three–dimensional distribution $f_0(\boldsymbol{v}_e)$

$$\frac{\partial}{\partial \boldsymbol{v}_i} H(\boldsymbol{v}_i) = \frac{\partial}{\partial \boldsymbol{v}_i} \int \mathrm{d}\boldsymbol{v}_e \frac{f_0(\boldsymbol{v}_e)}{|\boldsymbol{v}_e - \boldsymbol{v}_i|} = \int \mathrm{d}\boldsymbol{v}_e \frac{\boldsymbol{v}_e - \boldsymbol{v}_i}{|\boldsymbol{v}_e - \boldsymbol{v}_i|^3} f_0(\boldsymbol{v}_e) . \tag{2.25}$$

We will first consider a target in thermal equilibrium described by the Maxwell distribution

$$f_0(\boldsymbol{v}_e) = \frac{1}{(2\pi v_{\mathrm{th}}^2)^{3/2}} \exp\left(-\frac{v_e^2}{2 v_{\mathrm{th}}^2} \right) , \tag{2.26}$$

where the thermal velocity v_{th} is related to the temperature T by $m v_{\mathrm{th}}^2 = k_{\mathrm{B}} T$ with the Boltzmann constant k_{B}. Performing a multipole expansion in (2.25) we obtain

$$\frac{\mathrm{d}\mathcal{E}_i}{\mathrm{d}l} = \frac{4\pi Z^2 n_e e^4}{m} \Lambda_{\mathrm{th}} \hat{\boldsymbol{v}}_i \cdot \frac{\partial}{\partial \boldsymbol{v}_i} \left[\frac{1}{v_i} \mathrm{erf}\left(\frac{v_i}{\sqrt{2} v_{\mathrm{th}}} \right) \right] \tag{2.27}$$

$$\sim -\frac{4\pi Z^2 n_e e^4}{m} \Lambda_{\mathrm{th}} \begin{cases} \frac{1}{3}\sqrt{\frac{2}{\pi}} \frac{v_i}{v_{\mathrm{th}}^3}, & v_i \ll v_{\mathrm{th}} \\ \frac{1}{v_i^2}, & v_i \gg v_{\mathrm{th}} \end{cases} .$$

This confirms the gross estimate (2.2) shown in Fig. 2.1, which was obtained in the dielectric picture. In connection with electron cooling of ion beams we will employ anisotropic velocity distributions like

$$f_0(\boldsymbol{v}_e) = \frac{1}{2\pi v_{\mathrm{th}\perp}^2} \frac{1}{(2\pi v_{\mathrm{th}\|}^2)^{1/2}} \exp\left(-\frac{v_{e\perp}^2}{2 v_{\mathrm{th}\perp}^2} - \frac{v_{e\|}^2}{2 v_{\mathrm{th}\|}^2} \right) . \tag{2.28}$$

Now we have to solve a problem analoguous to the electric field of an axially symmetric ellipsoidal Gaussian charge distribution. This can be done in a straightforward manner [86] by representing $|\boldsymbol{v}_e - \boldsymbol{v}_i|^{-1}$ in cylindrical coordinates [68, Sect. 13]. The stopping force $\mathcal{F}$ on the ion, defined by

$$\frac{\mathrm{d}\mathcal{E}_i}{\mathrm{d}l} = \mathcal{F} \cdot \hat{\boldsymbol{v}}_i \tag{2.29}$$

is not more antiparallel to $\hat{\boldsymbol{v}}_i$ unless $v_i \gg \max(v_{\mathrm{th}\perp}, v_{\mathrm{th}\|})$. Assuming $v_{\mathrm{th}\perp} \gg v_{\mathrm{th}\|}$ and expanding yields three regimes:

$$\mathcal{F}_{\parallel} = -\frac{4\pi Z^2 n_e e^4}{m} \Lambda_{\text{th}} \begin{cases} \frac{v_{i\parallel}}{v_i^3}, & v_{i\parallel} \gg v_{\text{th}\perp} \\ \frac{v_{i\parallel}}{v_{i\parallel} v_{\text{th}\perp}^2}, & v_{\text{th}\parallel} \ll v_{i\parallel} \ll v_{\text{th}\perp} \\ \sqrt{\frac{2}{\pi}} \frac{v_{i\parallel}}{v_{\text{th}\parallel} v_{\text{th}\perp}^2}, & v_{i\parallel} \ll v_{\text{th}\parallel} \ll v_{\text{th}\perp} \end{cases}, \tag{2.30}$$

$$\mathcal{F}_{\perp} = -\frac{4\pi Z^2 n_e e^4}{m} \Lambda_{\text{th}} \begin{cases} \frac{v_{i\perp}}{v_i^3}, & v_{i\perp} \gg v_{\text{th}\perp} \\ \sqrt{\frac{\pi}{8}} \frac{v_{i\perp}}{v_{\text{th}\perp}^3}, & v_{i\perp} \ll v_{\text{th}\perp} \end{cases}. \tag{2.31}$$

While these expressions agree with [106] they differ in some prefactors from results reported in [85] and in [26]. There occurs now an intermediate velocity regime $v_{\text{th}\parallel} \ll v_{i\parallel} \ll v_{\text{th}\perp}$ in which $\mathcal{F}_{\parallel}$ is constant. The numerical evaluation of (2.25) shows however that this regime is more narrow than the limits in (2.30) suggest.

2.2 Challenges Imposed by the Magnetic Field

The influence of a magnetic field $\boldsymbol{B} = B\boldsymbol{b}$ on the energy loss cannot be easily described, in fact simple physical arguments lead to apparently contradictory conclusions. In a very strong magnetic field the electrons are restricted to move along the field lines like beads on a wire [85], their transverse motion being quenched.

1. Now consider an electron velocity distribution which is flattened in the direction of the magnetic field like (2.28) with $v_{\text{th}\parallel} \ll v_{\text{th}\perp}$. Then the simple dielectric result shown in Fig. 2.1 which has been confirmed in the binary collision model must be modified as shown in Fig. 2.4. The transverse motion of the electrons is quenched, so it is the smaller velocity $v_{\text{th}\parallel}$ which divides the linear and the inverse square regimes of the energy loss, the intermediate regime (2.30) (solid curve) being absent. The v_i^{-2} dependence of the drag (dashed curve) occurs now down to $v_{\text{th}\parallel}$. It has therefore been argued (in careful terms) that the magnetic field tends to increase the energy loss in case of such an anisotropic velocity distribution [120].

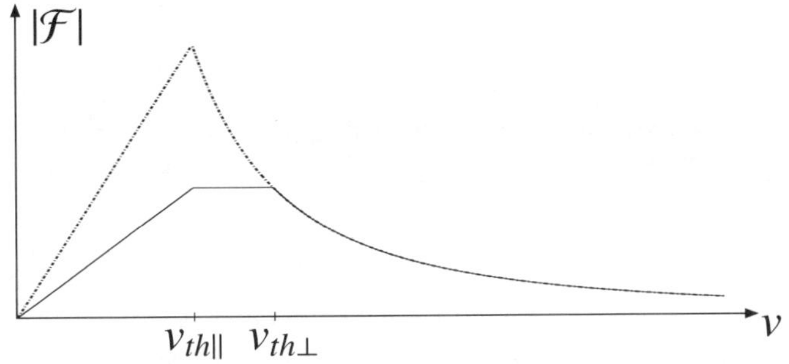

Fig. 2.4. Schematic dielectric model for the stopping force in strongly magnetized electrons with a flattened velocity distribution.

2. Consider, on the other hand, collisions of electrons with an ion moving along the field lines ($v_{i\perp} = 0$). The process is completely symmetric, the energy gained in approaching the ion is again lost on the way out. So in this situation the magnetic field suppresses the energy loss completely [35]. We note that this argument is only valid for the attractive interaction of electrons with positively charged ions. For the repulsive case, e.g. electrons and antiprotons, there will be reflections for small impact parameters. Of course such violent changes in the trajectories cannot be treated as perturbations of the helical motion of the electrons. We will show in Sect. 6.2.3 in explicit examples that the energy loss in the repulsive case is larger than for particles which attract each other.

Thus for reliable results on the influence of the magnetic field on the energy loss the scattering must be studied on a deeper level. It is useful to distinguish three regimes of trajectories shown in Fig. 2.5, depending on the relative size of the distance of closest approach, the cyclotron radius and the pitch of the helix.

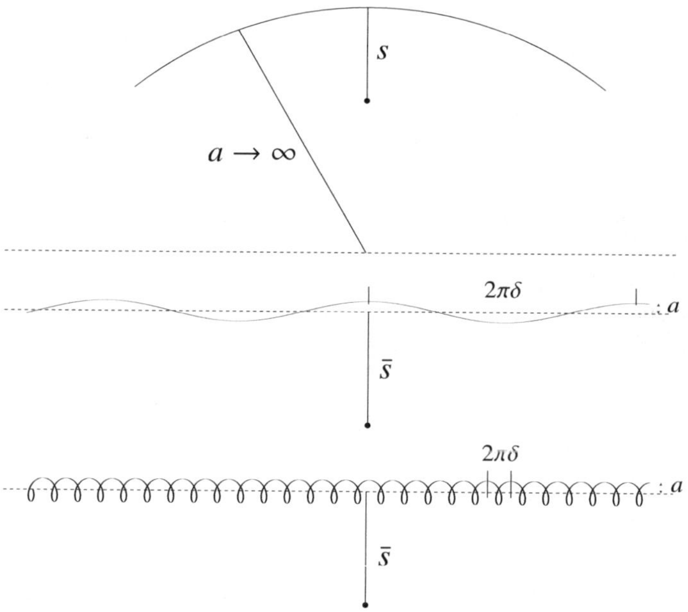

Fig. 2.5. Schematic trajectories of relative motion. Top: Weak field, $s \ll a$. Center: Stretched helices in case of a strong field, $\bar{s} \gg a$, $\bar{s} \ll \delta$. Bottom: Tight helices in case of a strong field, $\bar{s} \gg a$, $\bar{s} \gg \delta$. After [124].

1. If the cyclotron radius is much larger than the distance of closest approach the influence of the magnetic field is weak. As shown in Sect. 3.9 the energy loss in this Coulomb regime can be described by (2.8).

2. In a magnetic field $\boldsymbol{B}$ the unperturbed trajectory of the electrons is a helix with the cyclotron radius $a = v_{e\perp}/\Omega_e$. Its guiding center moves with a velocity $\bar{\boldsymbol{v}}_r = v_{e\|}\boldsymbol{b} - \boldsymbol{v}_i$ relative to the ion. Its impact parameter is $\bar{\boldsymbol{s}}$ and the pitch of the helix is $2\pi\delta$ with $\delta = \bar{v}_r/\Omega_e$. Here $v_{e\|} = \boldsymbol{v}_e \cdot \boldsymbol{b}$ is the component of the asymptotic electron velocity $\boldsymbol{v}_e$ parallel to the magnetic field and $v_{e\perp} = |\boldsymbol{v}_e - v_{e\|}\boldsymbol{b}|$ is its magnitude transverse to the magnetic field. (The subscript zero has again been dropped). We distinguish now two subcases:
   a. If the helices are stretched so that their pitch is larger than the impact parameter $\bar{s} \ll \delta$, an inspection of Fig. 2.5 shows that these trajectories are quite similar to those of the Coulomb regime in the vicinity of the scattering center. This suggests that the energy loss can be obtained by replacing the relative velocity $\boldsymbol{v}_r$ occuring in (2.8) by the relative velocity of the guiding center. Thus
   $$\langle \Delta E_i \rangle_{\hat{\bar{s}}} = \frac{2Z^2 e^4}{m\bar{v}_r^4(\bar{s}_{90}^2 + \bar{s}^2)} (\bar{\boldsymbol{v}}_r \cdot \boldsymbol{v}_i) \tag{2.32}$$
   which has been confirmed by explicit calculation [124].
   b. More interesting is the case of tight helices, where $\bar{s} \gg \delta$. The force $-\boldsymbol{F}_1$ on the ion (2.16) must be modified as the electrons are only allowed to move in the direction $\boldsymbol{b}$ of the magnetic field
   $$\boldsymbol{F}_1(\boldsymbol{r}_0(t), \boldsymbol{b}\,(\boldsymbol{b}\cdot\boldsymbol{r}_1(t)) = \boldsymbol{b}(\boldsymbol{b}\cdot\boldsymbol{r}_1)\frac{F(r_0)}{r_0} + \frac{\boldsymbol{r}_0(\boldsymbol{r}_0\cdot\boldsymbol{b})(\boldsymbol{b}\cdot\boldsymbol{r}_1)}{r_0}\frac{\partial}{\partial r_0}\left(\frac{F(r_0)}{r_0}\right). \tag{2.33}$$
   In the second–order energy transfer
   $$\left\langle \Delta E_i^{(2)} \right\rangle_{\hat{\bar{s}}} = \left\langle -\int_{-\infty}^{\infty} \mathrm{d}t' \boldsymbol{v}_i \cdot \boldsymbol{F}_1(\boldsymbol{r}_0(t'), \boldsymbol{b}(\boldsymbol{b}\cdot\boldsymbol{r}_1(t'))) \right\rangle_{\hat{\bar{s}}} \tag{2.34}$$
   only the longitudinal component of the force survives the $\hat{\bar{s}}$–averaging. The $\hat{\bar{s}}$–averages are
   $$v_{i\|}\langle \boldsymbol{b}\cdot\boldsymbol{r}_0(t)\rangle_{\hat{\bar{s}}} = v_{i\|}\bar{v}_{r\|}(t+t_0)\,,$$
   $$\langle(\boldsymbol{r}_0(t)\cdot\boldsymbol{v}_i)(\boldsymbol{r}_0(t)\cdot\boldsymbol{b})(\boldsymbol{r}_0(t')\cdot\boldsymbol{b})\rangle_{\hat{\bar{s}}} = \frac{1}{2}\bar{s}^2\bar{v}_{r\|}v_{e\|}\frac{\bar{v}_{r\perp}^2}{\bar{v}_r^2}((t'+t_0)+(t+t_0))$$
   $$+\frac{1}{2}\bar{s}^2\boldsymbol{v}_i\cdot\bar{\boldsymbol{v}}_r\frac{\bar{v}_{r\perp}^2}{\bar{v}_r^2}(t+t_0) + \boldsymbol{v}_i\cdot\bar{\boldsymbol{v}}_r\bar{v}_{r\|}^2(t+t_0)^2(t'+t_0)\,. \tag{2.35}$$
   Straightforward integration of (2.34) with (2.33) and (2.27) yields [124]
   $$\left\langle \Delta E_i^{(2)} \right\rangle_{\hat{\bar{s}}} = \frac{Z^2 e^4}{m\bar{v}_r^6\bar{s}^2}\bar{v}_{r\perp}^2(v_{e\|}^2 - v_i^2) \tag{2.36}$$
   $$= \frac{Z^2 e^4}{m\bar{v}_r^6\bar{s}^2}\bar{v}_{r\perp}^2\left[2\bar{\boldsymbol{v}}_{r\|} + \bar{\boldsymbol{v}}_{r\perp}\left(1 - \frac{\bar{v}_{r\|}^2}{\bar{v}_{r\perp}^2}\right)\right]\cdot\boldsymbol{v}_i\,.$$
   Here $\boldsymbol{v}_{i\|}$ and $\boldsymbol{v}_{i\perp}$ are the velocities of the ion parallel and transverse to the magnetic field and similarly for the components of the relative velocity.

As was already noted above this energy transfer vanishes for symmetry reasons unless the ion has a velocity component transverse to the magnetic field. For the special case $v_{e\parallel} = 0$ this was previously derived in [120]. Similar expressions with the unexplicable replacements $2\bar{\boldsymbol{v}}_{r\parallel} \to 3\bar{\boldsymbol{v}}_{r\parallel}$ and $1 - \bar{v}_{r\parallel}^2/\bar{v}_{r\perp}^2 \to 1 - 2\bar{v}_{r\parallel}^2/\bar{v}_{r\perp}^2$ are used in [35, 85]. For the energy loss (2.9) an integration with respect to the impact parameter has to be done. The integrand can be approximated piecewise by (2.8) for $\bar{s} < a$ and by (2.32) ($\bar{s} < \delta$) or (2.36) ($\bar{s} > \delta$) for $\bar{s} > a$. This leads to Coulomb logarithms involving a and δ as additional parameters. As these are velocity dependent this leads to expressions for the energy loss [35,85,124] which are not very transparent. Moreover, the impact parameters $\boldsymbol{s}$ of the actual trajectory and $\bar{\boldsymbol{s}}$ of the guiding center ought to be distinguished carefully; we return to this question in Sect. 3.5. The subsequent integration with respect to the electron velocity distribution reveals deeper lying, more serious problems. These become obvious if one attempts to calculate the $\boldsymbol{v}_e$-averaged energy loss along the lines presented in Sect. 2.1. But now the energy loss involves derivatives of the potential

$$\int \mathrm{d}\boldsymbol{v}_e \frac{f_0(\boldsymbol{v}_e)}{\bar{v}_r} = \int \mathrm{d}v_{e\parallel} \frac{f_0(v_{e\parallel})}{|\bar{\boldsymbol{v}}_{e\parallel} - \boldsymbol{v}_i|} \tag{2.37}$$

of a one–dimensional source. In contrast to the previous three–dimensional case there is no volume element which suppresses the singularity of the integrand in the limit $\bar{v}_r \to 0$. This indicates a failure of the second–order perturbation treatment which is not surprising in view of the asymmetry a strong magnetic field introduces with respect to the sign of the interaction. As mentioned above reflections with large velocity transfer will increasingly occur in the repulsive case for $\bar{v}_r \to 0$. In the attractive case obvious unphysical results are avoided if a self–cutting Coulomb logarithm like (2.11) is left under the velocity integral [86]. The order in which the integrations for the energy loss are done in ordinary and in velocity space matters very much for strong magnetic fields. We will discuss this in more detail in Sect. 3.10 and again when comparing the BC with the LR in Sect. 4.6.

The cut–offs in the treatment presented in this Chapter are required because of singularity of the Coulomb interaction at the origin and its infinite range. However, there are good physical reasons to use interactions of finite range (screening) and which are regularized near the origin to account for the quantum indeterminacy. But for these interactions the calculation of the second–order energy transfer along the lines used in this Chapter cannot be easily done. We postpone such a treatment to Chap. 3.

2.3 Classical-Trajectory-Monte-Carlo (CTMC) Simulations

A fully numerical treatment is required for applications beyond the perturbative regime and for checking the validity of the perturbative approach outlined above. This can be done on different levels of approximation. We first consider the case of binary ion–electron collisions with an effective interaction $U(r)$, like e.g. the interaction potentials (3.2), (3.3), which allow to incorporate both screening and quantum effects in the binary collision picture. The non–perturbative treatment of the dielectric approach based on a nonlinear Vlasov-Poisson description will be discussed in Sect. 2.5.

In the absence of a magnetic field, the energy transfer in a binary collision with an arbitrary interaction $U(r)$ depending on the relative distance between ion and electron can be easily calculated by standard numerical techniques. As the angular momentum l and the energy of relative motion E are conserved in such collisions the azimuthal angle θ and the radial distance r are related by [48]

$$\theta(r) = \theta_{\max} - \int_r^\infty \frac{l\,\mathrm{d}r'}{r'\sqrt{2mr'^2[E-U(r)]-l^2}} \,. \tag{2.38}$$

This yields after numerical integration and inversion the trajectory $r(\theta)$, and from a subsequent numerical integration of expression (2.3) (with $F(r) = -\partial U(r)/\partial r$ and $r = r(\theta)$) the energy transfer ΔE_i.

But in the presence of a magnetic field the numerical evaluation of the BC energy loss is very complicated. Neither the energy of relative motion nor the angular momentum are conserved and we have to deal with a chaotic system. Under these conditions, the energy transfer can be successfully investigated by Classical-Trajectory-Monte-Carlo (CTMC) simulations [135, 136, 138]. In the CTMC method [2] the trajectories for the relative motion between the ion and an electron are calculated by a numerical integration of the equations of motion (i.e. equation (3.16) for the case $M \gg m$ and $\boldsymbol{v}_i$ = const, see Sect. 3.2), starting with initial conditions for the guiding center velocity $\bar{\boldsymbol{v}}_r = v_{e\|}\boldsymbol{b} - \boldsymbol{v}_i$ and the transverse electron motion $\boldsymbol{v}_{e\perp}$. The initial positions are chosen to correspond to a certain impact parameter $\bar{\boldsymbol{s}}$ and are located outside the interaction zone, which is – employing a screened interaction like (3.2) – defined as a sphere of several screening lengths λ about the ion. The numerical calculation stops after the electron has left this interaction zone, that is, when the collision is completed. Deducing the velocity changes from the initial and final velocities $\bar{\boldsymbol{v}}_r, \boldsymbol{v}_{e\perp}$ yields the energy transfer $\Delta E_i(\bar{\boldsymbol{v}}_r, \boldsymbol{v}_{e\perp}, \bar{\boldsymbol{s}})$, see (3.32). The required accuracy is achieved by using a modified Velocity–Verlet algorithm which has been specifically designed for particle propagation in a (strong) magnetic field [121], and by adapting continuously the actual time–step by monitoring the constant of motion K (3.18). The resulting relative deviations of K are of the order of $10^{-6} - 10^{-5}$. For binary collisions in an infinitely strong magnetic field, which are modeled by a purely one dimensional motion of the electron along the field lines, a high precision Bulirsch–Stoer method with adaptive step size [108] is applied. The desired average over the initial orientation of the perpendicular electron velocity $\boldsymbol{v}_{e\perp}$ and the impact parameter $\bar{\boldsymbol{s}}$ is performed by a Monte–Carlo sampling [21, 42] of a large number of trajectories with different initial values. The actual number of computed trajectories is adjusted by monitoring the convergence of the averaging procedure. Around $10^5 - 10^6$ trajectories are typically needed for the energy transfer $\langle\Delta E_i\rangle(\bar{v}_{r\|}, v_{e\perp}, v_{i\perp})$ for one set of velocities $\bar{v}_{r\|}, v_{e\perp}, v_{i\perp}$ and at a given magnetic field, i.e. for an ion with velocity $\boldsymbol{v}_i$ in a monochromatic electron beam. Calculating $\langle\Delta E_i\rangle$ for a sufficiently large number of electron velocities $v_{e\|}, v_{e\perp}$ a final numerical integration over the electron velocity distribution $f_0(v_{e\|}, v_{e\perp})$ can be performed. Results of the CTMC calculation will be presented and discussed in Sects. 3.7, 3.10 and Chap. 6.

2.4 Dielectric Treatment (DT), Vlasov–Poisson Equation, Linear Response (LR)

The dielectric treatment describes the force on a charged particle moving through a plasma in a continuum picture as the interaction with its polarization cloud. Correlation effects between the plasma particles which go beyond a mean–field interaction are neglected. This approximation holds for weakly coupled plasmas with a large number of particles in the Debye sphere $N_D = 4\pi n\lambda_D^3 \gg 1$, where n is the plasma density and λ_D the Debye length. A plasma of different species ν is then treated as a superposition of continuous, polarizable fluids which are described by the single particle phase–space densities $f_\nu(\boldsymbol{r}, \boldsymbol{v}, t)$. Their time evolution is determined by the Vlasov–Poisson equations

$$\frac{\partial f_\nu}{\partial t} + \boldsymbol{v} \cdot \frac{\partial f_\nu}{\partial \boldsymbol{r}} + c_\nu \Omega_\nu \left[\boldsymbol{v} \times \boldsymbol{b}\right] \cdot \frac{\partial f_\nu}{\partial \boldsymbol{v}} - \frac{q_\nu}{m_\nu} \frac{\partial \phi}{\partial \boldsymbol{r}} \cdot \frac{\partial f_\nu}{\partial \boldsymbol{v}} = 0 \tag{2.39}$$

and

$$\varepsilon_0 \nabla^2 \phi(\boldsymbol{r}, t) = -\rho_i(\boldsymbol{r}, t) - \sum_\nu q_\nu \int \mathrm{d}\boldsymbol{v} f_\nu(\boldsymbol{r}, \boldsymbol{v}, t). \tag{2.40}$$

Here $c_\nu = |q_\nu| / q_\nu$, $\Omega_\nu = |q_\nu| B/m_\nu$, and ϕ is the self–consistent electrostatic potential given by the solution of the Poisson equation (2.40), where $\rho_i(\boldsymbol{r}, t) = Ze\delta(\boldsymbol{r} - \boldsymbol{r}_i(t))$ is the charge density of the projectile ion (charge $q = Ze$, mass M) moving with a non–relativistic velocity $\boldsymbol{v}_i(t) = \dot{\boldsymbol{r}}_i(t)$.

For a sufficiently heavy projectile, implying a constant velocity $\boldsymbol{v}_i$ and a charge density $\rho_i(\boldsymbol{r}, t) = Ze\delta(\boldsymbol{r} - \boldsymbol{v}_i t)$, the energy loss (2.1) of the charged particle is directly given by the created electric field and related force $\mathcal{F} = -Ze\nabla\phi$ at the particle's position, i.e.

$$\frac{\mathrm{d}\mathcal{E}_i}{\mathrm{d}t} = \boldsymbol{v}_i \cdot \mathcal{F} = -Ze\boldsymbol{v}_i \cdot \nabla\, \phi(\boldsymbol{r})|_{\boldsymbol{r}=\boldsymbol{v}_i t} \,. \tag{2.41}$$

Due to the assumption of constant velocity $\boldsymbol{v}_i$, it was presumed here that $\phi, \mathcal{F}$ take – after a short transient period – stationary values in the moving frame where the ion is at rest. For calculating the force $\mathcal{F}$ the Vlasov–Poisson system, equations (2.39) and (2.40) has to be solved with appropriate boundary conditions. This can be done by fully numerical Particle-In-Cell (PIC) or test-particle simulations and will be discussed in Sect. 2.5. If the projectile only exerts a weak perturbation on the target plasma, approximate solutions of the Vlasov–Poisson system are possible and usually more appropriate. A detailed discussion of this perturbative dielectric treatment will be given in Chap. 4. The basic ideas and procedures are outlined in the following.

To that end we consider the rather general case of an anisotropic two-component electron-ion plasma ($\nu = e, i$) with $q_e = -e$ and $q_i = Z_i e$, densities $n_i, n_e = Z_i n_i$ and two different temperatures $T_{\nu\|}$, $T_{\nu\perp}$ of the plasma particles. For each plasma species we define an average temperature $\bar{T}_\nu = \frac{1}{3}T_{\nu\|} + \frac{2}{3}T_{\nu\perp}$. The axis $\boldsymbol{b}$ defined by the external magnetic field coincides with the symmetry axis of the velocity distribution with temperature $T_{\nu\|}$. The strength of the coupling between the projectile ion moving with velocity $\boldsymbol{v}_i$ and the plasma is given by the coupling parameter

$$\mathcal{Z} = \frac{g}{\left(1 + v_i^2/\bar{v}_{\text{th}\nu}^2\right)^{3/2}} . \tag{2.42}$$

Here $\bar{v}_{\text{th}\nu} = (k_\text{B}\bar{T}_\nu/m_\nu)^{1/2}$ is the average thermal velocity of the plasma particle ν with $m_\nu = m$ and $m_\nu = m_i$ for the electrons and ions, respectively, $g = |Z|/N_\text{D}$. The derivation of (2.42) is discussed in detail in [134]. The parameter $\mathcal{Z}$ characterizes the ion–target coupling, where $\mathcal{Z} \ll 1$ corresponds to weak, linear coupling and $\mathcal{Z} \gtrsim 1$ to strong, nonlinear coupling. Equation (2.42) was originally derived for nonmagnetized targets, but is expected to essentially characterize as well the coupling with magnetized plasmas.

For a sufficiently small perturbation ($\mathcal{Z} \ll 1$) the Vlasov equation (2.39) may be linearized about the unperturbed equilibrium distribution $n_\nu f_{0\nu}$, with $f_{0\nu}$ given by the two–temperature Maxwell distribution

$$f_{0\nu}\left(v_\|, v_\perp\right) = \frac{1}{(2\pi)^{3/2}\, v_{\text{th}\nu\perp}^2 v_{\text{th}\nu\|}} \exp\left(-\frac{v_\perp^2}{2v_{\text{th}\nu\perp}^2}\right) \exp\left(-\frac{v_\|^2}{2v_{\text{th}\nu\|}^2}\right), \tag{2.43}$$

where $\langle v_\|^2\rangle = v_{\text{th}\|\nu}^2 = k_\text{B}T_{\|\nu}/m_\nu$, $\langle v_\perp^2\rangle = 2v_{\text{th}\perp\nu}^2 = 2k_\text{B}T_{\perp\nu}/m_\nu$.

The linearized Vlasov–Poisson equations with $n_\nu f_{1\nu} = f_\nu - n_\nu f_{0\nu}$ then read

$$\frac{\partial f_{1\nu}}{\partial t} + \boldsymbol{v} \cdot \frac{\partial f_{1\nu}}{\partial \boldsymbol{r}} + c_\nu \Omega_\nu \left[\boldsymbol{v} \times \boldsymbol{b}\right] \cdot \frac{\partial f_{1\nu}}{\partial \boldsymbol{v}} = \frac{q_\nu}{m_\nu} \frac{\partial \phi}{\partial \boldsymbol{r}} \cdot \frac{\partial f_{0\nu}}{\partial \boldsymbol{v}} , \tag{2.44}$$

and

$$\varepsilon_0 \nabla^2 \phi = -\rho_i\left(\boldsymbol{r}, t\right) - \sum_\nu n_\nu q_\nu \int \mathrm{d}\boldsymbol{v} f_{1\nu}(\boldsymbol{r}, \boldsymbol{v}, t) . \tag{2.45}$$

By solving equations (2.44) and (2.45) in space–time Fourier transforms (see Appendix A for details), we obtain the electrostatic potential

$$\begin{aligned} \phi\left(\boldsymbol{r}, t\right) = {} & \frac{Ze}{(2\pi)^4\, \varepsilon_0} \int \mathrm{d}\boldsymbol{k} \int_{-\infty}^{\infty} \frac{\mathrm{d}\omega}{k^2 \varepsilon\left(\boldsymbol{k}, \omega\right)} \exp\left(\mathrm{i}\boldsymbol{k} \cdot \boldsymbol{r} - \mathrm{i}\omega t\right) \\ & \times \int_{-\infty}^{\infty} \mathrm{d}t' \exp\left(\mathrm{i}\omega t' - \mathrm{i}\boldsymbol{k} \cdot \boldsymbol{r}_i\left(t'\right)\right) , \end{aligned} \tag{2.46}$$

which provides the dynamic linear response of the plasma to the motion of the projectile ion in the presence of the external magnetic field. The dielectric response function $\varepsilon\left(\boldsymbol{k}, \omega\right)$ is expressed by the susceptibilities $\chi_\nu^{(0)}(\boldsymbol{k}, \omega)$ of the magnetized plasma particles with $\nu = e, i$,

$$\varepsilon(\boldsymbol{k},\omega) = 1 + \sum_{\nu} u_C(k)\chi_{\nu}^{(0)}(\boldsymbol{k},\omega) , \tag{2.47}$$

where $(q_\nu^2/4\pi\varepsilon_0)u(k)$ is the Fourier transformed two–body interaction potential, in case of the repulsive Coulomb potential for electron and ion subsystems $u_C(k) = 1/2\pi^2k^2$ (see (3.5)). The dielectric function (DF) of a homogeneous, magnetized and anisotropic Maxwellian plasma is given by (see Appendix A for details)

$$\begin{aligned}
\varepsilon(\boldsymbol{k},\omega) &= 1 + \frac{1}{k^2\lambda_{\mathrm{D}\|}^2}(\mathcal{G} + \mathrm{i}\,\mathcal{F}) \qquad (2.48)\\
&= 1 + \sum_{\nu}\frac{1}{k^2\lambda_{\mathrm{D}\nu\|}^2}\Bigg\{1 + \mathrm{i}\zeta_\nu\sqrt{2}\int_0^\infty \mathrm{d}t \exp\left[\mathrm{i}\zeta_\nu t\sqrt{2} - X_\nu(t)\right]\\
&+ \frac{kv_{\mathrm{th}\nu\|}\sqrt{2}}{\Omega_\nu}(1-\tau_\nu)\sin^2\beta\int_0^\infty \mathrm{d}t \sin\left(\frac{\Omega_\nu t\sqrt{2}}{kv_{\mathrm{th}\nu\|}}\right)\exp\left[\mathrm{i}\zeta_\nu t\sqrt{2} - X_\nu(t)\right]\Bigg\}
\end{aligned}$$

with

$$X_\nu(t) = t^2\cos^2\beta + k^2a_\nu^2\sin^2\beta\left[1 - \cos\left(\frac{\Omega_\nu t\sqrt{2}}{kv_{\mathrm{th}\nu\|}}\right)\right] , \tag{2.49}$$

where $\lambda_{\mathrm{D}\nu\|} = v_{\mathrm{th}\nu\|}/\omega_{p\nu}$ and $\lambda_{\mathrm{D}\|}^{-2} = \sum_\nu \lambda_{\mathrm{D}\nu\|}^{-2}$, $\omega_{p\nu} = (n_\nu q_\nu^2/m_\nu\varepsilon_0)^{1/2}$ is the plasma frequency of the species ν ($\omega_p^2 = \sum_\nu \omega_{p\nu}^2$ is the plasma frequency of electron–ion plasma), $\zeta_\nu = \omega/kv_{\mathrm{th}\nu\|}$, $\tau_\nu = T_{\nu\perp}/T_{\nu\|}$, $a_\nu = v_{\mathrm{th}\nu\perp}/\Omega_\nu$ and β is the angle between the wave vector $\boldsymbol{k}$ and the magnetic field $\boldsymbol{b}$. In equation (2.48) we have introduced the real and imaginary parts, $\mathcal{G}$ and $\mathcal{F}$, of the dispersion function for the electron–ion magnetized plasma, respectively.

As shown in Appendix A, equations (2.48) and (2.49) are identical with the Bessel function representation of $\varepsilon(\boldsymbol{k},\omega)$ derived e.g. by Ichimaru [67]. Equations (2.48) and (2.49) are, however, more convenient when studying the weak and strong magnetic field limits in Sects. 4.2 and 4.3.

The stopping power (SP) S of an ion is defined as the energy loss of the ion in a unit length path due to interaction with the plasma. From (2.46) it is straightforward to calculate the electric field $\boldsymbol{E} = -\nabla\phi$ and the stopping force acting on the ion

$$\begin{aligned}
\boldsymbol{\mathcal{F}}(t) &= -\,Ze\nabla\phi(\boldsymbol{r},t)|_{\boldsymbol{r}=\boldsymbol{r}_i(t)} \qquad (2.50)\\
&= -\frac{2\mathrm{i}Z^2e^2}{(2\pi)^3}\int \mathrm{d}\boldsymbol{k}\int_{-\infty}^{\infty}\mathrm{d}\omega\frac{\boldsymbol{k}}{k^2\varepsilon(\boldsymbol{k},\omega)}\Xi_i(\boldsymbol{k},\omega,t)\int_{-\infty}^{\infty}\mathrm{d}t'\,\Xi_i^*(\boldsymbol{k},\omega,t') ,
\end{aligned}$$

where $e^2 = e^2/4\pi\varepsilon_0$,

$$\Xi_i(\boldsymbol{k},\omega,t) = \exp\left(\mathrm{i}\boldsymbol{k}\cdot\boldsymbol{r}_i(t) - \mathrm{i}\omega t\right) . \tag{2.51}$$

In (2.50) the asterix denotes the complex conjugate. Then, the stopping power of the projectile ion becomes

$$\begin{aligned}
S &= -\frac{\mathrm{d}\mathcal{E}_i}{\mathrm{d}l} = -\frac{\boldsymbol{v}_i(t)\cdot\boldsymbol{\mathcal{F}}(t)}{v_i(t)} \qquad (2.52)\\
&= \frac{2\mathrm{i}Z^2e^2}{(2\pi)^3 v_i(t)}\int \mathrm{d}\boldsymbol{k}\int_{-\infty}^{\infty}\mathrm{d}\omega\frac{\boldsymbol{k}\cdot\boldsymbol{v}_i(t)}{k^2\varepsilon(\boldsymbol{k},\omega)}\Xi_i(\boldsymbol{k},\omega,t)\int_{-\infty}^{\infty}\mathrm{d}t'\,\Xi_i^*(\boldsymbol{k},\omega,t') .
\end{aligned}$$

We now consider some particular cases of this expression. First the case of a heavy ion which moves in the plasma with constant velocity $\boldsymbol{v}_i$ and $\boldsymbol{r}_i(t) = \boldsymbol{v}_i t$. Then expression (2.52) together (2.51) yields

$$S = \frac{2Z^2 e^2}{(2\pi)^2 v_i} \int d\boldsymbol{k} \frac{\boldsymbol{k} \cdot \boldsymbol{v}_i}{k^2} \mathrm{Im} \frac{-1}{\varepsilon(\boldsymbol{k}, \boldsymbol{k} \cdot \boldsymbol{v}_i)} . \tag{2.53}$$

Equation (2.53) assumes that the projectile ion mass is infinitely large, $M \to \infty$, and the magnetic field has no influence on the trajectory of the ion. More precisely this requires that the cyclotron radius of an ion must be larger than any characteristic length scale, in particular the Debye screening length λ_{D}.

For the light projectile ions, however, we take into account the magnetic field effect on the ion trajectory. In this case the ion velocity and coordinate can be represented as

$$\boldsymbol{v}_i(t) = v_{i\|} \boldsymbol{b} + v_{i\perp} \left[\boldsymbol{u}_p \cos(\Omega_c t) - c_Z [\boldsymbol{b} \times \boldsymbol{u}_p] \sin(\Omega_c t) \right] , \tag{2.54}$$

$$\boldsymbol{r}_i(t) = v_{i\|} t \boldsymbol{b} + a_c \left[\boldsymbol{u}_p \sin(\Omega_c t) + c_Z [\boldsymbol{b} \times \boldsymbol{u}_p] \cos(\Omega_c t) \right] , \tag{2.55}$$

where $c_Z = |Z|/Z$, $\boldsymbol{u}_p = (\cos\varphi_p, \sin\varphi_p)$ is an arbitrary unit vector perpendicular to $\boldsymbol{b}$ (see Sect. 3.5). The vector $\boldsymbol{u}_p$ fixes the phase φ_p of the incoming projectile ion. Here $v_{i\|}$, $v_{i\perp}$, $\Omega_c = |Z| e/M$ and $a_c = v_{i\perp}/\Omega_c$ are the longitudinal and transverse velocities as well as the cyclotron frequency and cyclotron radius of the projectile ion, respectively. Note that $\boldsymbol{v}_i^2(t) = v_{i\|}^2 + v_{i\perp}^2 = v_i^2$ is constant. Using (A.5) the SP (2.52) for a light ion yields

$$S = \frac{2iZ^2 e^2}{(2\pi)^2 v_i} \int d\boldsymbol{k} \frac{\boldsymbol{k} \cdot \boldsymbol{v}_i(t)}{k^2} \sum_{m,n=-\infty}^{\infty} \frac{J_n(k_\perp a_c) J_m(k_\perp a_c)}{\varepsilon(\boldsymbol{k}, n\Omega_c + k_\| v_{i\|})}$$
$$\times \exp\left\{ i(m-n) \left[\Omega_c t + c_Z(\theta - \varphi_p) \right] \right\} , \tag{2.56}$$

where $k_\|$ and $\boldsymbol{k}_\perp$ are the components of $\boldsymbol{k}$ along and perpendicular to the magnetic field, respectively. Here θ is the azimuthal angle of $\boldsymbol{k}_\perp$ and J_n are the Bessel functions of the first kind and of the nth order. From expression (2.56) is seen that the SP is periodic function of time due to projectile ion cyclotron motion. For practical applications expression (2.56) should be averaged over a cyclotron period $2\pi/\Omega_c$ of the ion. The result reads

$$\langle S \rangle = \frac{2Z^2 e^2}{(2\pi)^2 v_i} \sum_{n=-\infty}^{\infty} \int d\boldsymbol{k} \frac{k_\| v_{i\|} + n\Omega_c}{k^2} J_n^2(k_\perp a_c) \mathrm{Im} \frac{-1}{\varepsilon(\boldsymbol{k}, n\Omega_c + k_\| v_{i\|})} . \tag{2.57}$$

The expression (2.57) differs significantly from (2.53) because the SP $\langle S \rangle$ involves besides the cyclotron motion of the plasma particles in addition the cyclotron motion of the projectile ion. We discuss this point in details in Sect. 4.5.3.

In the case of a bare Coulomb interaction between the projectile ion and plasma particles the cut–off parameters $k_{\min} = 1/s_{\max}$ and $k_{\max} = 1/s_{\min}$ (where $s_{\min}$ is the effective minimum impact parameter) must be introduced in (2.53) and (2.57) to avoid the logarithmic divergence at small and large k. The divergence at large k corresponds to the incapability of the linearized Vlasov theory to treat close encounters between the projectile ion and the plasma particles properly, while $k_{\min}$ accounts for screening. Ignoring the collisions of the projectile with the plasma ions for $s_{\min}$ we use the effective minimum impact parameter excluding hard Coulomb collisions with a scattering angle larger than 90^0 (see Sect. 2.1 as well as the discussion below (2.6))

$$s_{\min} = \frac{|Z|e^2}{mv_r^2}, \qquad s_{\max} = \frac{v_r}{\omega_p}. \tag{2.58}$$

The cut–off $s_{\max}$ describes the dynamic screening at high relative electron-ion velocities v_r (see, e.g., [134] for more detail).

In the linear response (LR) treatment cut–offs and impact parameters are used which are averaged with respect to the electron velocity distribution function. With the averaged relative velocity $\langle v_r \rangle \simeq (v_i^2 + \bar{v}_{the}^2)^{1/2}$, they read

$$\langle s_{\min} \rangle = \frac{1}{k_{\max}} = \frac{|Z|e^2}{m(v_i^2 + \bar{v}_{the}^2)}, \qquad \langle s_{\max} \rangle = \frac{1}{k_{\min}} = \frac{(v_i^2 + \bar{v}_{the}^2)^{1/2}}{\omega_p}. \tag{2.59}$$

The cut–off parameters (2.58) and (2.59) are well known (see, e.g., [85, 107, 120, 134, 136, 138]) for stopping power calculations without magnetic field. In particular, the minimum impact parameter, $s_{\min}$, is provided by the Rutherford scattering formula [68] (see also (2.6) and (2.8)). However, in the presence of a magnetic field, the cut–off $s_{\min}$ must be deduced by a comparison of the LR and the full non-perturbative BC treatments. The cut–off $s_{\max}$ or its velocity averaged form in (2.59) incorporates the static and dynamic screening effects at low– and high–velocity limits, respectively. In particular, in high–velocity limit $s_{\max} \simeq \langle s_{\max} \rangle \simeq v_i/\omega_p$ which is precisely the wavelength of the plasma wakewaves (divided by 2π) excited by a swift projectile ion in unmagnetized plasma [9, 37, 50, 99, 104, 105]. It turns out that this dynamic screening length is essentially valid also for the case of magnetized plasma [89, 97, 113, 129], where, however, the magnetic field may play an important role in an intermediate regimes discussed, e.g., in [97] and Sect. 4.5.1.

Introducing the maximum impact parameter the SP (2.53) of the heavy projectile ion becomes

$$S = \frac{2Z^2 e^2 \lambda_{D\|}^2}{\pi^2} \int_0^{k_{\max}} k^3 \mathrm{d}k \int_0^1 \mathrm{d}\mu \int_0^{\pi} \mathrm{d}\phi \frac{\mathcal{F}\cos\Theta}{(k^2\lambda_{D\|}^2 + \mathcal{G})^2 + \mathcal{F}^2}, \tag{2.60}$$

where β is the angle between $\boldsymbol{k}$ and $\boldsymbol{b}$, $\mu = \cos\beta$, Θ is the angle between $\boldsymbol{k}$ and $\boldsymbol{v}_i$, $\zeta_v = \boldsymbol{k} \cdot \boldsymbol{v}_i / k v_{\mathrm{th}v\|} = (v_i/v_{\mathrm{th}v\|}) \cos\Theta$. In passing from (2.53) to (2.60) the $\boldsymbol{k}$–integral has been represented in a spherical coordinate system (k, β, ϕ) with the axial axis along $\boldsymbol{b}$. In addition we have made a transformation $\phi \to \phi + \pi$. Thus, in this system

$$\cos\Theta = \mu\cos\alpha - \sqrt{1-\mu^2}\sin\alpha\cos\phi, \tag{2.61}$$

where α is the angle between $\boldsymbol{v}_i$ and $\boldsymbol{b}$. The functions $\mathcal{G}$ and $\mathcal{F}$ were defined in (2.48).

Let us briefly discuss the question of the isotropisation of the plasma temperature during ion beam–plasma interaction. A two temperature description of the plasma is valid only when the ion beam–plasma interaction time is less than the relaxation time between the two temperatures, $T_{\nu\parallel}$ and $T_{\nu\perp}$. For an estimate we consider the field-free case, because the external magnetic field suppresses the relaxation between the transverse and longitudinal temperatures [66, 67] during the time of flight of the ion beam through plasma. The interaction between identical plasma particles, e.g. from the species ν, essentially accounts for the relaxation between $T_{\nu\parallel}$ and $T_{\nu\perp}$. In addition since the ions are heavier than the electrons we consider the relaxation time only for electrons, $\Delta\tau_{\rm rel} = \Delta\tau_{e,\rm rel} \ll \Delta\tau_{i,\rm rel}$.

The problem of a temperature relaxation in an anisotropic plasma with and without an external magnetic field was considered by Ichimaru and Rosenbluth [66] (see also [67]). Within the dominant–term approximation the relaxation time $\Delta\tau_{\rm rel}$ for the plasma without magnetic field is given by

$$\frac{1}{\Delta\tau_{\rm rel}} = \frac{8}{15}\sqrt{\frac{\pi}{m}}\frac{n_e e^4}{(k_{\rm B}T_{\rm eff})^{3/2}}\ln(N_D)\ , \tag{2.62}$$

where $\ln(N_D)$ is the Coulomb logarithm and the effective electron temperature $T_{\rm eff}$ is defined through

$$\begin{aligned}\frac{1}{T_{\rm eff}^{3/2}} &= \frac{15}{2}\int_0^1 \frac{\mu^2(1-\mu^2)\mathrm{d}\mu}{[\mu^2 T_\parallel + (1-\mu^2)T_\perp]^{3/2}} \\ &= \frac{5\sqrt{3}}{12\bar{T}^{3/2}}\frac{(1+2\tau)^{3/2}}{(\tau-1)^2}\left[\frac{\tau+2}{\sqrt{|\tau-1|}}p_0(\tau) - 3\right],\end{aligned} \tag{2.63}$$

$$p_0(\tau) = \begin{cases} \ln\frac{1+\sqrt{1-\tau}}{\sqrt{\tau}}, & \tau < 1 \\ \arctan\sqrt{\tau-1}, & \tau > 1 \end{cases} . \tag{2.64}$$

Here we drop the electronic index for simplicity. The relaxation time calculated from (2.62) are of the order of 10^{-6} s, 0.5×10^{-5} s and 10^{-3} s for averaged temperatures $\bar{T} = 10^{-2}$ eV, $\bar{T} = 0.1$ eV and $\bar{T} = 1$ eV, respectively, for anisotropies $\tau \simeq 0.01 - 100$. The interaction time (for instance, for inertial confinement fusion (ICF) or for electron cooling) is about $10^{-7} - 10^{-8}$ s. Therefore, ion beam–plasma interaction time can be very small compared to the plasma relaxation time.

In Chap. 4 we analyze the general expressions (2.53) and (2.57) for the SP in details. In (2.47) and (2.48) the susceptibility of the ions is neglected for the energy loss of the projectile ion in an electron plasma. In such case we drop the electronic index for the sake of simplicity assuming that all parameters refer to the electrons.

2.5 Particle-In-Cell (PIC) Simulations

Numerical solutions of (2.39) and (2.40) are unavoidable in the regime of nonlinear coupling $\mathcal{Z} \gtrsim 1$, that is, e.g. for highly charged, slow ions. A widely used and very efficient method for this task are particle simulations where the phase–space distribution is represented by a swarm of pseudo-particles. Here we focus on the case of a heavy projectile ($M \gg m$) moving with a constant velocity $\boldsymbol{v}_i$ through a magnetized electron plasma. Ions in the target plasma are only taken into account as a homogenous, neutralizing background, if appropriate. After transforming to the ion rest frame, i.e. to the relative positions and velocities, $\boldsymbol{r} = \boldsymbol{r}_e - \boldsymbol{r}_i = \boldsymbol{r}_e - \boldsymbol{v}_i t$, $\boldsymbol{v} = \boldsymbol{v}_e - \boldsymbol{v}_i$, see Sect. 3.2, the nonlinear Vlasov–Poisson equations (2.39), (2.40) describing the evolution of the transformed electronic phase–space density $f(\boldsymbol{r}, \boldsymbol{v}, t)$ (suppressing the index e), then take the form

$$\frac{\partial f}{\partial t} + \boldsymbol{v} \cdot \frac{\partial f}{\partial \boldsymbol{r}} - \Omega_e \left[(\boldsymbol{v} + \boldsymbol{v}_i) \times \boldsymbol{b}\right] \cdot \frac{\partial f}{\partial \boldsymbol{v}} + \frac{1}{m} \frac{\partial}{\partial \boldsymbol{r}} \left(\frac{Z\not{e}^2}{r} + e\phi_p \right) \cdot \frac{\partial f}{\partial \boldsymbol{v}} = 0 , \qquad (2.65)$$

together with the Poisson–equation for the polarization $\phi_p = \phi - \phi_i$,

$$\varepsilon_0 \nabla^2 \phi_p(\boldsymbol{r}, t) = e \left(\int f(\boldsymbol{r}, \boldsymbol{v}, t) \mathrm{d}\boldsymbol{v} - n_e \right) , \qquad (2.66)$$

where $e\phi_i = Z\not{e}^2/r$ is the potential of the ion and n_e is the electron density in absence of the ion. The phase–space density $f(\boldsymbol{r}, \boldsymbol{v}, t)$ for N electrons is here normalized as

$$\int f(\boldsymbol{r}, \boldsymbol{v}, t) \mathrm{d}\boldsymbol{v} \mathrm{d}\boldsymbol{r} = \int n(\boldsymbol{r}, t) \mathrm{d}\boldsymbol{r} = N , \qquad (2.67)$$

where $n(\boldsymbol{r}, t)$ is the electron density when the ion is present.

These coupled equations are now solved numerically by the particle–in–cell (PIC) technique [22, 60] representing the phase-space distribution of N real electrons through

$$f(\boldsymbol{r}, \boldsymbol{v}, t) = \frac{N}{N_\mathrm{T}} \sum_\alpha w(\boldsymbol{r} - \boldsymbol{r}_\alpha(t))\, w_v(\boldsymbol{v} - \boldsymbol{v}_\alpha(t)) \qquad (2.68)$$

by a swarm of N_T pseudo-particles with positions $\boldsymbol{r}_\alpha(t)$ and velocities $\boldsymbol{v}_\alpha(t)$, $\alpha = 1, ..., N_\mathrm{T}$. The pseudo-particles have a mass $m_\mathrm{T} = mN/N_\mathrm{T}$ and charge $e_\mathrm{T} = eN/N_\mathrm{T}$ which usually differ from the values for physical particles, but the physical charge to mass ratio $e_\mathrm{T}/m_\mathrm{T} = e/m$ is the same. The w and w_v in (2.68) are smooth and well concentrated, normalized packets in phase space. For a time evolution of $f(\boldsymbol{r}, \boldsymbol{v}, t)$ given by (2.68) according to (2.65) the pseudo-particles now follow the Newtonian equations of motion

$$\dot{\boldsymbol{r}}_\alpha = \boldsymbol{v}_\alpha , \quad \dot{\boldsymbol{v}}_\alpha + \Omega_e[\boldsymbol{v}_\alpha \times \boldsymbol{b}] = -\Omega_e[\boldsymbol{v}_i \times \boldsymbol{b}] + \frac{1}{m} \frac{\partial}{\partial \boldsymbol{r}} \left(\frac{Z\not{e}^2}{r} + e\phi_p \right) \qquad (2.69)$$

in the potential ϕ_p produced from their own charges and the field of the charged projectile. The propagation of the Vlasov equation thus depends linearly on the number of pseudo-particles. The Poisson equation (2.66) for the polarization ϕ_p is usually solved on a spatial grid, where the electronic charge density at a grid point $\boldsymbol{r_m}$ is given by equation (2.68) and the pseudo-particle positions $\{\boldsymbol{r}_\alpha\}$ through

$$\rho_m = \rho_e(\boldsymbol{r_m}, t) = e\frac{N}{N_\mathrm{T}} \sum_\alpha w(\boldsymbol{r_m} - \boldsymbol{r}_\alpha(t)) . \tag{2.70}$$

There are various methods available for solving (2.66). A particularly efficient one employs fast Fourier transformation [108] to obtain the electrostatic potential for a given charge density. The number of pseudo-particles must be adapted to the spatial grid to achieve a sufficiently smooth representation of the electron density on the grid. Typically, a few pseudo-particles per grid cell are sufficient. Their number can be enhanced, if necessary, to suppress the amount of unavoidable fluctuations or noise. Two basically different scenarios are conceivable. In the one case, one pseudo–particle stands as a "macro particle" for many physical electrons. This is the genuine PIC scheme. In the other case, one physical electron is represented by many pseudo-particles. This is the regime of test–particle simulations. The choice of the appropriate scenario essentially depends on the involved physical length scales and the required spatial resolution. This determines the grid and, in turn, the required number of pseudo-particles and the numerical expense. For the actual case of energy loss calculations a spatial resolution of the order of some fraction of the screening length is needed, which typically enforces the test–particle scheme. For a given, affordable number of test–particles the required resolution at small distances limits, however, the size of the simulation box and sets an upper limit on the wave length of the dynamic response of the plasma. Thus a high spatial resolution runs into a conflict with the dynamic response at long wavelengths, like the excitation of plasma waves and the dynamic screening, which yields important contributions to energy loss at high projectile velocities. This usually limits energy loss calculations to a low velocity range up to $v_i \approx 10 v_\mathrm{th}$.

In the actual numerical treatment the electron density $\{\rho_{\boldsymbol{m}}\}$ (2.70) is sampled on a 3D spatial grid from the pseudo–particles contained in a cubic simulation box taking for the function w a product of triangular shaped weight functions in each dimension. For a detailed discussion about commonly used weight functions and other possible shapes and frequently chosen w, see e.g. [60]. The Poisson–equation is solved in Fourier space, i.e. by $\varepsilon_0 \phi_p(\boldsymbol{k}) = e^2 \rho_e(\boldsymbol{k})/k^2$, using fast Fourier transformation to switch forth and back simultaneously with the particle propagation in every time step. The forces on the particles are retrieved from the grid using the same weight functions w, see again [60] for details. As only the calculation of the electronic contribution $\nabla\phi_p$ to the force on the pseudo–particles involves the grid, no restriction of the resolution at small distances occurs in the ion–pseudo–particle interaction. The particle propagation given by (2.69) is performed using the same techniques and algorithms as described and discussed for propagating the electron trajectories in the CTMC treatment outlined in Sect. 2.3. The CTMC scheme and the PIC method just differ by the dynamic polarization contribution related to the Poisson equation (2.66). Replacing $Ze^2/r + e\phi_p(\boldsymbol{r})$ with a static ion–electron interaction $U(r)$, the phase–space dynamics and particle propagation given by (2.65) and

(2.69) become identical to the CTMC treatment with a subsequent average over the electron distribution. In the PIC scheme the energy loss is also derived from the velocity change as in the CTMC method, now with respect to the velocity of a pseudo-particle when it enters and leaves the simulation box. As an alternative and for a consistency check also the averaged force on the ion, which is calculated anyway for the equations of motion (2.69), is recorded.

For the case of unmagnetized electrons more details on the test–particle simulation technique and some subtleties, e.g. for treating problems arising from the Coulomb singularity, and a comprehensive discussion of results for the energy loss are reported in [134, 136, 139]. The more recent simulations of the energy loss in the presence of a magnetic field essentially employ the same numerical techniques, but larger spatial grids (up to 72×72×72) and up to 2.4×10^6 pseudo-particles. In contrast to earlier simulations no periodic boundary conditions for the pseudo-particle positions are used. Instead particles leaving the simulation box are replaced by new incoming particles generated at the side walls of the box with a velocity randomly sampled from the distribution of the incoming particle current $\boldsymbol{j} = n_e \boldsymbol{v} f_0(\boldsymbol{v})$ associated with the desired thermal distribution of the electrons $f_0(\boldsymbol{v})$, e.g. (2.43). Some results of this PIC/test–particle simulations will be given at the end of Sect. 4.6.

3 Binary Collision Model

3.1 Introductory Remarks

In the presence of an external magnetic field $\boldsymbol{B}$ even the nonrelativistic problem of two charged particles cannot be solved in a closed form as the relative motion and the motion of the center of mass are coupled to each other. There exists no closed solution of this problem that is uniformly valid for any strength of the magnetic field and the Coulomb force between the particles. The classical limit of a hydrogen or Rydberg atom in a strong magnetic field also falls in this category (see, e.g., [56] and references therein) but in contrast to the free-free transitions (scattering) the total energy is negative there.

Numerical calculations have been performed for binary collisions (BC) between magnetized electrons [47, 116] and for collisions between magnetized electrons and ions [135, 136, 138]. While the total energy W of the particles interacting in a magnetic field is a constant of motion, the relative and center of mass energies are not conserved separately. In addition, the presence of the magnetic field breaks the rotational symmetry of the system and as a consequence only the component of the angular momentum $\boldsymbol{L}$ parallel to the magnetic field $L_{\|}$ is a constant of motion. A different situation arises for the BC between an electron and uniformly moving heavy ion. As an ion is much heavier than an electron, its uniform motion is only weakly perturbed by collisions with the electrons and the magnetic field. In this case $L_{\|}$ is not conserved but there exists a conserved generalized energy K [124, 138] involving the energy of relative motion and a magnetic term. The apparently simple problem of charged particle interaction in a magnetic field is in fact a problem of considerable complexity as the cyclotron orbital motion can produce a chaotic dynamics [54, 64, 114]. Some examples of the chaotic scattering events are discussed in Sect. 3.7 using Classical-Trajectory-Monte-Carlo (CTMC) simulations.

In this Chapter we consider the BC between electrons and heavy ions treating the interaction with the ion as a perturbation to the helical motion of the magnetized electrons. This has been done previously in first order of the ion charge Z and for an ion at rest [46]. For applications in plasma physics (e.g., for calculation of the ion energy loss in a magnetized plasma) this is somewhat unsatisfactory as one has to calculate the angular averaged energy transfer which vanishes in first–order perturbation theory due to symmetry reasons. The ion energy change receives contribution only from higher orders, see Sect. 3.5. Indeed, the transport phenomena are of order $O(Z^2)$ in the ion charge. Here, the second–order energy transfer is calculated with the help of an improved BC treatment which is uniformly valid for any strength of the magnetic field and does not require an early specification of the interaction potential [93]. This allows a physically transparent introduction of cut–offs at small and large distances.

3.2 Equations of Motion

We consider two point charges, an electron with mass m and charge $-e$, and an ion with mass M and charge Ze moving in a homogeneous magnetic field $\boldsymbol{B} = B\boldsymbol{b}$. These particles interact with a potential $U(\boldsymbol{r})$ where $\boldsymbol{r}$ is the relative coordinate of the colliding particles. The bare Coulomb interaction is $U_C(\boldsymbol{r}) = -Z\not{e}^2 u_C(r)$ with

$$u_C(r) = \frac{1}{r}\,, \tag{3.1}$$

where $\not{e}^2 = e^2/(4\pi\varepsilon_0)$ with the permittivity of the vacuum ε_0. In plasma applications the infinite range of this potential is modified by screening, e.g. $U_D(\boldsymbol{r}) = -Z\not{e}^2 u_D(r)$ with

$$u_D(r) = u_C(r)\,\mathrm{e}^{-r/\lambda} = \frac{1}{r}\,\mathrm{e}^{-r/\lambda}\,, \tag{3.2}$$

with a screening length λ which can be chosen as the Debye screening length λ_D, see, for example [4, Chap. 1.1]. The quantum uncertainty principle prevents particles from falling into the center of these potentials. In a classical picture this can be achieved by regularization at the origin, for example [36, 69] $U_R(\boldsymbol{r}) = -Z\not{e}^2 u_R(r)$ with

$$u_R(r) = \frac{1 - \mathrm{e}^{-r/\lambda\!\!\!^-}}{r}\,\mathrm{e}^{-r/\lambda}\,, \tag{3.3}$$

where $\lambda\!\!\!^-$ is a parameter, which may be related to the de Broglie wavelength. For later purposes we note the Fourier transforms

$$u(\boldsymbol{k}) = \frac{1}{(2\pi)^3}\int \mathrm{d}\boldsymbol{r}\,\mathrm{e}^{-\mathrm{i}\boldsymbol{k}\cdot\boldsymbol{r}}u(\boldsymbol{r}) \tag{3.4}$$

of these potentials

$$u_C(\boldsymbol{k}) = \frac{1}{2\pi^2 k^2}\,, \tag{3.5}$$

$$u_D(\boldsymbol{k}) = \frac{1}{2\pi^2(k^2 + \lambda^{-2})}\,, \tag{3.6}$$

$$u_R(\boldsymbol{k}) = \frac{1}{2\pi^2}\left(\frac{1}{k^2 + \lambda^{-2}} - \frac{1}{k^2 + \lambda'^{-2}}\right) \tag{3.7}$$

with $\lambda'^{-1} = \lambda^{-1} + \lambda\!\!\!^-{}^{-1}$. Introducing the vector potential $\boldsymbol{A}$ through $\boldsymbol{B} = \nabla \times \boldsymbol{A}$ the Lagrangian for the electron–ion system is

$$L = \frac{m}{2}\,\boldsymbol{v}_e^2 + \frac{M}{2}\,\boldsymbol{v}_i^2 - e\boldsymbol{A}(\boldsymbol{r}_e)\cdot\boldsymbol{v}_e + Ze\boldsymbol{A}(\boldsymbol{r}_i)\cdot\boldsymbol{v}_i - U(\boldsymbol{r})\,. \tag{3.8}$$

Here $\boldsymbol{r}_e, \boldsymbol{r}_i$ and $\boldsymbol{v}_e, \boldsymbol{v}_i$ are the position and velocity vectors of the electron and the ion, respectively. This Lagrangian cannot in general be separated into parts describing the relative motion and the motion of the center of mass (CM). With

$$\boldsymbol{R} = \frac{m\boldsymbol{r}_e + M\boldsymbol{r}_i}{m+M}, \quad \boldsymbol{V} = \frac{m\boldsymbol{v}_e + M\boldsymbol{v}_i}{m+M}, \qquad (3.9)$$
$$\boldsymbol{r} = \boldsymbol{r}_e - \boldsymbol{r}_i, \quad \boldsymbol{v} = \boldsymbol{v}_e - \boldsymbol{v}_i,$$

the reduced mass $1/\mu = 1/m + 1/M$ and $\boldsymbol{A}(\boldsymbol{r}) = (\boldsymbol{B} \times \boldsymbol{r})/2$ the Langrangian is

$$L = \frac{m+M}{2}\boldsymbol{V}^2 + \frac{\mu}{2}\boldsymbol{v}^2 - U(\boldsymbol{r}) + \frac{Z-1}{2}e\,(\boldsymbol{B} \times \boldsymbol{R})\boldsymbol{V} \qquad (3.10)$$
$$+ \frac{\mu^2}{2}\left(\frac{Ze}{M^2} - \frac{e}{m^2}\right)(\boldsymbol{B} \times \boldsymbol{r}) \cdot \boldsymbol{v} - \frac{\mu}{2}\left(\frac{Ze}{M} + \frac{e}{m}\right)[(\boldsymbol{B} \times \boldsymbol{R}) \cdot \boldsymbol{v} + (\boldsymbol{B} \times \boldsymbol{r}) \cdot \boldsymbol{V}] .$$

The coupled equations of motion are

$$\dot{\boldsymbol{v}}(t) + \Omega_4(\boldsymbol{v}(t) \times \boldsymbol{b}) = -\Omega_3(\boldsymbol{V} \times \boldsymbol{b}) + \frac{1}{\mu}\boldsymbol{F}(\boldsymbol{r}(t)) , \qquad (3.11)$$
$$\dot{\boldsymbol{V}}(t) - \Omega_1(\boldsymbol{V}(t) \times \boldsymbol{b}) = -\Omega_2(\boldsymbol{v}(t) \times \boldsymbol{b}) , \qquad (3.12)$$

where $\boldsymbol{F}(\boldsymbol{r}) = -\partial U/\partial \boldsymbol{r}$ is the force exerted by the ion on the electron. The frequencies $\Omega_i, i = 1 \ldots 4$ are expressed in terms of the electron cyclotron frequency $\Omega_e = eB/m$

$$\Omega_1 = \frac{m(Z-1)}{M+m}\Omega_e , \quad \Omega_2 = \frac{m(M+Zm)}{(M+m)^2}\Omega_e , \qquad (3.13)$$
$$\Omega_3 = \left(1 + \frac{Zm}{M}\right)\Omega_e , \quad \Omega_4 = \frac{\mu}{m}\left(1 - \frac{Zm^2}{M^2}\right)\Omega_e . \qquad (3.14)$$

As the relative and the CM motion are coupled, their energies are not separately conserved but the total energy

$$W = \frac{M+m}{2}\boldsymbol{V}^2 + \frac{\mu}{2}\boldsymbol{v}^2 + U(\boldsymbol{r}) = \text{const} \qquad (3.15)$$

is a constant of motion. For interactions between electrons and heavy ions the equations of motion can be simplified using $M \gg m$. In this limit $\mu \to m$, $\Omega_1, \Omega_2 \to 0$ and $\Omega_3, \Omega_4 \to \Omega_e$. Then (3.12) leads to $\boldsymbol{V} \to \boldsymbol{v}_i$ =const. and the relative motion is described by

$$\dot{\boldsymbol{v}}(t) + \Omega_e(\boldsymbol{v}(t) \times \boldsymbol{b}) = -\Omega_e(\boldsymbol{v}_i \times \boldsymbol{b}) + \frac{1}{m}F(\boldsymbol{r}(t)) . \qquad (3.16)$$

Dropping unimportant constants the Lagrangian (3.10) becomes in this limit a Lagrangian for the relative motion in which the uniformly moving ion acts as a time–dependent perturbation to the electron motion

$$L = \frac{m}{2}\boldsymbol{v}^2 - U(\boldsymbol{r}) - \frac{e}{2}(\boldsymbol{B} \times \boldsymbol{r}) \cdot \boldsymbol{v} - \frac{e}{2}((\boldsymbol{B} \times \boldsymbol{v}_i t) \cdot \boldsymbol{v} + (\boldsymbol{B} \times \boldsymbol{r}) \cdot \boldsymbol{v}_i) . \qquad (3.17)$$

Because of the explicit time dependence it is not the energy $E = \frac{m}{2}v^2 + U(r)$ of relative motion that is conserved, but rather the quantity

$$K(\boldsymbol{v}, \boldsymbol{r}) = \frac{m}{2}\,\boldsymbol{v}^2 + U(\boldsymbol{r}) + m\Omega_e \boldsymbol{r} \cdot (\boldsymbol{v}_i \times \boldsymbol{b})\,, \tag{3.18}$$

which can easily be proved by inserting (3.16) into the total time derivative of K. In contrast to the unmagnetized case there is a transfer of relative energy in an electron–ion collision which is proportional to $\delta r_\perp v_{i\perp}$, where $\delta r_\perp$ and $v_{i\perp}$ are the components of the relative coordinate transfer and the ion velocity transverse to the magnetic field, respectively.

3.3 Energy Loss and Velocity Transfer

The rate at which the energy of an ion in a collision with an electron at time t changes is given by

$$\frac{\mathrm{d}E_i(t)}{\mathrm{d}t} = -\boldsymbol{v}_i \cdot \boldsymbol{F}(\boldsymbol{r}(t))\,, \tag{3.19}$$

as $-\boldsymbol{F}$ is the force exerted by the electron on the ion. Integration with respect to time yields the energy transfer itself

$$\delta E_i(t) = -\int_{-\infty}^{t} \mathrm{d}\tau \boldsymbol{v}_i \cdot \boldsymbol{F}(\boldsymbol{r}(\tau))\,, \tag{3.20}$$

which after completion of the collision becomes

$$\Delta E_i = \delta E_i(t \to \infty) = -\int_{-\infty}^{\infty} \mathrm{d}\tau \boldsymbol{v}_i \cdot \boldsymbol{F}(\boldsymbol{r}(\tau))\,. \tag{3.21}$$

Here one has to integrate the equations of motion for the relative trajectories. The limit $M \gg m$ leading to (3.13) implies that the change in the ion energy is calculated under the assumption of a constant ion velocity. Alternatively this energy transfer can be expressed by the velocity transferred to the electrons during the collision. For that purpose we substitute $\boldsymbol{v}_i = \boldsymbol{v}_e(t) - \boldsymbol{v}(t)$ into (3.21) and split the electron velocity into two terms $\boldsymbol{v}_e(t) = \boldsymbol{v}_{e0}(t) + \delta\boldsymbol{v}(t)$, where $\boldsymbol{v}_{e0}(t)$ describes the helical motion in the magnetic field

$$\dot{\boldsymbol{v}}_{e0} + \Omega_e(\boldsymbol{v}_{e0} \times \boldsymbol{b}) = 0 \tag{3.22}$$

and $\delta\boldsymbol{v}(t)$ the velocity transfer due to the collision with the ion

$$\delta\dot{\boldsymbol{v}} + \Omega_e(\delta\boldsymbol{v} \times \boldsymbol{b}) = \frac{1}{m}\boldsymbol{F}(\boldsymbol{r}(t))\,. \tag{3.23}$$

This yields

$$\delta E_i(t) = -\int_{-\infty}^{t} \mathrm{d}\tau \boldsymbol{v}_{e0}(\tau) \cdot \boldsymbol{F}(\boldsymbol{r}(\tau)) - \int_{-\infty}^{t} \mathrm{d}\tau \delta\boldsymbol{v}(\tau) \cdot \boldsymbol{F}(\boldsymbol{r}(\tau)) - U(\boldsymbol{r}(t))\,. \tag{3.24}$$

The time integrals can be done with the help of the derivative of the scalar product $\boldsymbol{v}_{e0}(t) \cdot \delta\boldsymbol{v}(t)$. Using the equations of motion (3.22) and (3.23) we obtain

$$\frac{\mathrm{d}}{\mathrm{d}t}\left(\boldsymbol{v}_{e0}(t) \cdot \delta\boldsymbol{v}(t)\right) = \frac{1}{m}\boldsymbol{v}_{e0}(t) \cdot \boldsymbol{F}(\boldsymbol{r}(t)) \,, \tag{3.25}$$

which yields

$$\int_{-\infty}^{t} \mathrm{d}\tau \boldsymbol{v}_{e0}(\tau) \cdot \boldsymbol{F}(\boldsymbol{r}(\tau)) = m\boldsymbol{v}_{e0}(t) \cdot \delta\boldsymbol{v}(t) \,. \tag{3.26}$$

Similarly from (3.23)

$$\frac{\mathrm{d}}{\mathrm{d}t}(\delta\boldsymbol{v}(t))^2 = \frac{2}{m}\delta\boldsymbol{v}(t) \cdot \boldsymbol{F}(\boldsymbol{r}(t)) \tag{3.27}$$

which yields

$$\int_{-\infty}^{t} \mathrm{d}\tau \delta\boldsymbol{v}(\tau) \cdot \boldsymbol{F}(\boldsymbol{r}(\tau)) = \frac{m}{2}[\delta\boldsymbol{v}(t)]^2 \,. \tag{3.28}$$

Thus

$$\delta E_i(t) = -U(\boldsymbol{r}(t)) - m\boldsymbol{v}_{e0}(t) \cdot \delta\boldsymbol{v}(t) - \frac{m}{2}[\delta\boldsymbol{v}(t)]^2 \,. \tag{3.29}$$

The last two terms in this equation represent the change in the electron energy due to the collision

$$\delta E_e(t) = \frac{m}{2}\left((\boldsymbol{v}_{e0}(t) + \delta\boldsymbol{v}(t))^2 - \boldsymbol{v}_{e0}(t)\right) = m\boldsymbol{v}_{e0}(t) \cdot \delta\boldsymbol{v}(t) + \frac{m}{2}[\delta\boldsymbol{v}(t)]^2 \,. \tag{3.30}$$

This shows energy conservation

$$\delta E_i(t) + \delta E_e(t) + U(\boldsymbol{r}(t)) = 0 \,. \tag{3.31}$$

As $U(\boldsymbol{r}(t \to \infty)) = 0$ the energy change of the ion can also be calculated from the velocity transfer $\Delta\boldsymbol{v} = \delta\boldsymbol{v}(t \to \infty)$ with the help of

$$\Delta E_i = -\delta E_e(t \to \infty) = -m\left(\boldsymbol{v}_{e0}(t) \cdot \Delta\boldsymbol{v} + \frac{1}{2}(\Delta\boldsymbol{v})^2\right) \,. \tag{3.32}$$

This method has been adopted in [124]. In this approach the potential $U(\boldsymbol{r})$ has to be specified at an early stage. In Sect. 3.4 we will show that (3.21) allows for a more general formulation in which the cut–off at large distances (3.2) and the regularization at small distances (3.3) suggested by screening and quantum diffraction, respectively, can be treated easily.

3.4 General Interactions, no Magnetic Field

In the previous perturbation theory for binary collisions explicit use was made of the Coulomb interaction. We now seek a more general formulation. As before we start from the expansions

$$\boldsymbol{r}(t) = \boldsymbol{r}_0(t) + \boldsymbol{r}_1(t) + \boldsymbol{r}_2(t) + \dots\,, \tag{3.33}$$

$$\boldsymbol{v}(t) = \boldsymbol{v}_0(t) + \boldsymbol{v}_1(t) + \boldsymbol{v}_2(t) + \dots\,, \tag{3.34}$$

$$\boldsymbol{F}\left(\boldsymbol{r}(t)\right) = \boldsymbol{F}_0\left(\boldsymbol{r}_0(t)\right) + \boldsymbol{F}_1\left(\boldsymbol{r}_0(t), \boldsymbol{r}_1(t)\right) + \dots \tag{3.35}$$

with

$$\boldsymbol{F}_1\left(\boldsymbol{r}_0(t), \boldsymbol{r}_1(t)\right) = \left(\boldsymbol{r}_1(t) \cdot \frac{\partial}{\partial \boldsymbol{r}}\right) \boldsymbol{F}(\boldsymbol{r})\Big|_{\boldsymbol{r}=\boldsymbol{r}_0(t)}\,. \tag{3.36}$$

It turns out to be useful to Fourier transform with respect to space

$$\boldsymbol{F}_0\left(\boldsymbol{r}_0(t)\right) = -\mathrm{i} \int \mathrm{d}\boldsymbol{k} U(\boldsymbol{k}) \boldsymbol{k} \exp\left(\mathrm{i}\boldsymbol{k} \cdot \boldsymbol{r}_0(t)\right)\,, \tag{3.37}$$

$$\boldsymbol{F}_1\left(\boldsymbol{r}_0(t), \boldsymbol{r}_1(t)\right) = \int \mathrm{d}\boldsymbol{k} U(\boldsymbol{k}) \boldsymbol{k} \left(\boldsymbol{k} \cdot \boldsymbol{r}_1(t)\right) \exp(\mathrm{i}\boldsymbol{k} \cdot \boldsymbol{r}_0(t))\,. \tag{3.38}$$

For pedagogical purposes we treat the case without magnetic field first. This will allow to calculate the energy transfer ΔE_i for potentials which are more general than Coulomb. In particular the use of screened and regularized interaction potentials like (3.2) and (3.3) obviates cut–offs when integrating with respect to the impact parameter. As before the zero–order motion is described by the linear trajectory

$$\boldsymbol{r}_0(t) = \boldsymbol{R}_0 + \boldsymbol{v}_r t = \boldsymbol{s} + \boldsymbol{v}_r(t + t_0) \tag{3.39}$$

with the constant relative velocity $\boldsymbol{v}_r$ and the impact parameter $\boldsymbol{s} \perp \boldsymbol{v}_r$, see Fig. 2.3. The first–order energy transfer is

$$\Delta E_i^{(1)} = \mathrm{i} \int \mathrm{d}\boldsymbol{k} U(\boldsymbol{k}) \left(\boldsymbol{k} \cdot \boldsymbol{v}_i\right) \int_{-\infty}^{\infty} \mathrm{d}t\, \mathrm{e}^{\mathrm{i}\boldsymbol{k} \cdot \boldsymbol{r}_0(t)} = 2\pi\mathrm{i} \int \mathrm{d}\boldsymbol{k} U(\boldsymbol{k}) \left(\boldsymbol{k} \cdot \boldsymbol{v}_i\right) \mathrm{e}^{\mathrm{i}\boldsymbol{k} \cdot \boldsymbol{R}_0} \delta\left(\boldsymbol{k} \cdot \boldsymbol{v}_r\right)\,. \tag{3.40}$$

Now the δ-function enforces $\boldsymbol{k}$ to lie in the plane transverse to $\boldsymbol{v}_r$. Thus

$$\begin{aligned} \Delta E_i^{(1)} &= 2\pi\mathrm{i} \int \mathrm{d}\boldsymbol{k} U(\boldsymbol{k}) \left(\boldsymbol{k} \cdot \boldsymbol{v}_i\right) \mathrm{e}^{\mathrm{i}ks \sin\vartheta_s} \delta\left(\boldsymbol{k} \cdot \boldsymbol{v}_r\right) \\ &= 2\pi\mathrm{i} \int \mathrm{d}\boldsymbol{k} U(\boldsymbol{k}) \left(\boldsymbol{k} \cdot \boldsymbol{v}_i\right) \delta\left(\boldsymbol{k} \cdot \boldsymbol{v}_r\right) \sum_{n=-\infty}^{\infty} J_n(ks) \mathrm{e}^{\mathrm{i}n\vartheta_s}\,, \end{aligned} \tag{3.41}$$

where the exponential has been expanded into Bessel functions [1, Chap. 9] (see also (A.5)). After averaging with respect to ϑ_s there remains an odd integrand as $U(\boldsymbol{k})$ is even for real interactions. Therefore $\langle \Delta E_i^{(1)} \rangle_{\hat{s}} = 0$.

With the first–order trajectory correction (2.17)

$$\begin{aligned} \boldsymbol{r}_1(t) &= \frac{1}{m} \int_{-\infty}^{t} \mathrm{d}t' \left(t - t'\right) \boldsymbol{F}_0\left(\boldsymbol{r}_0(t')\right) \\ &= -\frac{\mathrm{i}}{m} \int \mathrm{d}\boldsymbol{k} U(\boldsymbol{k}) \boldsymbol{k} \int_{-\infty}^{t} \mathrm{d}t' \left(t - t'\right) \mathrm{e}^{\mathrm{i}\boldsymbol{k} \cdot \boldsymbol{r}_0(t')} \end{aligned} \tag{3.42}$$

the second–order energy transfer is

$$\begin{aligned}\Delta E_i^{(2)} &= -\boldsymbol{v}_i \cdot \int_{-\infty}^{\infty} \mathrm{d}t \boldsymbol{F}_1\left(\boldsymbol{r}_0(t), \boldsymbol{r}_1(t)\right) = \\ &= -\int_{-\infty}^{\infty} \mathrm{d}t \int \mathrm{d}\boldsymbol{k} U(\boldsymbol{k})\,(\boldsymbol{k}\cdot\boldsymbol{v}_i)\,(\boldsymbol{k}\cdot\boldsymbol{r}_1(t))\,\mathrm{e}^{\mathrm{i}\boldsymbol{k}\cdot\boldsymbol{r}_0(t))} \\ &= \frac{\mathrm{i}}{m}\int \mathrm{d}\boldsymbol{k} \int \mathrm{d}\boldsymbol{k}' U(\boldsymbol{k})U(\boldsymbol{k}')\,(\boldsymbol{k}\cdot\boldsymbol{v}_i)\,(\boldsymbol{k}\cdot\boldsymbol{k}')\,\mathrm{e}^{\mathrm{i}(\boldsymbol{k}+\boldsymbol{k}')\cdot\boldsymbol{s}} \\ &\times \int_{-\infty}^{\infty} \mathrm{d}t\,\mathrm{e}^{\mathrm{i}\boldsymbol{k}\cdot\boldsymbol{v}_r(t+t_0)} \int_{-\infty}^{t} \mathrm{d}t'(t-t')\mathrm{e}^{\mathrm{i}\boldsymbol{k}'\cdot\boldsymbol{v}_r(t'+t_0)}\,.\end{aligned} \tag{3.43}$$

For real potentials symmetrization shows that $\Delta E_i^{(2)}$ is real. The time integrals can be expressed as

$$\int_{-\infty}^{\infty} \mathrm{d}\tau\,\mathrm{e}^{\mathrm{i}(\omega+\omega')\tau} \int_0^{\infty} \mathrm{d}\tau\tau'\,\mathrm{e}^{-\mathrm{i}\omega'\tau'} = 2\pi\,\delta(\omega+\omega')\frac{\partial}{\partial\omega'}\,\frac{1}{\omega'-\mathrm{i}0}\,. \tag{3.44}$$

Here and in the following we introduce cylindrical coordinates oriented along the relative velocity $\boldsymbol{n}_r = \boldsymbol{v}_r/v_r$, i.e.

$$\boldsymbol{C} = C_{\parallel}^{(r)}\boldsymbol{n}_r + \boldsymbol{C}_{\perp}^{(r)}\,. \tag{3.45}$$

for any vector $\boldsymbol{C}$. In this coordinate system we have

$$\mathrm{e}^{\mathrm{i}(\boldsymbol{k}+\boldsymbol{k}')\cdot\boldsymbol{s}} = \mathrm{e}^{\mathrm{i}(\boldsymbol{k}_{\perp}^{(r)}+\boldsymbol{k'}_{\perp}^{(r)})\cdot\boldsymbol{s}} = \exp\left(\mathrm{i}s\left|\boldsymbol{k}_{\perp}^{(r)}+\boldsymbol{k'}_{\perp}^{(r)}\right|\sin\vartheta_s\right)\,. \tag{3.46}$$

Expanding as above in (3.41) and averaging with respect to ϑ_s yields

$$\begin{aligned}\left\langle\Delta E_i^{(2)}\right\rangle_{\hat{s}} &= -\frac{2\pi^2}{mv_r^3}\int \mathrm{d}\boldsymbol{k}\int \mathrm{d}\boldsymbol{k}' U(\boldsymbol{k})U(\boldsymbol{k}')J_0\left(s\left|\boldsymbol{k}_{\perp}^{(r)}+\boldsymbol{k'}_{\perp}^{(r)}\right|\right) \\ &\times (\boldsymbol{k}\cdot\boldsymbol{v}_i)(\boldsymbol{k}\cdot\boldsymbol{k}')\delta\left(\boldsymbol{k}_{\parallel}^{(r)}+\boldsymbol{k'}_{\parallel}^{(r)}\right)\delta'\left(\boldsymbol{k'}_{\parallel}^{(r)}\right)\,,\end{aligned} \tag{3.47}$$

where we have used

$$\frac{1}{\omega'-\mathrm{i}0} = \frac{P}{\omega'} + \pi\mathrm{i}\delta(\omega') \tag{3.48}$$

and the reality of $\Delta E_i^{(2)}$. The integrals with respect to the parallel components are done with the help of

$$\begin{aligned}&\int \mathrm{d}x \int \mathrm{d}x' f(x,x')\delta(x+x')\delta'(x'-x_0) \\ &= -\left.\frac{\partial f(-x_0,x')}{\partial x'}\right|_{x'=x_0} + \left.\frac{\partial f(x,x_0)}{\partial x}\right|_{x=-x_0}\,.\end{aligned} \tag{3.49}$$

The result is

$$\begin{aligned}\left\langle\Delta E_i^{(2)}\right\rangle_{\hat{s}} &= -\frac{2\pi^2}{mv_r^3}\int \mathrm{d}^2\boldsymbol{k}_{\perp}^{(r)}\int \mathrm{d}^2\boldsymbol{k}_{\perp}^{\prime(r)} J_0\left(s\left|\boldsymbol{k}_{\perp}^{(r)}+\boldsymbol{k'}_{\perp}^{(r)}\right|\right) \\ \times\Big[&U(0,\boldsymbol{k}_{\perp}^{(r)})U(0,\boldsymbol{k}_{\perp}^{\prime(r)})(\boldsymbol{v}_i\cdot\boldsymbol{n}_r)\left(\boldsymbol{k}_{\perp}^{(r)}\cdot\boldsymbol{k}_{\perp}^{\prime(r)}\right) + \left(\boldsymbol{v}_{i\perp}^{(r)}\cdot\boldsymbol{k}_{\perp}^{(r)}\right)\left(\boldsymbol{k}_{\perp}^{(r)}\cdot\boldsymbol{k}_{\perp}^{\prime(r)}\right) \\ &\times\left(\left(\frac{\partial U(\boldsymbol{k})}{\partial k_{\parallel}^{(r)}}\right)_{k_{\parallel}^{(r)}=0} U(0,\boldsymbol{k}_{\perp}^{\prime(r)}) - U(0,\boldsymbol{k}_{\perp}^{(r)})\left(\frac{\partial U(\boldsymbol{k}')}{\partial k_{\parallel}^{\prime(r)}}\right)_{k_{\parallel}^{\prime(r)}=0}\right)\Big]\,.\end{aligned} \tag{3.50}$$

In the transversal plane we have

$$\begin{aligned} \boldsymbol{k}_{\perp}^{(r)} \cdot \boldsymbol{v}_{i\perp}^{(r)} &= k_{\perp}^{(r)} v_{i\perp}^{(r)} \cos\vartheta \,, \\ \boldsymbol{k}_{\perp}^{\prime(r)} \cdot \boldsymbol{v}_{i\perp}^{(r)} &= k_{\perp}^{\prime(r)} v_{i\perp}^{(r)} \cos\vartheta' \,, \\ \boldsymbol{k}_{\perp}^{(r)} \cdot \boldsymbol{k}_{\perp}^{\prime(r)} &= k_{\perp}^{(r)} k_{\perp}^{\prime(r)} \cos(\vartheta - \vartheta') \,, \end{aligned} \tag{3.51}$$

where ϑ and ϑ' are azimuthal angles of $\boldsymbol{k}_{\perp}^{(r)}$ and $\boldsymbol{k}'^{(r)}_{\perp}$, respectively. For axially symmetric potentials the integration with respect to these angles can be done after using the addition theorem [1]

$$J_0\left(s\left|\boldsymbol{k}_{\perp}^{(r)} + \boldsymbol{k}'^{(r)}_{\perp}\right|\right) = J_0\left(k_{\perp}^{(r)} s\right) J_0\left(k_{\perp}^{'(r)} s\right) \tag{3.52}$$

$$+2 \sum_{n=1}^{\infty} J_n\left(k_{\perp}^{(r)} s\right) J_n\left(k_{\perp}^{'(r)} s\right) \cos\left(n(\pi - \vartheta + \vartheta')\right) \,.$$

Inspection shows that all terms involving derivatives of the potential yield odd powers of $\cos\vartheta$ or $\cos\vartheta'$ and vanish upon integration. There remains

$$\left\langle \Delta E_i^{(2)} \right\rangle_{\hat{s}} = \frac{8\pi^4}{m v_r^4} (\boldsymbol{v}_r \cdot \boldsymbol{v}_i) T_{1,2}^2(0, s) \,, \tag{3.53}$$

where the transforms

$$T_{\mu,\nu}(k_{\|}, s) = \int_0^{\infty} \mathrm{d}k_{\perp} k_{\perp}^{\nu} U(k_{\|}, k_{\perp}) J_{\mu}(k_{\perp} s) \tag{3.54}$$

have been introduced. For the potentials (3.5)-(3.7) further progress is achieved with the help of the integrals [52]

$$\int_0^{\infty} \mathrm{d}x \frac{x^{\nu+1} J_{\nu}(ax)}{x^2 + b^2} = b^{\nu} K_{\nu}(ab) \,. \tag{3.55}$$

There results in the screened case (3.6)

$$T_{1,2}^{(D)}(0, s) = -\frac{Z e^2}{2\pi^2 \lambda} K_1(s/\lambda) \tag{3.56}$$

and for the regularized and screened potential (3.7)

$$T_{1,2}^{(R)}(0, s) = -\frac{Z e^2}{2\pi^2} \left[\frac{1}{\lambda} K_1(s/\lambda) - \frac{1}{\lambda'} K_1(s/\lambda')\right] . \tag{3.57}$$

We investigate the asymptotic behavior of this amplitude in the limit $s \to 0$. As $K_1(z) \sim z^{-1} + z \ln(z/2)$

$$T_{1,2}^{(R)}(0, s) \sim -\frac{Z e^2}{2\pi^2} s \left[\frac{1}{\lambda^2} \ln(s/2\lambda) - \frac{1}{\lambda'^2} \ln(s/2\lambda')\right] \tag{3.58}$$

which causes no harm when integrating over the impact parameters in (2.9). This is not so for the screened potential. Insertion of (3.56) into (3.53) yields

$$\left\langle \Delta E_i^{(2)} \right\rangle_{\hat{s}} = \frac{2Z^2 e^4}{m v_r^4 s^2} (\boldsymbol{v}_r \cdot \boldsymbol{v}_i) \left[\rho K_1(\rho)\right]^2 \tag{3.59}$$

with $\rho = s/\lambda$. Of course the finite screening length λ prevents problems for large impact parameters. Indeed, all modified Bessel functions vanish exponentially for large arguments. The Coulomb case (2.21) is recovered in the limit $\lambda \to \infty$.

3.5 Binary Collisions (BC) in a Magnetic Field

We are now prepared to include the magnetic field in our perturbative treatment of binary collisions. The unperturbed helical relative motion is described by $\boldsymbol{v}_0(t) = \boldsymbol{v}_e(t) - \boldsymbol{v}_i$

$$\dot{\boldsymbol{v}}_0(t) + \Omega_e(\boldsymbol{v}_0(t) \times \boldsymbol{b}) = -\Omega_e(\boldsymbol{v}_i \times \boldsymbol{b}) \,. \tag{3.60}$$

This equation is projected in turn on $\boldsymbol{b}$, on a unit vector $\boldsymbol{u}$ transverse to the magnetic field and on $\boldsymbol{b} \times \boldsymbol{u}$. This yields

$$\boldsymbol{v}_0(t) = \bar{\boldsymbol{v}}_r + v_{e\perp} \left[\boldsymbol{u} \cos(\Omega_e t) + (\boldsymbol{b} \times \boldsymbol{u}) \sin(\Omega_e t)\right] \,, \tag{3.61}$$

where $\bar{\boldsymbol{v}}_r = v_{e\|}\boldsymbol{b} - \boldsymbol{v}_i$ is the relative velocity of the electron guiding center, $v_{e\|}$ is the component of the unperturbed electron velocity $\boldsymbol{v}_e(t)$ parallel to the magnetic field and $v_{e\perp}$ its magnitude transverse to $\boldsymbol{b}$. A further integration gives the unperturbed relative coordinate

$$\boldsymbol{r}_0(t) = \bar{\boldsymbol{R}}_0 + \bar{\boldsymbol{v}}_r t + a\left[\boldsymbol{u} \sin(\Omega_e t) - (\boldsymbol{b} \times \boldsymbol{u}) \cos(\Omega_e t)\right] \tag{3.62}$$

where $a = v_{e\perp}/\Omega_e$ is the cyclotron radius. It should be noted that the vectors $\boldsymbol{u}$ and $\bar{\boldsymbol{R}}_0$ are independent and defined by the initial conditions. The first–order corrections involve the zero–order force. Inserting (3.62) into (3.37) we obtain

$$\boldsymbol{F}_0\left(\boldsymbol{r}_0(t)\right) = -\mathrm{i} \int \mathrm{d}\boldsymbol{k} U(\boldsymbol{k}) \boldsymbol{k}\, \mathrm{e}^{\mathrm{i}\boldsymbol{k}\cdot\left\{\bar{\boldsymbol{R}}_0 + \bar{\boldsymbol{v}}_r t + a[\boldsymbol{u}\sin(\Omega_e t) - (\boldsymbol{b}\times\boldsymbol{u})\cos(\Omega_e t)]\right\}} \,. \tag{3.63}$$

We introduce a coordinate system oriented along $\boldsymbol{b}$ so that $\boldsymbol{u} = (\cos\varphi, \sin\varphi, 0)$, $(\boldsymbol{b} \times \boldsymbol{u}) = (-\sin\varphi, \cos\varphi, 0)$ and $\boldsymbol{k} = (k_\perp \cos\theta, k_\perp \sin\theta, k_\|)$ and expand as in (3.41)

$$\boldsymbol{F}_0\left(\boldsymbol{r}_0(t)\right) = -\mathrm{i} \int \mathrm{d}\boldsymbol{k} U(\boldsymbol{k}) \boldsymbol{k}\, \mathrm{e}^{\mathrm{i}\boldsymbol{k}\cdot(\bar{\boldsymbol{R}}_0 + \bar{\boldsymbol{v}}_r t)} \sum_{n=-\infty}^{\infty} J_n\left(k_\perp a\right) \mathrm{e}^{\mathrm{i}n(\Omega_e t + \varphi - \theta)} \,. \tag{3.64}$$

Insertion into (3.21) yields the first–order energy transfer

$$\Delta E_i^{(1)} = 2\pi\mathrm{i} \int \mathrm{d}\boldsymbol{k} U(\boldsymbol{k}) \left(\boldsymbol{k} \cdot \boldsymbol{v}_i\right) \mathrm{e}^{\mathrm{i}\boldsymbol{k}\cdot\bar{\boldsymbol{R}}_0} \sum_{n=-\infty}^{\infty} J_n\left(k_\perp a\right) \mathrm{e}^{\mathrm{i}n(\varphi - \theta)} \delta\left(\boldsymbol{k} \cdot \bar{\boldsymbol{v}}_r + n\Omega_e\right) \,. \tag{3.65}$$

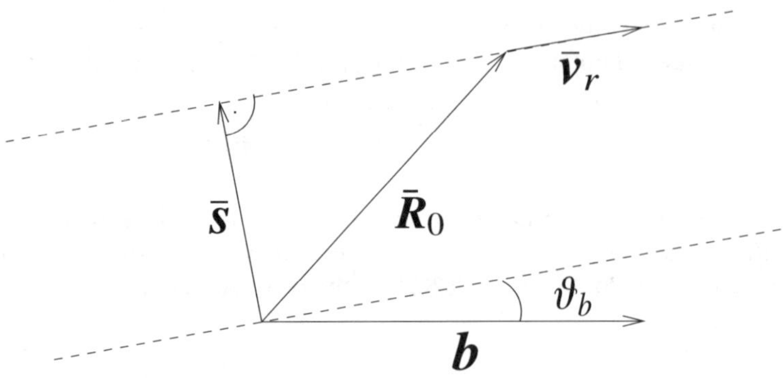

Fig. 3.1. The direction $\boldsymbol{b}$ of the magnetic field lies in general not in the common plane of the vectors $\bar{\boldsymbol{v}}_r$, $\bar{\boldsymbol{R}}_0$ and $\bar{\boldsymbol{s}}$.

Now this energy change is averaged with respect to the initial phase φ of the electrons as well as with respect to the azimuthal angle $\vartheta_{\bar{s}}$ of the impact parameter $\bar{\boldsymbol{s}}$. This vector points from the ion to the closest point of the electron's guiding center, it is the component of the initial coordinate $\bar{\boldsymbol{R}}_0$ perpendicular to $\bar{\boldsymbol{v}}_r$, see figure 3.1.

Averaging with respect to φ yields

$$\left\langle \Delta E_i^{(1)} \right\rangle_\varphi = 2\pi \mathrm{i} \int \mathrm{d}\boldsymbol{k} U(\boldsymbol{k}) \, (\boldsymbol{k} \cdot \boldsymbol{v}_i) \, \mathrm{e}^{\mathrm{i}\boldsymbol{k} \cdot \bar{\boldsymbol{R}}_0} J_0 \, (k_\perp a) \, \delta \, (\boldsymbol{k} \cdot \bar{\boldsymbol{v}}_r) \; . \tag{3.66}$$

As in (3.41) a further average with respect to $\vartheta_{\bar{s}}$ vanishes as the integrand is odd for a real interaction. The equation of motion for the velocity transfer $\delta\boldsymbol{v}(t) = \boldsymbol{v}(t) - \boldsymbol{v}_0(t)$ is obtained from (3.16)

$$\delta\dot{\boldsymbol{v}}(t) + \Omega_e \, (\delta\boldsymbol{v}(t) \times \boldsymbol{b}) = \frac{1}{m} \boldsymbol{F}(\boldsymbol{r}(t)) \tag{3.67}$$

with the electron–ion relative coordinate $\boldsymbol{r}(t) = \boldsymbol{r}_e(t) - \boldsymbol{v}_i t$. Expanding according to (3.33)-(3.38) the equation for the first–order velocity transfer is given by

$$\dot{\boldsymbol{v}}_1(t) + \Omega_e(\boldsymbol{v}_1(t) \times \boldsymbol{b}) = \frac{1}{m} \boldsymbol{F}_0 \, (\boldsymbol{r}_0(t)) \; . \tag{3.68}$$

This inhomogeneous equation can be solved with the help of a tensorial Green's function $G_{ij}(t - t')$

$$v_{1i}(t) = \frac{1}{m} \int_{-\infty}^{\infty} \mathrm{d}t' G_{ij}(t - t') F_{0j} \, (\boldsymbol{r}_0(t')) \; . \tag{3.69}$$

Being a solution of

$$\dot{G}_{ij}(t - t') + \Omega_e \varepsilon_{ikl} b_l G_{kj}(t - t') = \delta_{ij} \delta(t - t') \tag{3.70}$$

the Green's function must be a function of $\boldsymbol{b}$. Furthermore it consists of a symmetric and an antisymmetric part. Inserting the ansatz

$$G_{ij}(t-t') = b_i b_j u(t-t') + \delta_{ij} v(t-t') + \varepsilon_{ikj} b_k w(t-t') \tag{3.71}$$

yields

$$b_i b_j \dot{u} + \delta_{ij}\dot{v} + \varepsilon_{ikj} b_k \dot{w} + \Omega_e \varepsilon_{ijl} b_l v + \Omega_e(\delta_{ij} - b_i b_j) w = \delta_{ij}\delta(t-t') \,. \tag{3.72}$$

These equations can be decoupled by taking in turn the trace, by multiplying with b_i and by multiplying with ε_{ijm}. There results

$$G_{ij}(t-t') = \Theta(t-t')\Big[b_i b_j (1 - \cos(\Omega_e(t-t'))) + \delta_{ij}\cos(\Omega_e(t-t')) + \varepsilon_{ikj} b_k \sin(\Omega_e(t-t'))\Big] \,. \tag{3.73}$$

Inserting into (3.69) yields the first–order velocity correction

$$\begin{aligned} \boldsymbol{v}_1(t) = \frac{1}{m}\int_{-\infty}^{t} \mathrm{d}t' \, [&\cos(\Omega_e(t-t'))\boldsymbol{F}_0(\boldsymbol{r}_0(t')) + (1-\cos(\Omega_e(t-t'))) \\ &\times (\boldsymbol{b}\cdot\boldsymbol{F}_0(\boldsymbol{r}_0(t')))\,\boldsymbol{b} + \sin(\Omega_e(t-t'))\,(\boldsymbol{b}\times\boldsymbol{F}_0(\boldsymbol{r}_0(t')))] \end{aligned} \tag{3.74}$$

which can be written as

$$\boldsymbol{v}_1(t) = \boldsymbol{b}V_{\|}(t) - \mathrm{Re}\,[\boldsymbol{b}(\boldsymbol{b}\cdot\boldsymbol{V}_\perp(t)) - \boldsymbol{V}_\perp(t) + \mathrm{i}(\boldsymbol{b}\times\boldsymbol{V}_\perp(t))] \tag{3.75}$$

with

$$V_{\|}(t) = \frac{1}{m}\int_{-\infty}^{t} \mathrm{d}t' \, (\boldsymbol{b}\cdot\boldsymbol{F}_0(\boldsymbol{r}_0(t'))) \tag{3.76}$$

and

$$\boldsymbol{V}_\perp(t) = \frac{1}{m}\mathrm{e}^{\mathrm{i}\Omega_e t}\int_{-\infty}^{t} \mathrm{d}t' \mathrm{e}^{-\mathrm{i}\Omega_e t'}\boldsymbol{F}_0(\boldsymbol{r}_0(t')) \,. \tag{3.77}$$

A further integration yields the first–order coordinate correction

$$\boldsymbol{r}_1(t) = \boldsymbol{b}R_{\|}(t) - \mathrm{Re}\,[\boldsymbol{b}(\boldsymbol{b}\cdot\boldsymbol{R}_\perp(t)) - \boldsymbol{R}_\perp(t) + \mathrm{i}(\boldsymbol{b}\times\boldsymbol{R}_\perp(t))] \tag{3.78}$$

with

$$R_{\|}(t) = \int_{-\infty}^{t} \mathrm{d}t' V_{\|}(t') \tag{3.79}$$

and

$$\boldsymbol{R}_\perp(t) = \int_{-\infty}^{t} \mathrm{d}t' \boldsymbol{V}_\perp(t') \,. \tag{3.80}$$

Here it has been assumed that all corrections vanish for $t \to -\infty$. For an investigation of the asymptotic behavior of these corrections we insert, for example, (3.76) into (3.79) and perform a partial integration

$$\begin{aligned} R_{\|}(t) &= \frac{1}{m}\int_{-\infty}^{t} \mathrm{d}t' \int_{-\infty}^{t'} \mathrm{d}t'' \, (\boldsymbol{b}\cdot\boldsymbol{F}_0(\boldsymbol{r}_0(t''))) \\ &= \frac{1}{m} t' \int_{-\infty}^{t'} \mathrm{d}t'' \, (\boldsymbol{b}\cdot\boldsymbol{F}_0(\boldsymbol{r}_0(t'')))\Big|_{-\infty}^{t} - \frac{1}{m}\int_{-\infty}^{t} \mathrm{d}t' t' \, (\boldsymbol{b}\cdot\boldsymbol{F}_0(\boldsymbol{r}_0(t'))) \\ &= tV_{\|}(t) - tV_{\|}(t)|_{t\to-\infty} - \frac{1}{m}\int_{-\infty}^{t} \mathrm{d}t' t' \, (\boldsymbol{b}\cdot\boldsymbol{F}_0(\boldsymbol{r}_0(t'))) \,. \end{aligned} \tag{3.81}$$

The first–order corrections involve the zero–order force. Inserting (3.37) into (3.74) we obtain

$$\begin{aligned} \boldsymbol{v}_1(t) = &-\frac{\mathrm{i}}{2m}\int \mathrm{d}\boldsymbol{k}'U(\boldsymbol{k}')\int_{-\infty}^{t}\mathrm{d}t'\left[2(\boldsymbol{k}'\cdot\boldsymbol{b})\boldsymbol{b} + \mathrm{e}^{\mathrm{i}\Omega_e(t-t')}(\boldsymbol{k}'-(\boldsymbol{k}'\cdot\boldsymbol{b})\boldsymbol{b}+\mathrm{i}(\boldsymbol{k}'\times\boldsymbol{b}))\right.\\ &\left.+\ \mathrm{e}^{\mathrm{i}\Omega_e(t-t')}(\boldsymbol{k}'-(\boldsymbol{k}'\cdot\boldsymbol{b})\boldsymbol{b}-\mathrm{i}(\boldsymbol{k}'\times\boldsymbol{b}))\right]\mathrm{e}^{\mathrm{i}\boldsymbol{k}'\cdot\boldsymbol{r}_0(t')}\,. \end{aligned} \tag{3.82}$$

Another integration yields the first–order trajectory correction

$$\begin{aligned} \boldsymbol{r}_1(t) = &-\frac{\mathrm{i}}{2m}\int \mathrm{d}\boldsymbol{k}'U(\boldsymbol{k}')\int_{-\infty}^{t}\mathrm{d}t'\int_{-\infty}^{t'}\mathrm{d}t''\mathrm{e}^{\mathrm{i}\boldsymbol{k}'\cdot\boldsymbol{r}_0(t'')}\left\{2(\boldsymbol{k}'\cdot\boldsymbol{b})\boldsymbol{b}\right. \\ &+\ \mathrm{e}^{\mathrm{i}\Omega_e(t-t'')}\left[\boldsymbol{k}'-(\boldsymbol{k}'\cdot\boldsymbol{b})\boldsymbol{b}+\mathrm{i}(\boldsymbol{k}'\times\boldsymbol{b})\right]+\mathrm{e}^{-\mathrm{i}\Omega_e(t-t'')}\left.\left[\boldsymbol{k}'-(\boldsymbol{k}'\cdot\boldsymbol{b})\boldsymbol{b}-\mathrm{i}(\boldsymbol{k}'\times\boldsymbol{b})\right]\right\}\,. \end{aligned} \tag{3.83}$$

The double time integrals are done by partial integration according to

$$\begin{aligned} \int_{-\infty}^{t}\mathrm{d}t'\int_{-\infty}^{t'}\mathrm{d}t''f(t'') &= (t'-t)\int_{-\infty}^{t'}\mathrm{d}t''f(t'')\Big|_{-\infty}^{t} - \int_{-\infty}^{t}\mathrm{d}t'(t'-t)f(t') = \\ &= \int_{-\infty}^{t}\mathrm{d}t'(t-t')f(t') \end{aligned} \tag{3.84}$$

and similarly

$$\int_{-\infty}^{t}\mathrm{d}t'\int_{-\infty}^{t'}\mathrm{d}t''\mathrm{e}^{\pm\mathrm{i}\Omega_e(t-t'')}f(t'') = \pm\frac{1}{\mathrm{i}\Omega_e}\int_{-\infty}^{t}\mathrm{d}t'\mathrm{e}^{\pm\mathrm{i}\Omega_e(t-t')}f(t')\,, \tag{3.85}$$

so that

$$\begin{aligned} \boldsymbol{r}_1(t) = &-\frac{\mathrm{i}}{m}\int \mathrm{d}\boldsymbol{k}'U(\boldsymbol{k}')\int_{-\infty}^{t}\mathrm{d}t'\left[(t-t')(\boldsymbol{k}'\cdot\boldsymbol{b})\boldsymbol{b}+\frac{1}{2\mathrm{i}\Omega_e}(\boldsymbol{k}'-(\boldsymbol{k}'\cdot\boldsymbol{b})\boldsymbol{b}\right. \\ &+\ \mathrm{i}(\boldsymbol{k}'\times\boldsymbol{b}))\left(\mathrm{e}^{\mathrm{i}\Omega_e(t-t')}-1\right)-\frac{1}{2\mathrm{i}\Omega_e}(\boldsymbol{k}'-(\boldsymbol{k}'\cdot\boldsymbol{b})\boldsymbol{b} \\ &-\ \left.\mathrm{i}(\boldsymbol{k}'\times\boldsymbol{b}))\left(\mathrm{e}^{-\mathrm{i}\Omega_e(t-t')}-1\right)\right]\mathrm{e}^{\mathrm{i}\boldsymbol{k}'\cdot\boldsymbol{r}_0(t')}\,. \end{aligned} \tag{3.86}$$

Insertion into (3.38) yields the second–order energy transfer

$$\begin{aligned} \Delta E_i^{(2)} &= -\int_{-\infty}^{\infty}\mathrm{d}t\int \mathrm{d}\boldsymbol{k}\ U(\boldsymbol{k})\,(\boldsymbol{k}\cdot\boldsymbol{v}_i)\,(\boldsymbol{k}\cdot\boldsymbol{r}_i(t))\,\mathrm{e}^{\mathrm{i}\boldsymbol{k}\cdot\boldsymbol{r}_0(t)} \\ &= -\frac{\mathrm{i}}{m}\int \mathrm{d}\boldsymbol{k}\int \mathrm{d}\boldsymbol{k}'\ U(\boldsymbol{k})U(\boldsymbol{k}')\,(\boldsymbol{k}\cdot\boldsymbol{v}_i)\left[g_0(\boldsymbol{k},\boldsymbol{k}')\psi_0(\boldsymbol{k},\boldsymbol{k}')\right. \\ &\qquad \left.+g_-(\boldsymbol{k},\boldsymbol{k}')\psi_+(\boldsymbol{k},\boldsymbol{k}')+g_+(\boldsymbol{k},\boldsymbol{k}')\psi_-(\boldsymbol{k},\boldsymbol{k}')\right] \end{aligned} \tag{3.87}$$

with the functions

$$g_0(\boldsymbol{k},\boldsymbol{k}') = -(\boldsymbol{k}\cdot\boldsymbol{b})(\boldsymbol{k}'\cdot\boldsymbol{b})\,, \tag{3.88}$$
$$g_\pm(\boldsymbol{k},\boldsymbol{k}') = g_1(\boldsymbol{k},\boldsymbol{k}')\pm\mathrm{i}g_2(\boldsymbol{k},\boldsymbol{k}') = (\boldsymbol{k}\cdot\boldsymbol{b})(\boldsymbol{k}'\cdot\boldsymbol{b})-\boldsymbol{k}\cdot\boldsymbol{k}'\pm\mathrm{i}\boldsymbol{k}\cdot(\boldsymbol{k}'\times\boldsymbol{b})$$

and the time integrals

$$\psi_0(\boldsymbol{k},\boldsymbol{k}') = \int_{-\infty}^{\infty} \mathrm{d}t\, \mathrm{e}^{\mathrm{i}\boldsymbol{k}\cdot\boldsymbol{r}_0(t)} \int_{-\infty}^{t} \mathrm{d}t'(t-t')\mathrm{e}^{\mathrm{i}\boldsymbol{k}'\cdot\boldsymbol{r}_0(t')} \tag{3.89}$$
$$= \int_{-\infty}^{\infty} \mathrm{d}t\, \mathrm{e}^{\mathrm{i}\boldsymbol{k}\cdot\boldsymbol{r}_0(t)} \int_{0}^{\infty} \mathrm{d}\tau\tau\mathrm{e}^{\mathrm{i}\boldsymbol{k}'\cdot\boldsymbol{r}_0(t-\tau)}\,,$$

$$\psi_\pm(\boldsymbol{k},\boldsymbol{k}') = \frac{1}{\pm 2\mathrm{i}\Omega_e}\int_{-\infty}^{\infty} \mathrm{d}t\, \mathrm{e}^{\mathrm{i}\boldsymbol{k}\cdot\boldsymbol{r}_0(t)} \int_{-\infty}^{t} \mathrm{d}t'\left(\mathrm{e}^{\pm\mathrm{i}\Omega_e(t-t')}-1\right)\mathrm{e}^{\mathrm{i}\boldsymbol{k}'\cdot\boldsymbol{r}_0(t')} \tag{3.90}$$
$$= \frac{1}{\pm 2\mathrm{i}\Omega_e}\int_{-\infty}^{\infty} \mathrm{d}t\, \mathrm{e}^{\mathrm{i}\boldsymbol{k}\cdot\boldsymbol{r}_0(t)} \int_{0}^{\infty} \mathrm{d}\tau\left(\mathrm{e}^{\pm\mathrm{i}\Omega_e\tau}-1\right)\mathrm{e}^{\mathrm{i}\boldsymbol{k}'\cdot\boldsymbol{r}_0(t-\tau)}\,.$$

Here the helical trajectories (3.62) must be inserted. We will show below in Sect. 3.9 that the weak field limit of the energy transfer is obtained by subsequent expansion with respect to powers of Ω_e.

For magnetic fields of arbitrary strength we introduce $\boldsymbol{r}_{0\perp}(t) = \boldsymbol{r}_0(t) - \bar{\boldsymbol{R}}_0 - \bar{\boldsymbol{v}}_r t$ and expand into Bessel functions as in (3.41) and (3.64)

$$\mathrm{e}^{\mathrm{i}\boldsymbol{k}\cdot\boldsymbol{r}_{0\perp}(t)} = \mathrm{e}^{\mathrm{i}k_\perp a\sin(\varphi-\theta+\Omega_e t)} = \sum_{n=-\infty}^{\infty} J_n(k_\perp a)\,\mathrm{e}^{\mathrm{i}n(\varphi-\theta+\Omega_e t)} \tag{3.91}$$

and similarly for $\exp[\mathrm{i}\boldsymbol{k}'\cdot\boldsymbol{r}_0(t-\tau)]$. Then the second–order energy transfer is

$$\Delta E_i^{(2)} = -\frac{\mathrm{i}}{m}\int \mathrm{d}\boldsymbol{k}\int \mathrm{d}\boldsymbol{k}'\; U(\boldsymbol{k})U(\boldsymbol{k}')(\boldsymbol{k}\cdot\boldsymbol{v}_i)\mathrm{e}^{\mathrm{i}(\boldsymbol{k}+\boldsymbol{k}')\cdot\bar{\boldsymbol{R}}_0}$$
$$\times \sum_{n,m=-\infty}^{\infty} \mathrm{e}^{\mathrm{i}n(\varphi-\theta)+\mathrm{i}m(\varphi-\theta')} J_n(k_\perp a)J_m(k'_\perp a) \tag{3.92}$$
$$\times \int_{-\infty}^{\infty} \mathrm{d}t\, \mathrm{e}^{\mathrm{i}(\boldsymbol{k}+\boldsymbol{k}')\cdot\bar{\boldsymbol{v}}_r t}\mathrm{e}^{\mathrm{i}(n+m)\Omega_e t}\int_0^\infty \mathrm{d}\tau\, \mathrm{e}^{-\mathrm{i}(\boldsymbol{k}'\cdot\bar{\boldsymbol{v}}_r+m\Omega_e-\mathrm{i}0)\tau}$$
$$\times\left[\tau g_0(\boldsymbol{k},\boldsymbol{k}') + \frac{1}{\Omega_e}\sin(\Omega_e\tau)g_1(\boldsymbol{k},\boldsymbol{k}') + \frac{1}{\Omega_e}\left(1-\cos(\Omega_e\tau)\right)g_2(\boldsymbol{k},\boldsymbol{k}')\right].$$

After averaging with respect to φ the remaining series can be summed with the help of an addition theorem for Bessel functions [1]

$$\sum_{n=-\infty}^{\infty} \mathrm{e}^{\mathrm{i}n(\theta'-\theta)}\mathrm{e}^{\mathrm{i}n\Omega_e\tau} J_n(k_\perp a)J_{-n}(k'_\perp a) = J_0\left(K(\tau)a\right) \tag{3.93}$$

with

$$K^2(\tau) = k_\perp^2 + k_\perp'^2 + 2k_\perp k'_\perp\cos(\Omega_e\tau+\theta'-\theta)\,. \tag{3.94}$$

The t-integral in (3.92) can now be done

$$\langle\Delta E_i^{(2)}\rangle_\varphi = -\frac{2\pi\mathrm{i}}{m}\int \mathrm{d}\boldsymbol{k}\int \mathrm{d}\boldsymbol{k}'\; U(\boldsymbol{k})U(\boldsymbol{k}')(\boldsymbol{k}\cdot\boldsymbol{v}_i)\mathrm{e}^{\mathrm{i}(\boldsymbol{k}+\boldsymbol{k}')\cdot\bar{\boldsymbol{R}}_0}$$
$$\times\delta\left((\boldsymbol{k}+\boldsymbol{k}')\cdot\bar{\boldsymbol{v}}_r\right)\int_0^\infty \mathrm{d}\tau J_0(K(\tau)a)\mathrm{e}^{-\mathrm{i}(\boldsymbol{k}'\cdot\bar{\boldsymbol{v}}_r-\mathrm{i}0)\tau} \tag{3.95}$$
$$\times\left[\tau g_0(\boldsymbol{k},\boldsymbol{k}') + \frac{1}{\Omega_e}\sin(\Omega_e\tau)g_1(\boldsymbol{k},\boldsymbol{k}') + \frac{1}{\Omega_e}\left(1-\cos(\Omega_e\tau)\right)g_2(\boldsymbol{k},\boldsymbol{k}')\right].$$

Similar as in (3.40) the δ-function enforces $\boldsymbol{k}+\boldsymbol{k}'$ to lie in the plane transverse to $\bar{\boldsymbol{v}}_r$ so that

$$\mathrm{e}^{\mathrm{i}(\boldsymbol{k}+\boldsymbol{k}')\cdot\bar{\boldsymbol{R}}_0}\delta\left((\boldsymbol{k}+\boldsymbol{k}')\cdot\bar{\boldsymbol{v}}_r\right)=\mathrm{e}^{\mathrm{i}(\boldsymbol{k}+\boldsymbol{k}')\cdot\bar{\boldsymbol{s}}}\delta\left((\boldsymbol{k}+\boldsymbol{k}')\cdot\bar{\boldsymbol{v}}_r\right)\,, \tag{3.96}$$

$$\mathrm{e}^{\mathrm{i}\bar{s}|\boldsymbol{k}+\boldsymbol{k}'|\sin\vartheta_{\bar{s}}}\delta\left((\boldsymbol{k}+\boldsymbol{k}')\cdot\bar{\boldsymbol{v}}_r\right)=\sum_{n=-\infty}^{\infty}J_n\left(\left|\boldsymbol{k}+\boldsymbol{k}'\right|\bar{s}\right)\mathrm{e}^{\mathrm{i}n\vartheta_{\bar{s}}}\delta\left((\boldsymbol{k}+\boldsymbol{k}')\cdot\bar{\boldsymbol{v}}_r\right)\,, \tag{3.97}$$

where $\vartheta_{\bar{s}}$ is the azimuthal angle of $\bar{\boldsymbol{s}}$ in that plane. After averaging with respect to this angle there remains

$$\begin{aligned}\langle\Delta E_i^{(2)}\rangle_{\varphi,\hat{\bar{s}}}=&-\frac{2\pi\mathrm{i}}{m}\int\mathrm{d}\boldsymbol{k}\int\mathrm{d}\boldsymbol{k}'\;U(\boldsymbol{k})U(\boldsymbol{k}')(\boldsymbol{k}\cdot\boldsymbol{v}_i)J_0\left(\left|\boldsymbol{k}+\boldsymbol{k}'\right|\bar{s}\right)\\&\times\delta\left((\boldsymbol{k}+\boldsymbol{k}')\cdot\bar{\boldsymbol{v}}_r\right)\int_0^\infty\mathrm{d}\tau J_0\left(K(\tau)a\right)\mathrm{e}^{-\mathrm{i}(\boldsymbol{k}\cdot\bar{\boldsymbol{v}}_r-\mathrm{i}0)\tau}\\&\times\left[\tau g_0(\boldsymbol{k},\boldsymbol{k}')+\frac{1}{\Omega_e}\sin(\Omega_e\tau)g_1(\boldsymbol{k},\boldsymbol{k}')+\frac{1}{\Omega_e}\left(1-\cos(\Omega_e\tau)\right)g_2(\boldsymbol{k},\boldsymbol{k}')\right].\end{aligned} \tag{3.98}$$

Symmetrization according to $k\rightarrow-\boldsymbol{k},\boldsymbol{k}\rightarrow-\boldsymbol{k}'$ shows that for real potentials the energy transfer can be written as

$$\begin{aligned}\left\langle\Delta E_i^{(2)}\right\rangle_{\varphi,\hat{\bar{s}}}=&\frac{2\pi}{m}\int\mathrm{d}\boldsymbol{k}\int\mathrm{d}\boldsymbol{k}'U(\boldsymbol{k})U(\boldsymbol{k}')(\boldsymbol{k}\cdot\boldsymbol{v}_i)J_0\left(\left|\boldsymbol{k}+\boldsymbol{k}'\right|\bar{s}\right)\\&\times\delta\left((\boldsymbol{k}+\boldsymbol{k}')\cdot\bar{\boldsymbol{v}}_r\right)\,\mathrm{Im}\int_0^\infty\mathrm{d}\tau\mathrm{e}^{-\mathrm{i}(\boldsymbol{k}'\cdot\bar{\boldsymbol{v}}_r-\mathrm{i}0)\tau}J_0(aK(\tau))\\&\times\left[\tau g_0(\boldsymbol{k},\boldsymbol{k}')+\frac{1}{\Omega_e}\sin(\Omega_e\tau)g_1(\boldsymbol{k},\boldsymbol{k}')+\frac{1}{\Omega_e}(1-\cos(\Omega_e\tau))g_2(\boldsymbol{k},\boldsymbol{k}')\right].\end{aligned} \tag{3.99}$$

This integral representation of the second–order energy change is valid for any strength of the magnetic field. Besides the direction $\boldsymbol{b}$ of the magnetic field the direction $\bar{\boldsymbol{n}}_r=\bar{\boldsymbol{v}}_r/\bar{v}_r$ of the relative velocity is singled out in the arguments of the δ–function and integrand of the τ–integration. This prevents a closed evaluation of the integrals (3.99). However the limiting case of an ion motion parallel to a magnetic field of arbitrary strength and the case of an arbitrary motion in a magnetic field with small electron cyclotron radius a can be treated in a straightforward manner.

3.6 Parallel Ion Motion

Introducing cylindrical coordinates along $\boldsymbol{b}$, i.e. $\boldsymbol{v}_i = \boldsymbol{v}_{i\perp} + v_{i\|}\boldsymbol{b}$ etc. and setting $\boldsymbol{v}_{i\perp} = 0$, the energy change (3.99) is

$$\left\langle \Delta E_{i\|}^{(2)} \right\rangle = \frac{2\pi v_{i\|}}{m|\bar{v}_{r\|}|} \int \mathrm{d}\boldsymbol{k} \int \mathrm{d}\boldsymbol{k}' U(\boldsymbol{k})U(\boldsymbol{k}')k_\| J_0\left(\bar{s}\sqrt{k_\perp^2 + k_\perp'^2}\right)$$
$$\times \delta(k_\| + k_\|') \,\mathrm{Im} \int_0^\infty \mathrm{d}\tau \, \mathrm{e}^{\mathrm{i}(k_\| \bar{v}_{r\|} + \mathrm{i}0)\tau} \tag{3.100}$$
$$\times J_0\left(a\sqrt{k_\perp^2 + k_\perp'^2 + 2k_\perp k_\perp' \cos(\Omega_e \tau + \theta' - \theta)}\right)$$
$$\times \left[k_\|^2 \tau - \frac{k_\perp k_\perp'}{\Omega_e} \sin(\Omega_e \tau)\cos(\theta - \theta') - \frac{k_\perp k_\perp'}{\Omega_e}(1 - \cos(\Omega_e \tau))\sin(\theta - \theta')\right],$$

where the subscripts φ and $\hat{\boldsymbol{s}}$ have been dropped to simplify the notation. For the Bessel functions we use the addition theorem [1]

$$J_0\left(a\sqrt{k_\perp^2 + k_\perp'^2 + 2k_\perp k_\perp' \cos(\Omega_e \tau + \theta' - \theta)}\right) = J_0(ak_\perp)J_0(ak_\perp') +$$
$$+2\sum_{n=1} J_n(ak_\perp)J_n(ak_\perp')\cos(n(\pi + \Omega_e \tau - \theta + \theta')) \tag{3.101}$$

and similarly for $J_0\left(\bar{s}\sqrt{k_\perp^2 + k_\perp'^2}\right)$. Then the angular integrations with respect to θ and θ' are trivial for angle–independent interactions $U(k_\perp, k_\|)$. The integrals with respect to the parallel components $k_\|, k_\|'$ are done with the help of (3.49).

To continue the interaction must be specified. The potentials proposed in (3.5)-(3.7) have the common behavior $\propto (k_\perp^2 + \kappa^2)^{-2}$, where κ is some constant to be specified later. The remaining transversal integrals are then of the form [52]

$$\int_0^\infty \mathrm{d}k_\perp k_\perp \frac{J_n(k_\perp a)J_n(k_\perp \bar{s})}{k_\perp^2 + \kappa^2} = I_n(\xi)K_n(\eta) \tag{3.102}$$

and their derivatives with respect to κ. Here $\xi = \kappa \min(a, \bar{s})$ and $\eta = \kappa \max(a, \bar{s})$. In the following we consider the regularized screened potential (3.7) with

$$u_R(k_\perp, k_\|) = \frac{1}{2\pi^2}\left(\frac{1}{k_\perp^2 + \kappa^2} - \frac{1}{k_\perp^2 + \chi^2}\right), \tag{3.103}$$

where $\kappa^2 = k_\|^2 + \lambda^{-2}, \chi^2 = k_\|^2 + \lambda'^{-2}$ and $\lambda'^{-1} = \lambda^{-1} + \bar{\lambda}^{-1}$. Carrying out the calculation indicated above between (3.100) and (3.103) we obtain

$$\left\langle \Delta E_{i\|}^{(2)} \right\rangle = \frac{4Z^2 e^4 v_{i\|}}{m\bar{v}_{r\|}^3 \delta^2} \sum_{n=1}^{\infty} n^2 \Bigg\{ 3\left[u_n(\kappa_n a, \kappa_n \bar{s}) - u_n(\chi_n a, \chi_n \bar{s})\right]^2 \tag{3.104}$$
$$+ \frac{2n^2}{\delta^2}\left[u_n(\kappa_n a, \kappa_n \bar{s}) - u_n(\chi_n a, \chi_n \bar{s})\right]\left[\frac{1}{\kappa_n^2} t_n(\kappa_n a, \kappa_n \bar{s}) - \frac{1}{\chi_n^2} t_n(\chi_n a, \chi_n \bar{s})\right]$$
$$+ 2\delta^2\left[\kappa_n^2 q_n(\kappa_n a, \kappa_n \bar{s}) + \chi_n^2 q_n(\chi_n a, \chi_n \bar{s}) - 2\kappa_n \chi_n d_n(\kappa_n a, \kappa_n \bar{s}; \chi_n a, \chi_n \bar{s})\right]\Bigg\}.$$

Here $\delta = |\bar{v}_{r\|}|/\Omega_e$ is the pitch of the helix, divided by 2π,

$$\kappa_n^2 = \frac{n^2}{\delta^2} + \frac{1}{\lambda^2}\,, \quad \chi_n^2 = \frac{n^2}{\delta^2} + \frac{1}{\lambda'^2}\,, \tag{3.105}$$

$$\begin{aligned} u_n(x,y) &= I_n(\xi)K_n(\eta)\,, \\ t_n(x,y) &= \xi I_n'(\xi)K_n(\eta) + \eta K_n'(\eta)I_n(\xi)\,, \\ q_n(x,y) &= I_n(\xi)K_n(\eta)\left[\frac{1}{\xi}I_n'(\xi)K_n(\eta) + \frac{1}{\eta}K_n'(\eta)I_n(\xi)\right]\,, \end{aligned} \tag{3.106}$$

with $\xi = \min(x,y)$, $\eta = \max(x,y)$,

$$\begin{aligned} d_n(x,y;X,Y) = \frac{1}{4n}\big[&s_{n-1}(x,y)\,s_{n-1}(X,Y) - s_n(x,y)\,s_n(X,Y) \\ &+ s_{n-1}(y,x)\,s_{n-1}(Y,X) - s_n(y,x)\,s_n(Y,X)\big]\,, \end{aligned} \tag{3.107}$$

and

$$s_n(x,y) = \begin{cases} I_n(x)K_{n+1}(y)\,, & y > x \\ \frac{1}{2}[I_n(x)K_{n+1}(x) - I_{n+1}(x)K_n(x)]\,, & y = x \\ -I_{n+1}(y)K_n(x)\,, & y < x \end{cases}\,. \tag{3.108}$$

For a study of the convergence of the series (3.104) we note that in all terms the modified Bessel functions I_n carry the smaller argument $\propto \min(a, \bar{s})$, while the K_n depend on $\max(a, \bar{s})$. The uniform asymptotic expansions of these functions are [1]

$$\begin{aligned} I_n(nz) &\sim \frac{1}{\sqrt{2\pi n}}\frac{e^{n\varphi(z)}}{(1+z^2)^{1/4}}\left[1 - \frac{a_1(z)}{n(1+z^2)^{1/2}}\right]\,, \\ I_n'(nz) &\sim \frac{1}{\sqrt{2\pi n}}\frac{(1+z^2)^{1/4}}{z}e^{n\varphi(z)}\left[1 - \frac{b_1(z)}{n(1+z^2)^{1/2}}\right]\,, \\ K_n(nz) &\sim \sqrt{\frac{\pi}{2n}}\frac{e^{-n\varphi(z)}}{(1+z^2)^{1/4}}\left[1 + \frac{a_1(z)}{n(1+z^2)^{1/2}}\right]\,, \\ K_n'(nz) &\sim -\sqrt{\frac{\pi}{2n}}\frac{(1+z^2)^{1/4}}{z}e^{-n\varphi(z)}\left[1 + \frac{b_1(z)}{n(1+z^2)^{1/2}}\right] \end{aligned} \tag{3.109}$$

with

$$\varphi(z) = (1+z^2)^{1/2} - \ln\frac{1+(1+z^2)^{1/2}}{z} \tag{3.110}$$

and

$$a_1(z) = -\frac{1}{8} + \frac{5}{24(1+z^2)}\,, \quad b_1(z) = \frac{3}{8} - \frac{7}{24(1+z^2)}\,. \tag{3.111}$$

Therefore the case $\bar{s} = a$ is most critical for the convergence of (3.104). This is intuitively clear as the gyrating electron hits the ion on such a trajectory. This should not matter for the potential (3.3), (3.103) which has been regularized near the origin for exactly that purpose. Insertion of the expansions (3.109) shows indeed that the

nth member of the series are of the order $O(n^{-4})$, see Sect. 3.10, so the series converges even for $\bar{s} = a$. On the other hand, the energy transfer for the unregularized potentials (3.1), (3.2) diverges for $\bar{s} = a$.

In the limit $\lambda \to 0$ all functions u_n, t_n, q_n and d_n involving χ_n in their arguments tend to zero. There remains

$$\left\langle \Delta E_{i\|}^{(2)} \right\rangle = \frac{4Z^2 e^4 v_{i\|}}{m \bar{v}_{r\|}^3 \delta^2} \sum_{n=1}^{\infty} n^2 \left\{ 3u_n^2(\kappa_n a; \kappa_n \bar{s}) + \frac{2n^2}{\kappa_n^2 \delta^2} u_n(\kappa_n a; \kappa_n \bar{s}) t_n(\kappa_n a; \kappa_n \bar{s}) \right.$$
$$\left. + 2\kappa_n^2 \delta^2 q_n(\kappa_n a; \kappa_n \bar{s}) \right\} . \qquad (3.112)$$

Now insertion of the expansions (3.109) shows that at $n \to \infty$ the members of this series are independent of n for $\bar{s} = a$, hence the series diverges. In the Coulomb case (3.1), i.e. for $\lambda \to \infty$, $\kappa_n = n/\delta$, the resulting series are geometric and can be summed in closed form

$$\left\langle \Delta E_{i\|}^{(2)} \right\rangle = \frac{Z^2 e^4 v_{i\|}}{2m \bar{v}_{r\|}^3 \delta^2} \frac{(1+\xi^2)^{-1/2}(1+\eta^2)^{-1/2}}{\sinh \Psi(\xi,\eta)}$$
$$\times \left\{ [2 + \Xi_1(\xi,\eta)]\, \mathrm{e}^{-\Psi(\xi,\eta)} + \frac{\Xi_2(\xi,\eta)}{\sinh \Psi(\xi,\eta)} \right\} \qquad (3.113)$$

with $\xi = \delta^{-1}\min(a,\bar{s})$, $\eta = \delta^{-1}\max(a,\bar{s})$, $\Psi(\xi,\eta) = \varphi(\xi) - \varphi(\eta)$ and

$$\Xi_1(\xi,\eta) = \frac{5(1+\xi^2)^{3/2}}{6\xi^2(1+\eta^2)^{3/2}} + \frac{5(1+\eta^2)^{3/2}}{6\eta^2(1+\xi^2)^{3/2}} - \frac{1}{3}\left(\frac{1}{\xi^2} + \frac{1}{\eta^2}\right)$$
$$- \frac{(1+\eta^2)^{3/2}}{2\eta^2(1+\xi^2)^{1/2}} - \frac{(1+\xi^2)^{3/2}}{2\xi^2(1+\eta^2)^{1/2}} , \qquad (3.114)$$
$$\Xi_2(\xi,\eta) = \frac{(1+\xi^2)^{3/2}}{\xi^2} - \frac{(1+\eta^2)^{3/2}}{\eta^2} .$$

For the limit $|\bar{s} - a| \to 0$ the Taylor expansions

$$\Psi(\xi,\eta) \simeq (\xi - \eta) \left.\frac{\partial \Psi}{\partial \xi}\right|_{\xi=\eta} = -(\xi - \eta)\frac{(1+\eta^2)^{1/2}}{\eta} \qquad (3.115)$$

and

$$\frac{(1+\xi^2)^{3/2}}{\xi^2} \simeq \frac{(1+\eta^2)^{3/2}}{\eta^2} + (\xi - \eta)(1+\eta^2)^{1/2}\frac{\eta^2 - 2}{\eta^3} \qquad (3.116)$$

are used. This yields

$$\left\langle \Delta E_{i\|}^{(2)} \right\rangle \simeq \frac{Z^2 e^4 v_{i\|}}{m \bar{v}_{r\|}^3 \delta} \frac{1}{\eta(1+\eta^2)^{3/2}} \frac{1}{|\bar{s} - a|} \qquad (3.117)$$

which exhibits the divergence at $\bar{s} = a$. Note that in equation (3.117) $\eta = a/\delta = v_{e\perp}/|\bar{v}_{r\|}|$ at $\bar{s} = a$.

For later purposes we also note the limits of (3.112) for a small cyclotron radius with $a \ll \delta$,

$$\left\langle \Delta E_{i\|}^{(2)} \right\rangle = \frac{2Z^2 e^4 v_{i\|}}{m \bar{v}_{r\|}^3 \bar{s}^2} \left[(\kappa_1 \bar{s}) K_1(\kappa_1 \bar{s}) \right]^2 \left[1 + \frac{a^2}{\delta^2} \mathfrak{F}(\kappa_1 \delta, \kappa_1 \bar{s}, \kappa_2 \bar{s}) \right] \quad (3.118)$$

with

$$\mathfrak{F}(\zeta, \lambda, \mu) = \frac{3}{2} + \frac{\zeta^2}{2} \left\{ 1 + \frac{2K_1'(\lambda)}{\lambda K_1(\lambda)} + \left[\frac{\mu^2 K_2(\mu)}{\lambda^2 K_1(\lambda)} \right]^2 \right\} + \frac{1}{\zeta^2} \left[1 + \frac{\lambda K_1'(\lambda)}{K_1(\lambda)} \right] . \quad (3.119)$$

Because of the symmetry of (3.112) in respect to its arguments the limit of a small impact parameter $\bar{s} \ll \delta$, is given by (3.118), (3.119) with the roles of a and $\bar{s}$ interchanged.

3.7 Chaotic Scattering and Validity of the Perturbation Treatment

Having obtained the results for the second–order energy transfer in the case of parallel ion motion we interrupt the discussion of (3.99) in order to investigate the validity of the perturbation treatment by comparing with CTMC calculations.

For all the binary ion-electron collisions in a magnetic field as given by the equations of motion (3.16) and (3.17) with interactions (3.1)-(3.3) there are less integrals of motion than degrees of freedom. So the motion can be chaotic. Coulomb scattering in a magnetic field is in fact the paradigm of a chaotic system. This immediately raises the question whether a perturbative treatment as proposed in the previous sections can be applied here at all. We will address this issue in this section by showing some examples for the energy transfer obtained from a numerical solution of the equations of motion (3.16) as outlined for the CTMC simulations in Sect. 2.3 and by making some comparison with the perturbative treatment. To avoid all problems related to the Coulomb singularity at $r \to 0$ we employ the regularized screened potential $U_R(\boldsymbol{r})$ (3.3) throughout.

Starting point is thus the equation of relative motion for $M \gg m$, $\boldsymbol{v}_i = \text{const}$ (3.16) with the interaction $U_R(\boldsymbol{r})$ (3.3)

$$\dot{\boldsymbol{v}} = -\Omega_e [(\boldsymbol{v} + \boldsymbol{v}_i) \times \boldsymbol{b}] + \frac{Ze^2}{m} \frac{\partial}{\partial \boldsymbol{r}} \left[\frac{1 - e^{-r/\lambda}}{r} e^{-r/\lambda} \right] . \quad (3.120)$$

For the forthcoming discussion we put it in a more appropriate dimensionless form by scaling lengths in units of the screening length λ and velocities in units of a characteristic velocity v_s defined by

$$v_s^2 = \frac{|Z| e^2}{m\lambda} . \quad (3.121)$$

This velocity v_s gives a measure for the strength of the Coulomb interaction with respect to the (initial) kinetic energy of relative motion $mv_r^2/2$. For $v_r < v_s$ the kinetic energy is small compared to the characteristic potential energy $|Z|e^2/\lambda$ in a screened Coulomb potential and we expect to be in a non-perturbative regime. A perturbative treatment on the hand should be applicable for $v_r \gg v_s$.

The scaled version of equation (3.120) reads

$$\frac{\mathrm{d}\tilde{\boldsymbol{v}}}{\mathrm{d}\tilde{t}} = -\left(\frac{\lambda}{a_s}\right)[(\tilde{\boldsymbol{v}} + \tilde{\boldsymbol{v}}_i) \times \boldsymbol{b}] + \frac{\partial}{\partial \tilde{\boldsymbol{r}}}\left[\frac{1}{\tilde{r}}\left(1 - \mathrm{e}^{-\tilde{r}(\lambda/\bar{\lambda})}\right)\mathrm{e}^{-\tilde{r}}\right], \tag{3.122}$$

where $\tilde{\boldsymbol{v}} = \boldsymbol{v}/v_s, \tilde{\boldsymbol{r}} = \boldsymbol{r}/\lambda$, $\tilde{t} = t/t_s$ with the corresponding time scale $t_s = \lambda/v_s$, while $a_s = v_s/\Omega_e$ is the cyclotron radius of an electron with $v_{e\perp} = v_s$. In the scaled variables a specific collision depends on the two dimensionless parameters a_s/λ and $\bar{\lambda}/\lambda$, and the initial conditions. The parameter $a_s/\lambda \propto v_s/B$ represents a measure for the strength of the magnetic field compared to the strength of the Coulomb interaction (which is $\propto v_s^2$, see (3.121)). The ratio $\bar{\lambda}/\lambda$ describes the amount of softening of the screened interaction at $r \to 0$ with $U_R(r \to 0) \to |Z|e^2/\bar{\lambda}$. The initial conditions are the scaled position $\boldsymbol{r}(t \to -\infty)/\lambda$ and the scaled velocities $\tilde{\boldsymbol{v}}(t \to -\infty) = \boldsymbol{v}_r/v_s$ and $\boldsymbol{v}_i/v_s$.

Here we are going to discuss the energy transfer $\langle \Delta E_{i\|} \rangle$ for parallel ion motion $\boldsymbol{v}_{i\perp} = 0$. This specific case shows qualitatively the same features of chaotic motion as the general case, but allows some further simplification and an explicit comparison with the second–order expression for $\langle \Delta E_{i\|}^{(2)} \rangle$ (3.104) as derived in the previous Sect. 3.6. For parallel ion motion the only relevant initial conditions are the impact parameter $\bar{s}/\lambda$, the velocity $\bar{v}_r/v_s = \bar{v}_{r\|}/v_s$ of the guiding center, the initial transverse electron velocity $v_{e\perp}/v_s$, or equivalently the cyclotron radius $a/\lambda = v_{e\perp}/\Omega_e\lambda = (v_{e\perp}/v_s)(a_s/\lambda)$, and its initial phase φ (i.e. the direction of $\boldsymbol{v}_{e\perp}$). The longitudinal initial positions must be located outside the interaction region of a few λ, but is otherwise of no relevance. In contrast to the general case the energy of the relative motion is conserved here as the conserved quantity K (3.18) turns into $K = mv^2/2 + U_R(\boldsymbol{r}) = E$ for $\boldsymbol{v}_{i\perp} = 0$. Hence $v^2(t \to \infty) = v^2(t \to -\infty)$ which results (with $\boldsymbol{v}(t) = \boldsymbol{v}_{e0}(t) - \boldsymbol{v}_i + \delta\boldsymbol{v}(t)$, $\delta\boldsymbol{v}(t \to -\infty) = 0$ and $v_0^2 = (\boldsymbol{v}_{e0}(t) - \boldsymbol{v}_i)^2 =$ const for the unperturbed helical motion (3.22)) in $\Delta\boldsymbol{v}^2 = -2\Delta\boldsymbol{v} \cdot (\boldsymbol{v}_{e0}(t) - \boldsymbol{v}_i)$, where $\Delta\boldsymbol{v} = \delta\boldsymbol{v}(t \to \infty)$. Inserting this into expression (3.32) for the energy transfer we get

$$\frac{\Delta E_{i\|}}{mv_s^2} = -\frac{\boldsymbol{v}_i}{v_s} \cdot \frac{\Delta\boldsymbol{v}}{v_s} = -\frac{v_{i\|}}{v_s}\frac{\Delta v_\|}{v_s}. \tag{3.123}$$

The energy transfer for parallel ion motion is thus entirely given by the parallel velocity transfer and we are focusing on the scaled quantity

$$\frac{\Delta E_{i\|}}{mv_s v_{i\|}} = -\frac{\Delta v_\|}{v_s}\left(\frac{a_s}{\lambda}, \frac{\bar{\lambda}}{\lambda}, \frac{\bar{v}_{r\|}}{v_s}, \frac{a}{\lambda}, \varphi, \frac{\bar{s}}{\lambda}\right), \tag{3.124}$$

where $\Delta v_\| = \delta v_\|(t \to \infty) = \boldsymbol{v}(t \to \infty) \cdot \boldsymbol{b} - \bar{v}_{r\|}$.

Within the perturbative treatment the average of (3.124) over the initial phase φ of the helical electron motion is given by (3.104) derived for parallel ion motion and the regularized screened interaction U_R. In the scaled form of equation (3.124) and in terms of the scaled parameters and variables, equation (3.104) reads

$$\frac{\left\langle \Delta E_{i\|}^{(2)} \right\rangle}{m v_s v_{i\|}} = \left(\frac{v_s}{\bar{v}_{r\|}}\right)^5 \left(\frac{\lambda}{a_s}\right)^2 \sum_{n=1}^{\infty} 4n^2 \Big\{ 3\left[u_n(\kappa_n a, \kappa_n \bar{s}) - u_n(\chi_n a, \chi_n \bar{s})\right]^2 \tag{3.125}$$
$$+ 2n^2 \left[u_n(\kappa_n a, \kappa_n \bar{s}) - u_n(\chi_n a, \chi_n \bar{s})\right] \left[\frac{t_n(\kappa_n a, \kappa_n \bar{s})}{(\kappa_n \delta)^2} - \frac{t_n(\chi_n a, \chi_n \bar{s})}{(\chi_n \delta)^2}\right]$$
$$- 4(\kappa_n \delta)(\chi_n \delta) d_n(\kappa_n a, \kappa_n \bar{s}; \chi_n a, \chi_n \bar{s})$$
$$+ 2(\kappa_n \delta)^2 q_n(\kappa_n a, \kappa_n \bar{s}) + 2(\chi_n \delta)^2 q_n(\chi_n a, \chi_n \bar{s}) \Big\}$$

with $u_n(x, y)$, $t_n(x, y)$, $q_n(x, y)$ and $d_n(x, y; X, Y)$ as defined in (3.106)-(3.108) and

$$(\kappa_n a)^2 = \left[n^2 \left(\frac{v_s}{\bar{v}_{r\|}}\right)^2 \left(\frac{\lambda}{a_s}\right)^2 + 1\right] \left(\frac{a}{\lambda}\right)^2, \quad (\chi_n a)^2 = \left[n^2 \left(\frac{v_s}{\bar{v}_{r\|}}\right)^2 \left(\frac{\lambda}{a_s}\right)^2 + \varkappa^2\right] \left(\frac{a}{\lambda}\right)^2,$$
$$(\kappa_n \bar{s})^2 = \left[n^2 \left(\frac{v_s}{\bar{v}_{r\|}}\right)^2 \left(\frac{\lambda}{a_s}\right)^2 + 1\right] \left(\frac{\bar{s}}{\lambda}\right)^2, \quad (\chi_n \bar{s})^2 = \left[n^2 \left(\frac{v_s}{\bar{v}_{r\|}}\right)^2 \left(\frac{\lambda}{a_s}\right)^2 + \varkappa^2\right] \left(\frac{\bar{s}}{\lambda}\right)^2,$$
$$(\kappa_n \delta)^2 = n^2 + \left(\frac{\bar{v}_{r\|}}{v_s}\right)^2 \left(\frac{a_s}{\lambda}\right)^2, \qquad (\chi_n \delta)^2 = n^2 + \left(\frac{\bar{v}_{r\|}}{v_s}\right)^2 \left(\frac{a_s}{\lambda}\right)^2 \varkappa^2, \tag{3.126}$$

where $\varkappa = 1 + \lambda/\bar{\lambda}$ and we have used $\delta/\lambda = |\bar{v}_{r\|}|/(\Omega_e \lambda) = (|\bar{v}_{r\|}|/v_s)(a_s/\lambda)$.

As a first example we consider the particular case of a motion which is initially purely parallel (i.e. $v_{e\perp} = 0$ or $a/\lambda = 0$) at a finite magnetic field. From the energy conservation with $v_{\|}^2(t \to \infty) + v_{\perp}^2(t \to \infty) = v^2(t \to \infty) = v^2(t \to -\infty) = \bar{v}_{r\|}^2$, it here follows that the (negative, scaled) velocity transfer $-\Delta v_{\|}/\bar{v}_{r\|}$ can range from $0 \ldots 2$, where $-\Delta v_{\|}/\bar{v}_{r\|} = 2$ corresponds to a reversion of the initial motion, i.e. to a backscattering event. Fig. 3.2 shows the results for the velocity transfer for different parameter settings as found in the numerical CTMC-simulations outlined in Sect. 2.3. Each dot corresponds to one calculated trajectory with a randomly chosen impact parameter (as $v_{e\perp} = 0$ no initial phase must be considered). Each figure comprises 10^5 events. For regular motion, like in the integrable case of Rutherford scattering, where the velocity transfer $\Delta v_{\|}$ is a smooth function of the impact parameter $\bar{s}$, all dots will fall on one single curve. In the given situation this is the case for large $\bar{s}$ and/or $\bar{v}_{r\|}/v_s \gg 1$. But for low $\bar{s}$ and $\bar{v}_{r\|}$ chaotic motion takes place where the scattering events completely fill the kinematically allowed range with some 'islands' of regular motion, i.e. the velocity transfer $\Delta v_{\|}$ is very sensitive to small changes in $\bar{s}$ (see e.g. Fig. 3.2 top left for $\bar{s}/\lambda < 1$). This chaotic region is characterized by frequent very violent scattering events ('hard collisions') with $|\Delta v_{\|}/\bar{v}_{r\|}| \gtrsim 1$, like backscattering or large transfers of initial parallel relative motion to transverse motion of the electron. Scattering events with an almost complete transfer to transverse electron motion are, however, strongly suppressed (see the thinly dotted regions around $-\Delta v_{\|}/\bar{v}_{r\|} = 1$) as they are associated with a vanishing outgoing velocity $v_{\|}(t \to \infty) \to 0$, i.e. the electron would not reemerge outside the interaction zone. The region of chaotic motion shrinks strongly with increasing

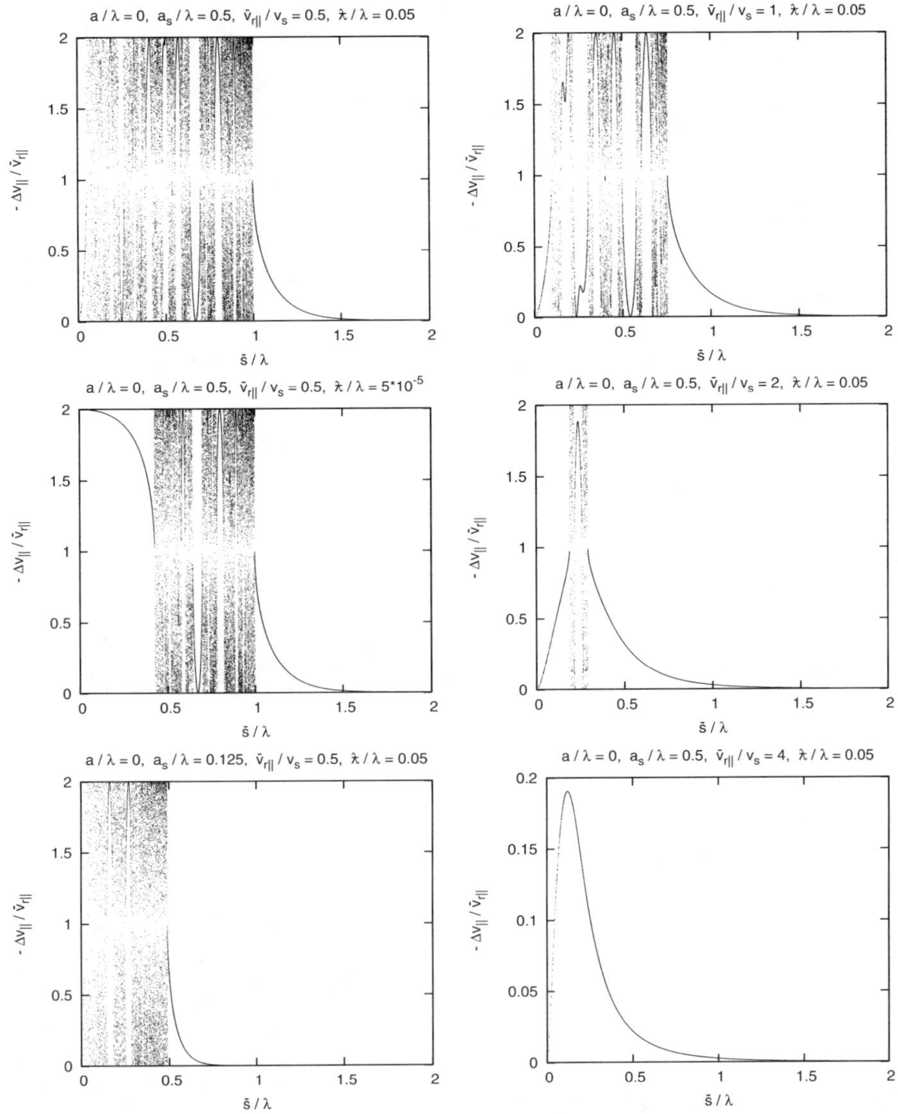

Fig. 3.2. Scaled velocity transfer (3.124) as function of the impact parameter $\bar{s}$ in units of the screening length λ for vanishing initial transverse electron velocity $v_{e\perp} = 0$ $(a/\lambda = 0)$ and different parameters $ƛ/\lambda$, a_s/λ and initial velocities $\bar{v}_{r\parallel}$ in units of v_s (3.121).

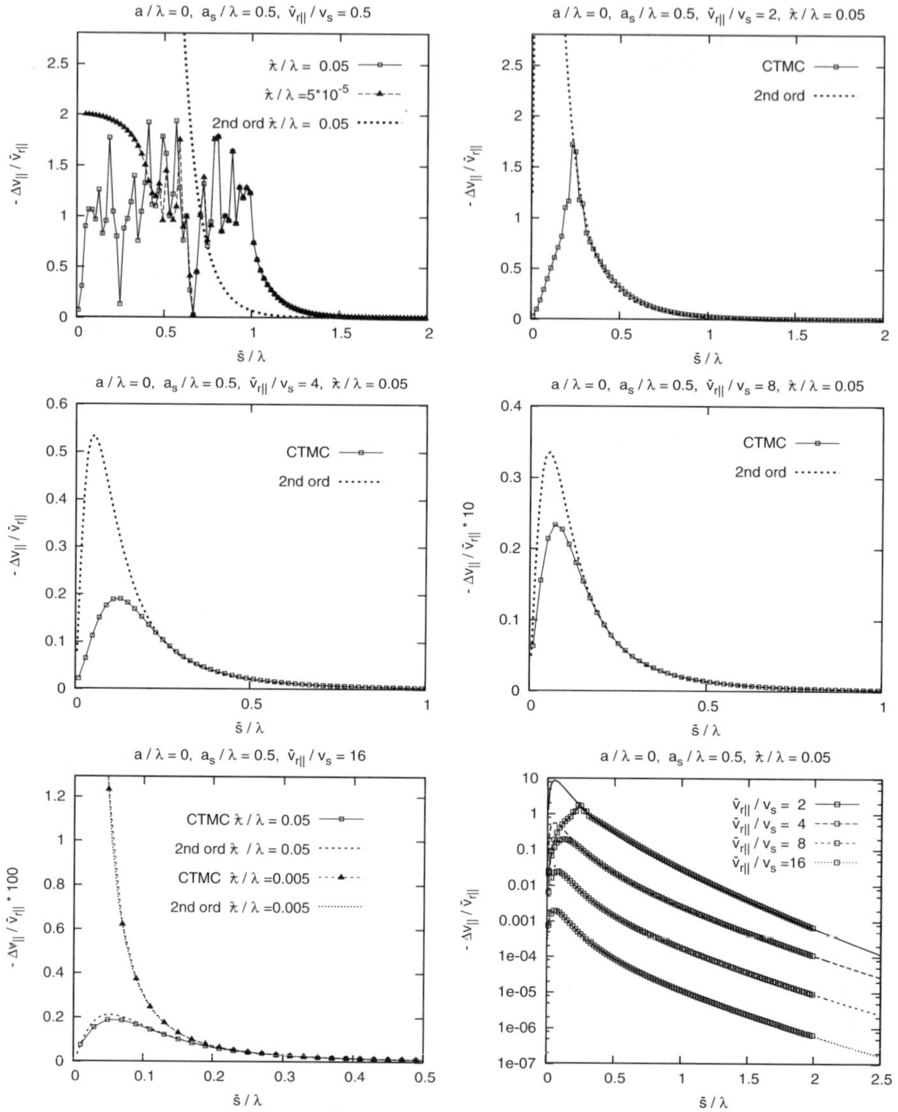

Fig. 3.3. Scaled velocity transfer (3.124) as in Fig. 3.2 for vanishing initial transverse electron velocity $v_{e\perp} = 0$ ($a/\lambda = 0$) and different parameters $\bar{\lambda}/\lambda$, a_s/λ and initial velocities $\bar{v}_{r\|}$. Compared is the velocity transfer from CTMC calculations averaged over impact parameter intervals $\Delta\bar{s}/\lambda = 0.02$ (symbols) with the second–order energy transfer (3.127). Note the different ranges, both of $\bar{s}$ and $\Delta v_{\|}$.

initial velocity $\bar{v}_{r\|}$ (see Fig. 3.2, going from top left to top right to down right). It even completely vanishes in the given example for $\bar{v}_{r\|}/v_s \gtrsim 4$ (Fig. 3.2 down right), where $|\Delta v_{\|}|$ gets small compared to $\bar{v}_{r\|}$ and $\Delta v_{\|}(\bar{s})$ behaves fully regular. Increasing the magnetic field (i.e. decreasing $a_s \propto 1/B$) at fixed $\bar{v}_{r\|}$ reduces the development of large transverse spatial excursions of the electron during the collision. The chaotic region then starts at lower impact parameters $\bar{s}$, as shown in Fig. 3.2 by comparing the case $a_s/\lambda = 0.5$ (top left) with $a_s/\lambda = 0.125$ (bottom left), that is, with a four times larger magnetic field. Rather interesting is also the sensitivity with respect to a variation of the regularization parameter $\lambda\!\!\!^{-}/\lambda$ of the interaction. Reducing $\lambda\!\!\!^{-}/\lambda$, i.e. approaching the (screened) Coulomb interaction, leads to an additional region of regular motion at very small impact parameters, see Fig. 3.2 center left compared to top left. Here the Coulomb force dominates the motion and the influence of the magnetic field gets negligible. And, as in the case of Rutherford scattering at small impact parameters $\bar{s}$, the velocity transfer tends to $-\Delta v_{\|}/\bar{v}_{r\|} = 2$ for $\bar{s} \to 0$, that is, backscattering. This is in contrast to the behavior seen for the 'softer' regularized potential ($\lambda\!\!\!^{-}/\lambda = 0.05$) where $\Delta v_{\|} \to 0$ for $\bar{s} \to 0$.

The numerical CTMC results are now compared to the predictions of the perturbative treatment (3.125) which reduces for the specific case discussed in Fig. 3.2 with $v_{e\perp} = 0$, i.e $a = 0, \kappa_n a = 0, \chi_n a = 0$ to

$$\left.\frac{\left\langle \Delta E_{i\|}^{(2)} \right\rangle}{m v_s v_{i\|}}\right|_{a=0} = \left(\frac{v_s}{\bar{v}_{r\|}}\right)^5 \left(\frac{\lambda}{a_s}\right)^2 2\left[(\kappa_1\delta)K_1(\kappa_1\bar{s}) - (\chi_1\delta)K_1(\chi_1\bar{s})\right]^2 . \qquad (3.127)$$

Some examples for different parameter settings are given in Fig. 3.3. While the CTMC results shown in Fig. 3.2 have been taken from 10^5 events for graphical reasons, we employed here 10^7 trajectories and averaged the impact parameters over intervals $\Delta\bar{s}/\lambda = 0.02$. For low initial velocities $\bar{v}_{r\|}$ the chaotic character of the scattering events is still present, despite the averaging procedure with respect to $\bar{s}$, see Fig. 3.3 top left. At $\bar{s} \to 0$ the regular behavior of the more Coulomb–like regularized potential (with $\lambda\!\!\!^{-}/\lambda = 5 \times 10^{-5}$) at small $\bar{s}$ can again nicely be seen. As expected, the perturbative treatment (3.127) reproduces the velocity transfer at large impact parameters. At small impact parameters it yields, however, a $|\Delta v_{\|}|$ which clearly exceeds the allowed upper limit given by energy conservation. This is still the case for the higher $\bar{v}_{r\|}/v_s = 2$ (Fig. 3.3 top right). But here the averaged CTMC results have a regular character (see Fig. 3.2 center right for comparison) and the agreement with the perturbative treatment is perfect for $\bar{s}/\lambda \gtrsim 0.3$. For increasing $\bar{v}_{r\|}$, where no chaotic regime at all has been found in the CTMC, the agreement at smaller impact parameters continuously improves, as shown in Fig. 3.3 center and down left.

As expected for a perturbative approach, expression (3.127) excellently describes the velocity transfer in cases where a small velocity transfer occurs and chaotic motion is completely absent, that is, for sufficiently large initial velocities $\bar{v}_{r\|} \gg 1$ or for large impact parameters (where $\Delta v_{\|}$ gets exponentially small, see Fig. 3.3 bottom right).

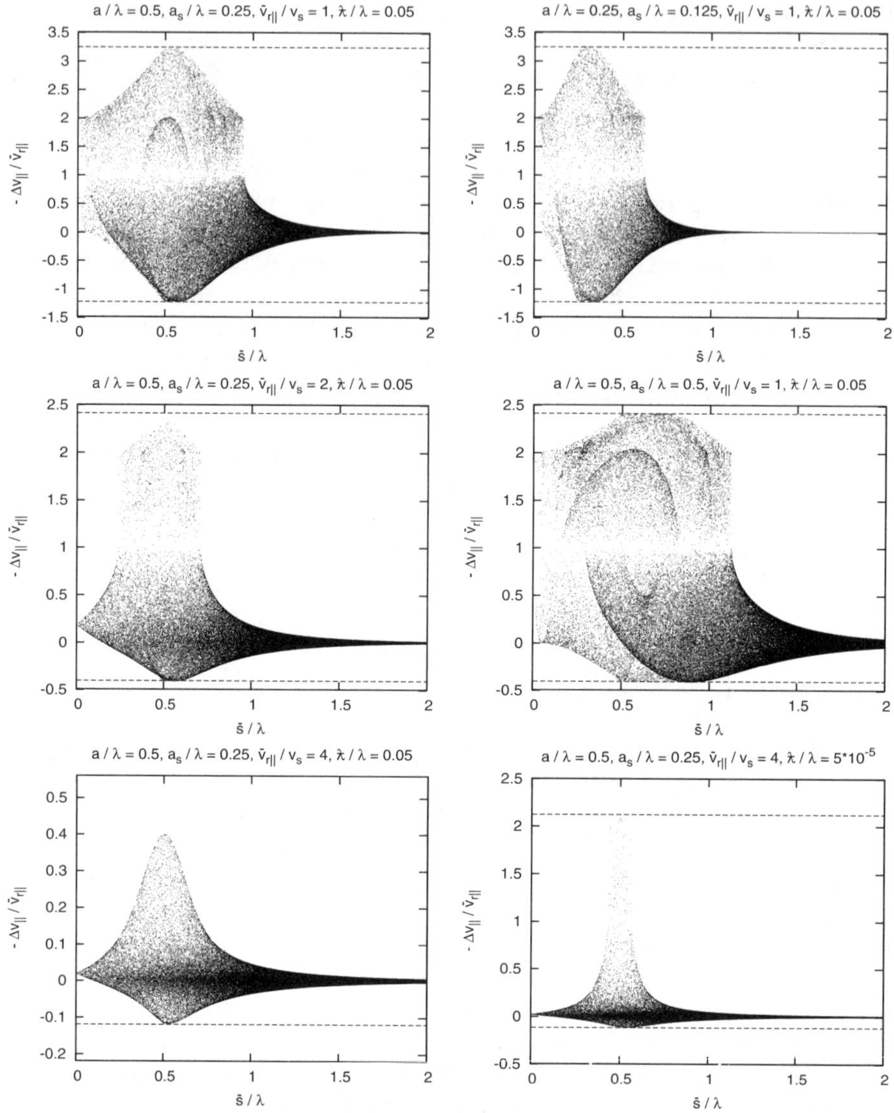

Fig. 3.4. Scaled velocity transfer (3.124) as function of the impact parameter $\bar{s}$ as in Fig. 3.2, but here for finite initial transverse electron velocities $v_{e\perp}/v_s = a/a_s$. Different values of $\Delta v_{\|}$ for the same given impact parameter $\bar{s}$ belong to different randomly chosen initial phases of the unperturbed helical motion of the electrons. The dashed lines indicate the lower and upper limits of the velocity transfer (3.128) due to energy conservation.

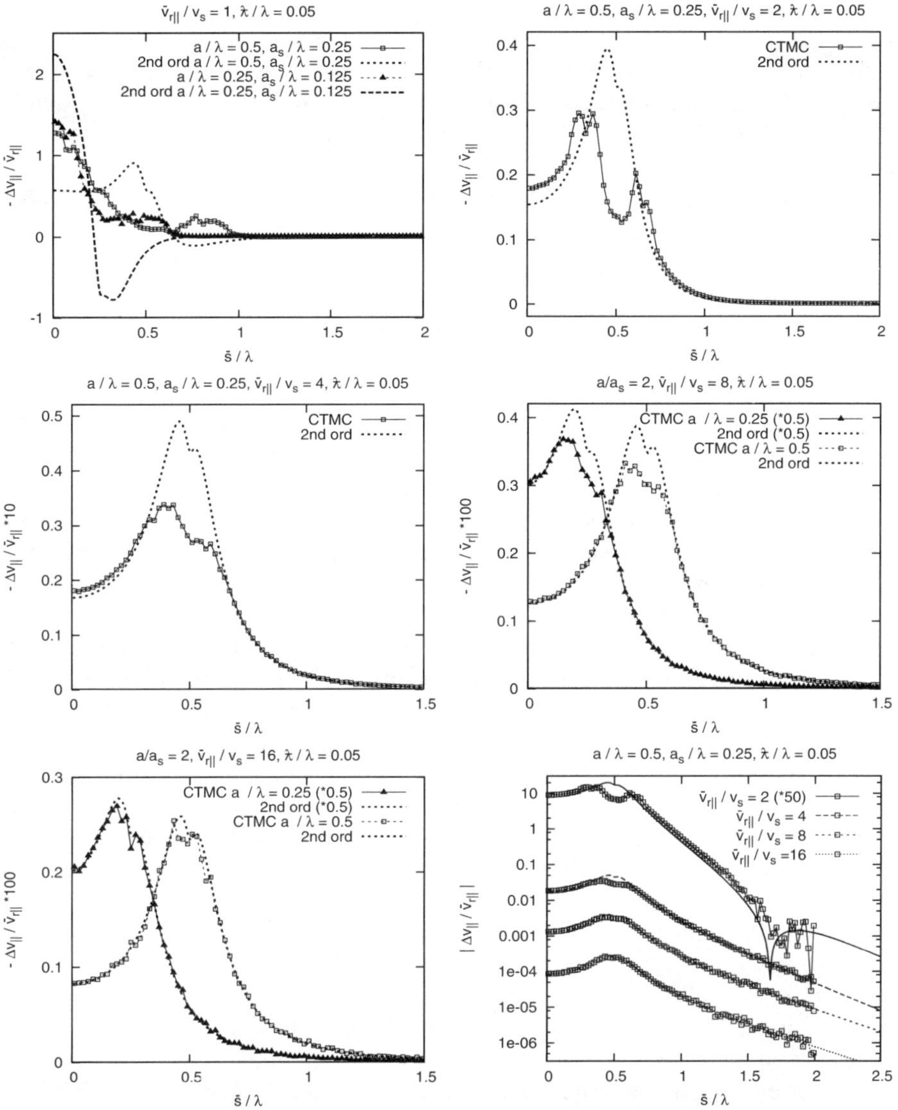

Fig. 3.5. Comparison of the scaled velocity transfer (3.124) from the CTMC calculations as in Fig. 3.4, now averaged over impact parameter intervals $\Delta\bar{s}/\lambda = 0.02$ and the initial phase φ, with the second–order energy transfer (3.125). Note again the different ranges, both of $\bar{s}$ and $\Delta v_{\parallel}$.

This turns out to be essentially true also for the more general case with $v_{e\perp} \neq 0$, where however the initial phase now plays an important role. Some examples for the velocity transfer obtained in the CTMC treatment for $v_{e\perp}/v_s = a/a_s \neq 0$ are shown in Fig. 3.4, taken again from 10^5 calculated trajectories for each parameter setting. The dots for an event with a given impact parameter depend here additionally on a randomly chosen initial phase φ of the unperturbed helical motion of the electrons. The allowed values for the scaled velocity transfer $-\Delta v_{\|}/\bar{v}_{r\|}$ again follow from energy conservation $v_{\|}^2(t \to \infty) + v_{\perp}^2(t \to \infty) = \bar{v}_{r\|}^2 + v_{e\perp}^2$ when setting $v_{\perp}(t \to \infty) = 0$ and are given by

$$1 - \left[1 + \left(\frac{v_{e\perp}}{\bar{v}_{r\|}}\right)^2\right]^{1/2} \leq -\frac{\Delta v_{\|}}{\bar{v}_{r\|}} \leq 1 + \left[1 + \left(\frac{v_{e\perp}}{\bar{v}_{r\|}}\right)^2\right]^{1/2} . \tag{3.128}$$

The lower and upper limits (3.128) for $-\Delta v_{\|}/\bar{v}_{r\|}$ are indicated as dashed lines in Fig. 3.4. Now a chaotic regime can be seen around $\bar{s} = a$, where all allowed velocity transfers are realized except those with vanishing $v_{\|}(t \to \infty)$ i.e. with $-\Delta v_{\|}/\bar{v}_{r\|} = 1$. These chaotic regions are larger and more pronounced for small $\bar{v}_{r\|}/v_s \lesssim 1$ but their size decreases with increasing $\bar{v}_{r\|}$, see the sequence from top left to bottom left in Fig. 3.4. It also shrinks while shifting towards lower $\bar{s}$ for increasing magnetic field, compare the plots on top of Fig. 3.4, but blows up for decreasing $v_{e\perp}/v_s = a/a_s$ at given $\bar{v}_{r\|}$ (top left versus center right in Fig. 3.4) as the total initial energy ($\propto \bar{v}_{r\|}^2 + v_{e\perp}^2$) gets smaller. The actual velocity transfer depends here sensitively both on the impact parameter $\bar{s}$ and the initial phase φ, where the latter dependence dominates. Simply considering an unperturbed helical motion, the closest and most violent ion–electron encounters will happen for impact parameters around the cyclotron radius a and initial phases φ such that the electron (almost) hits the ion. This will result in a large $|\Delta v_{\|}|$ while for a different φ, where the electron passes the ion in a distance of a or larger, $|\Delta v_{\|}|$ will be rather small. Here again the comparison of different regularized potentials as shown in the bottom panel of Fig. 3.4 is interesting. Whereas for the more Coulomb–like interaction ($\lambda\!\!\!^{-}/\lambda = 5 \times 10^{-5}$) scattering events with a velocity transfer out of the full range of allowed values (3.128) occur, only much 'weaker' collisions with much smaller $|\Delta v_{\|}|$ are observed for the 'softer' interaction with $\lambda\!\!\!^{-}/\lambda = 5 \times 10^{-2}$. In all cases, however, the large variation of $|\Delta v_{\|}|$ seen around $\bar{s} = a$ mainly results from the variation in the initial phase. It will thus be interesting to see how the φ (and $\bar{s}$)–averaged CTMC results will compare to the second–order perturbative expression (3.125). To that end the CTMC data for $\Delta v_{\|}(\bar{s}, \varphi)$ as obtained from 10^7 trajectories are averaged again over intervals $\Delta \bar{s}/\lambda = 0.02$, which automatically comprises an average over the initial phase φ.

Some results of this comparison are shown in Fig. 3.5. Even at low initial energies/velocities there are no more signatures for a chaotic character in the numerical data averaged with respect to $\bar{s}$ (at least not within the unavoidable numerical fluctuations). As in the previous case $v_{e\perp} = 0$ the CTMC results and the perturbative treatment disagree at small velocities and small impact parameters (Fig. 3.5 top left), but with increasing initial velocity $\bar{v}_{r\|}$ the agreement again continuously improves and is almost perfect over the full range of impact parameters at sufficiently large $\bar{v}_{r\|}/v_s \gg 1$ (Fig. 3.5 down left).

The general conclusion thus remains the same as previously. For large initial velocities $v_0/v_s = (\bar{v}_{r\|}^2 + v_{e\perp}^2)^{1/2}/v_s \gg 1$ and/or large impact parameters $\bar{s}$ occur only small velocity transfers $|\Delta v_\||/v_0 \ll 1$ and a chaotic behavior of the ion–electron collision is either absent or averages out. Then the non–perturbative numerical CTMC results agree excellently with the predictions of the perturbative treatment, i.e. expression (3.125). In other words the inherent chaotic character of the collision dynamics and the related problems for an analytical description of the energy/velocity transfer turn out to be of no relevance in parameter regimes where a perturbative treatment is applicable.

3.8 Binary Collision Model for Arbitrary Ion Motion in a Strong Field

After this discussion of the energy loss for an ion moving parallel to a magnetic field of arbitrary strength we return to the general case, where the ion velocity has a component transverse to the magnetic field. As mentioned above the integrals in (3.99) cannot be done in this case unless other simplifications are made. In the following we consider strong magnetic fields or, alternatively small electron transverse velocity $v_{e\perp}$ and expand $J_0(aK(\tau)) \simeq 1 - (aK(\tau))^2/4 + \ldots$. We obtain two contributions to the energy transfer, $\langle \Delta E_i^{(2)} \rangle_{\mathrm{I}}$ which is independent of the cyclotron radius and $\langle \Delta E_i^{(2)} \rangle_{\mathrm{II}}$ which is proportional to a^2. We split $\langle \Delta E_i^{(2)} \rangle_{\mathrm{I}}$ according to

$$\left\langle \Delta E_i^{(2)} \right\rangle_{\mathrm{I}} = \left\langle \Delta E_i^{(2)} \right\rangle_{\mathrm{I0}} + \left\langle \Delta E_i^{(2)} \right\rangle_{\mathrm{I1}} + \left\langle \Delta E_i^{(2)} \right\rangle_{\mathrm{I2}} \tag{3.129}$$

where

$$\left\langle \Delta E_i^{(2)} \right\rangle_{\mathrm{I}j} = \frac{2\pi}{m} \int \mathrm{d}\boldsymbol{k} \int \mathrm{d}\boldsymbol{k}'\, U(\boldsymbol{k}) U(\boldsymbol{k}')\, (\boldsymbol{k} \cdot \boldsymbol{v}_i)\, J_0\left(|\boldsymbol{k} + \boldsymbol{k}'|\bar{s}\right) \delta((\boldsymbol{k} + \boldsymbol{k}') \cdot \bar{\boldsymbol{v}}_r)$$
$$\times \operatorname{Im} \int_0^\infty \mathrm{d}\tau e^{\mathrm{i}(\boldsymbol{k}\cdot\bar{\boldsymbol{v}}_r + \mathrm{i}0)\tau} \begin{cases} \tau g_0(\boldsymbol{k}, \boldsymbol{k}')\,, & j = 0 \\ \frac{1}{\Omega_e} \sin(\Omega_e \tau) g_1(\boldsymbol{k}, \boldsymbol{k}')\,, & j = 1 \\ \frac{1}{\Omega_e}(1 - \cos(\Omega_e \tau)) g_2(\boldsymbol{k}, \boldsymbol{k}')\,, & j = 2 \end{cases} . \tag{3.130}$$

We will assume axially symmetric potentials $U(\boldsymbol{k}) = U(|k_\||, k_\perp)$. The integration is done by using cylindrical coordinates oriented along $\bar{\boldsymbol{n}}_r = \bar{\boldsymbol{v}}_r/\bar{v}_r$, i.e. any vector $\boldsymbol{C}$ will be represented as

$$\boldsymbol{C} = C_\|^{(r)} \bar{\boldsymbol{n}}_r + \boldsymbol{C}_\perp^{(r)} . \tag{3.131}$$

(Note that this orientation differs from the one adopted in Sect. 3.6). Inserting the function $g_0(\boldsymbol{k}, \boldsymbol{k}')$ we obtain the contribution

$$\left\langle \Delta E_i^{(2)} \right\rangle_{\mathrm{I0}} = \frac{2\pi^2}{m\bar{v}_r^3} \int \mathrm{d}\boldsymbol{k} \int \mathrm{d}\boldsymbol{k}'\, U(k_\|^{(r)}, k_\perp^{(r)}) U\left(k_\|^{\prime(r)}, k_\perp^{\prime(r)}\right) (\boldsymbol{k} \cdot \boldsymbol{v}_i) \tag{3.132}$$
$$\times J_0\left(\left|\boldsymbol{k} + \boldsymbol{k}'\right| \bar{s}\right) \delta\left(k_\|^{(r)} + k_\|^{\prime(r)}\right) (\boldsymbol{k} \cdot \boldsymbol{b})(\boldsymbol{k}' \cdot \boldsymbol{b}) \delta'\left(k_\|^{(r)}\right) ,$$

where

$$\mathrm{Im} \int_0^\infty \mathrm{e}^{-\mathrm{i}(\omega - \mathrm{i}0)\tau} \tau \mathrm{d}\tau = \pi \delta'(\omega) \tag{3.133}$$

has been used. The integration with respect to $k_\parallel^{(r)}$ and $k_\parallel^{\prime(r)}$ is done with the help of (3.49) with the result

$$\begin{aligned}
\left\langle \Delta E_i^{(2)} \right\rangle_{\mathrm{I0}} &= \frac{2\pi^2}{m\bar{v}_r^3} \int \mathrm{d}^2 \boldsymbol{k}_\perp^{(r)} \int \mathrm{d}^2 \boldsymbol{k}_\perp^{\prime(r)} J_0 \left(\left| \boldsymbol{k}_\perp^{(r)} + \boldsymbol{k}_\perp^{\prime(r)} \right| \bar{s} \right) \\
&\times \left\{ \left(\boldsymbol{k}_\perp^{(r)} \cdot \boldsymbol{b}_\perp^{(r)} \right) \left(\boldsymbol{v}_{i\perp}^{(r)} \cdot \boldsymbol{k}_\perp^{(r)} \right) U\left(0, k_\perp^{(r)}\right) \right. \\
&\times \left[b_\parallel^{(r)} U\left(0, k_\perp^{\prime(r)}\right) + \left(\boldsymbol{k}_\perp^{\prime(r)} \cdot \boldsymbol{b}_\perp^{(r)} \right) \left. \frac{\partial U\left(k_\parallel^{\prime(r)}, k_\perp^{\prime(r)}\right)}{\partial k_\parallel^{\prime(r)}} \right|_{k_\parallel^{\prime(r)}=0} \right] \\
&- \left(\boldsymbol{k}_\perp^{\prime(r)} \cdot \boldsymbol{b}_\perp^{(r)} \right) U\left(0, k_\perp^{\prime(r)}\right) \left[\left[v_{i\parallel}^{(r)} \left(\boldsymbol{k}_\perp^{(r)} \cdot \boldsymbol{b}_\perp^{(r)} \right) + \left(\boldsymbol{v}_{i\perp}^{(r)} \cdot \boldsymbol{k}_\perp^{(r)} \right) b_\parallel^{(r)} \right] U\left(0, k_\perp^{(r)}\right) \right. \\
&+ \left. \left. \left(\boldsymbol{v}_{i\perp}^{(r)} \cdot \boldsymbol{k}_\perp^{(r)} \right) \left(\boldsymbol{k}_\perp^{(r)} \cdot \boldsymbol{b}_\perp^{(r)} \right) \left. \frac{\partial U(k_\parallel^{(r)}, k_\perp^{(r)})}{\partial k_\parallel^{(r)}} \right|_{k_\parallel^{(r)}=0} \right] \right\} .
\end{aligned} \tag{3.134}$$

Now the integrations with respect to the azimuthal angles ϑ and ϑ' of $\boldsymbol{k}_\perp^{(r)}$ and $\boldsymbol{k}_\perp^{\prime(r)}$,

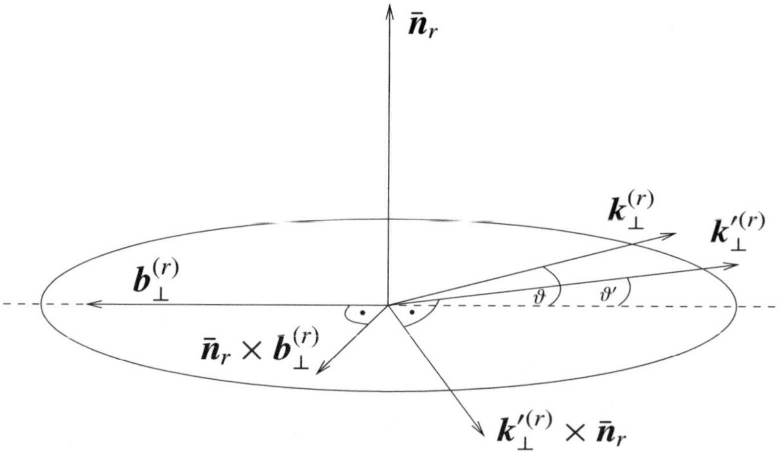

Fig. 3.6. Vectors in the plane tranverse to the relative motion of the guiding center, see (3.131).

respectively, can be done. Let ϑ_b be the angle between $\boldsymbol{b}$ and $\boldsymbol{v}_r$, see Fig. 3.1. Then according to Fig. 3.6

$$\begin{aligned}
\boldsymbol{k}_\perp^{(r)} \cdot \boldsymbol{b}_\perp^{(r)} &= -k_\perp^{(r)} \sin \vartheta_b \cos \vartheta , \\
\boldsymbol{k}_\perp^{\prime(r)} \cdot \boldsymbol{b}_\perp^{(r)} &= -k_\perp^{\prime(r)} \sin \vartheta_b \cos \vartheta' ,
\end{aligned} \tag{3.135}$$

and because of $\boldsymbol{v}_{i\perp}^{(r)} = \boldsymbol{v}_i - \boldsymbol{v}_{i\|}^{(r)} = v_{e\|}\boldsymbol{b} - \bar{\boldsymbol{v}}_r - \boldsymbol{v}_{i\|}^{(r)}$

$$\boldsymbol{k}_\perp^{(r)} \cdot \boldsymbol{v}_{i\perp}^{(r)} = -v_{e\|} k_\perp^{(r)} \sin\vartheta_b \cos\vartheta \,. \tag{3.136}$$

For the Bessel functions we use the addition theorem (3.52). The angular integrals are trivial as they involve powers of trigonometric functions. For symmetry reasons all terms involving odd powers of $\sin\vartheta_b = -v_{i\perp}/\bar{v}_r$ vanish. There remains

$$\begin{aligned}
\left\langle \Delta E_i^{(2)} \right\rangle_{\mathrm{IO}} &= \frac{(2\pi^2)^2}{m\bar{v}_r^3} \frac{v_{i\perp}^2}{\bar{v}_r^2} \int_0^\infty \mathrm{d}k_\perp^{(r)} k_\perp^{(r)} U\left(0, k_\perp^{(r)}\right) \int_0^\infty \mathrm{d}k_\perp^{\prime(r)} k_\perp^{\prime(r)} U\left(0, k_\perp^{\prime(r)}\right) \\
&\times \Bigg[v_{e\|}\, b_\|^{(r)} k_\perp^{(r)2} J_0\left(k_\perp^{(r)} \bar{s}\right) J_0\left(k_\perp^{\prime(r)} \bar{s}\right) \\
&\quad + \left(v_{i\|}^{(r)} + v_{e\|} b_\|^{(r)}\right) k_\perp^{(r)} k_\perp^{\prime(r)} J_1\left(k_\perp^{(r)} \bar{s}\right) J_1\left(k_\perp^{\prime(r)} \bar{s}\right) \Bigg] \\
&= \frac{(2\pi^2)^2}{m\bar{v}_r^3} \frac{v_{i\perp}^2}{\bar{v}_r^2} \left[v_{e\|} b_\|^{(r)} T_{0,3}(0,\bar{s}) T_{0,1}(0,\bar{s}) + \left(v_{i\|}^{(r)} + v_{e\|} b_\|^{(r)}\right) T_{1,2}^2(0,\bar{s}) \right]
\end{aligned} \tag{3.137}$$

with the transforms $T_{\mu,\nu}$ introduced in (3.54). Inserting $b_\|^{(r)} = \cos\vartheta_b = \bar{v}_{r\|}/\bar{v}_r = (v_{e\|} - v_{i\|})/\bar{v}_r$ and $v_{i\|}^{(r)} = v_{i\|} \cos\vartheta_b + v_{i\perp} \sin\vartheta_b = (v_{i\|}(v_{e\|} - v_{i\|}) - v_{i\perp}^2)/\bar{v}_r$ this contribution to the energy loss is

$$\left\langle \Delta E_i^{(2)} \right\rangle_{\mathrm{IO}} = \frac{(2\pi^2)^2}{m\bar{v}_r^6} v_{i\perp}^2 \left[v_{e\|} \left(v_{e\|} - v_{i\|}\right) T_{0,3}\left(0,\bar{s}\right) T_{0,1}\left(0,\bar{s}\right) + \left(v_{e\|}^2 - v_i^2\right) T_{1,2}^2(0,\bar{s}) \right] . \tag{3.138}$$

For further progress the transforms (3.54) of the interaction potentials must be done. From (3.55) and its derivatives with respect to a we obtain in the screened case (3.6)

$$\begin{aligned}
T_{0,1}^D(0,\bar{s}) &= -\frac{Z e^2}{2\pi^2} K_0(\bar{s}/\lambda) \,, \\
T_{1,2}^D(0,\bar{s}) &= -\frac{Z e^2}{2\pi^2 \lambda} K_1(\bar{s}/\lambda) \,, \\
T_{0,3}^D(0,\bar{s}) &= \frac{Z e^2}{2\pi^2 \lambda^2} K_0(\bar{s}/\lambda)
\end{aligned} \tag{3.139}$$

and for the regularized and screened potential (3.7)

$$\begin{aligned}
T_{0,1}^R(0,\bar{s}) &= -\frac{Z e^2}{2\pi^2} \left[K_0(\bar{s}/\lambda) - K_0(\bar{s}/\lambda')\right] \,, \\
T_{1,2}^R(0,\bar{s}) &= -\frac{Z e^2}{2\pi^2} \left[\frac{1}{\lambda} K_1(\bar{s}/\lambda) - \frac{1}{\lambda'} K_1(\bar{s}/\lambda')\right] \,, \\
T_{0,3}^R(0,\bar{s}) &= \frac{Z e^2}{2\pi^2} \left[\frac{1}{\lambda^2} K_0(\bar{s}/\lambda) - \frac{1}{\lambda'^2} K_0(\bar{s}/\lambda')\right] .
\end{aligned} \tag{3.140}$$

We investigate the asymptotic behavior of the amplitudes in the limit $\bar{s} \to 0$. As $K_0(z) \sim -\ln z$ and $K_1(z) \sim z^{-1} + z \ln \frac{z}{2}$

$$
\begin{aligned}
T^R_{0,1}(0,\bar{s}) &\sim -\frac{Z\not{e}^2}{2\pi^2}\ln\frac{\lambda}{\lambda'}\,,\\
T^R_{1,2}(0,\bar{s}) &\sim -\frac{Z\not{e}^2}{2\pi^2}\,\bar{s}\left(\frac{1}{\lambda^2}\ln\frac{\bar{s}}{2\lambda}-\frac{1}{\lambda'^2}\ln\frac{\bar{s}}{2\lambda'}\right)\,,\\
T^R_{0,3}(0,\bar{s}) &\sim \frac{Z\not{e}^2}{2\pi^2}\left(\frac{1}{\lambda^2}\ln\frac{\bar{s}}{\lambda}-\frac{1}{\lambda'^2}\ln\frac{\bar{s}}{\lambda'}\right)
\end{aligned}
\tag{3.141}
$$

the divergence is not worse than logarithmic and will cause no harm when integrating over the impact parameter in (2.9). This is not so in the case of the screened potential. Insertion of (3.139) into (3.138) yields

$$
\left\langle \Delta E_i^{(2)}\right\rangle_{\mathrm{I0}} = \frac{Z^2\not{e}^4 v_{i\perp}^2}{m\bar{v}_r^6\bar{s}^2}\left[\left(v_{e\|}^2-v_i^2\right)\rho^2K_1^2(\rho)-v_{e\|}(v_{e\|}-v_{i\|})\rho^2K_0^2(\rho)\right] \tag{3.142}
$$

with $\rho=\bar{s}/\lambda$, which behaves like $\bar{s}^{-2}$ for small impact parameters. On the other hand all amplitudes (3.139) and (3.140) vanish exponentially for large impact parameters because of the finite range of the potentials U_R and U_D. We obtain the Coulomb case by taking the limit $\lambda\to\infty$ in (3.142). This yields

$$
\left\langle \Delta E_i^{(2)}\right\rangle_{\mathrm{I0}} = \frac{Z^2\not{e}^4 v_{i\perp}^2}{m\bar{v}_r^6\bar{s}^2}(v_{e\|}^2-v_i^2) \tag{3.143}
$$

which is precisely the energy transfer (2.36) for tight helices.

We turn now to the next term in $\langle\Delta E_i^{(2)}\rangle_{\mathrm{I1}}$ in (3.130). It involves

$$
\frac{1}{\Omega_e}\,\mathrm{Im}\int_0^\infty \mathrm{d}\tau\, e^{\mathrm{i}(\boldsymbol{k}\cdot\bar{\boldsymbol{v}}_r+\mathrm{i}0)\tau}\sin(\Omega_e\tau)=\frac{\pi}{2\Omega_e}\left[\delta(\boldsymbol{k}\cdot\bar{\boldsymbol{v}}_r-\Omega_e)-\delta(\boldsymbol{k}\cdot\bar{\boldsymbol{v}}_r+\Omega_e)\right]\,. \tag{3.144}
$$

Then

$$
\begin{aligned}
\left\langle \Delta E_i^{(2)}\right\rangle_{\mathrm{I1}} &= \frac{2\pi^2}{m\Omega_e}\int \mathrm{d}\boldsymbol{k}\int \mathrm{d}\boldsymbol{k}'\,U(\boldsymbol{k})\,U(\boldsymbol{k}')\,(\boldsymbol{k}\cdot\boldsymbol{v}_i)\,J_0\left(|\boldsymbol{k}+\boldsymbol{k}'|\bar{s}\right)\\
&\quad\times\left[(\boldsymbol{k}\cdot\boldsymbol{b})(\boldsymbol{k}'\cdot\boldsymbol{b})-(\boldsymbol{k}\cdot\boldsymbol{k}')\right]\delta\left((\boldsymbol{k}+\boldsymbol{k}')\cdot\bar{\boldsymbol{v}}_r\right)\delta\left(\boldsymbol{k}\cdot\bar{\boldsymbol{v}}_r-\Omega_e\right)\,.
\end{aligned}
\tag{3.145}
$$

Using cylindrical coordinates the integrations are done as before between (3.134) and (3.137). With the help of the δ–functions we obtain

$$
\begin{aligned}
\left\langle \Delta E_i^{(2)}\right\rangle_{\mathrm{I1}} &= \frac{2\pi^2}{m\bar{v}_r^2\Omega_e}\int \mathrm{d}^2\boldsymbol{k}_\perp^{(r)}\int \mathrm{d}^2\boldsymbol{k}_\perp'^{(r)}\,U\left(\delta^{-1},k_\perp^{(r)}\right)U\left(\delta^{-1},k_\perp'^{(r)}\right)\\
&\quad\times J_0\left(|\boldsymbol{k}_\perp^{(r)}+\boldsymbol{k}_\perp'^{(r)}|\bar{s}\right)\left(\delta^{-1}v_{i\|}^{(r)}+\boldsymbol{k}_\perp^{(r)}\cdot\boldsymbol{v}_{i\perp}^{(r)}\right)\left[\delta^{-2}b_\perp^{(r)2}+\left(\boldsymbol{k}_\perp^{(r)}\cdot\boldsymbol{b}_\perp^{(r)}\right)\left(\boldsymbol{k}_\perp'^{(r)}\cdot\boldsymbol{b}_\perp^{(r)}\right)\right.\\
&\quad\left.-\left(\boldsymbol{k}_\perp^{(r)}\cdot\boldsymbol{k}_\perp'^{(r)}\right)+\delta^{-1}b_\|^{(r)}\left(\left(\boldsymbol{k}_\perp'^{(r)}-\boldsymbol{k}_\perp^{(r)}\right)\cdot\boldsymbol{b}_\perp^{(r)}\right)\right]\,.
\end{aligned}
\tag{3.146}
$$

The angular integrations yield

$$
\begin{aligned}
\left\langle \Delta E_i^{(2)}\right\rangle_{\mathrm{I1}} &= \frac{(2\pi^2)^2}{m\bar{v}_r^6}\left\{v_{i\perp}^2\left[2\delta^{-2}\left(v_{e\|}v_{i\|}-v_i^2\right)T_{0,1}^2\left(\delta^{-1},\bar{s}\right)\right.\right.\\
&\quad\left.-\,v_{e\|}\bar{v}_{r\|}T_{0,1}\left(\delta^{-1},\bar{s}\right)T_{0,3}\left(\delta^{-1},\bar{s}\right)\right]\\
&\quad\left.+\left[2\bar{v}_r^2\left(v_{e\|}v_{i\|}-v_i^2\right)-v_{i\perp}^2\left(v_{e\|}^2-v_i^2\right)\right]T_{1,2}^2\left(\delta^{-1},\bar{s}\right)\right\}\,.
\end{aligned}
\tag{3.147}
$$

For the potentials (3.6), (3.7) we obtain with the integrals (3.55)

$$\begin{aligned} T^D_{0,1}(\delta^{-1}, \bar{s}) &= -\frac{Z e^2}{2\pi^2} \, K_0(\kappa_1 \bar{s}) \, , \\ T^D_{1,2}(\delta^{-1}, \bar{s}) &= -\frac{Z e^2}{2\pi^2} \, \kappa_1 K_1(\kappa_1 \bar{s}) \, , \\ T^D_{0,3}(\delta^{-1}, \bar{s}) &= \frac{Z e^2}{2\pi^2} \, \kappa_1^2 K_0(\kappa_1 \bar{s}) \end{aligned} \tag{3.148}$$

in the screened case (3.2) and

$$\begin{aligned} T^R_{0,1}(\delta^{-1}, \bar{s}) &= -\frac{Z e^2}{2\pi^2} \, \left[K_0(\kappa_1 \bar{s}) - K_0(\chi_1 \bar{s})\right] \, , \\ T^R_{1,2}(\delta^{-1}, \bar{s}) &= -\frac{Z e^2}{2\pi^2} \, \left[\kappa_1 K_1(\kappa_1 \bar{s}) - \chi_1 K_1(\chi_1 \bar{s})\right] \, , \\ T^R_{0,3}(\delta^{-1}, \bar{s}) &= \frac{Z e^2}{2\pi^2} \left[\kappa_1^2 K_0(\kappa_1 \bar{s}) - \chi_1^2 K_0(\chi_1 \bar{s})\right] \end{aligned} \tag{3.149}$$

with $\kappa_1^2 = \delta^{-2} + \lambda^{-2}$ and $\chi_1^2 = \delta^{-2} + \lambda'^{-2}$ in the regularized case (3.3). As before, see (3.141), the behavior in the limit $\bar{s} \to 0$ is logarithmic for the regularized potential (3.7). For the screened potential (3.6) the insertion of (3.148) yields

$$\begin{aligned} \left\langle \Delta E_i^{(2)} \right\rangle_{\mathrm{I1}} = \frac{Z^2 e^4}{m \bar{v}_r^6 \bar{s}^2} \Big\{ & v_{i\perp}^2 \left[v_{e\|} \bar{v}_{r\|} + \frac{2}{(\kappa_1 \delta)^2} \left(v_{e\|} v_{i\|} - v_i^2 \right) \right] \rho_1^2 K_0^2(\rho_1) \\ & + \left[2\bar{v}_r^2 \left(v_{e\|} v_{i\|} - v_i^2 \right) - v_{i\perp}^2 \left(v_{e\|}^2 - v_i^2 \right) \right] \rho_1^2 K_1^2(\rho_1) \Big\} \end{aligned} \tag{3.150}$$

with $\rho_1 = \kappa_1 \bar{s}$. For parallel ion motion, $v_{i\perp} \to 0$, there remains a contribution to $\langle \Delta E_i^{(2)} \rangle_{\mathrm{I1}}$, which is equal to the leading term of the corresponding expansion of (3.118) in orders of a/δ.

Taking now the limit $\lambda \to \infty$ for the unscreened Coulomb interaction we see that $\langle \Delta E_i^{(2)} \rangle_{\mathrm{I1}}$ vanishes exponentially for $\bar{s} > \delta$. Hence the energy transfer is given by the tight helix term (3.143). For the case $\bar{s} < \delta$ and $\rho_1 = 0$ we obtain

$$\left\langle \Delta E_i^{(2)} \right\rangle_{\mathrm{I0}} + \left\langle \Delta E_i^{(2)} \right\rangle_{\mathrm{I1}} = \left(\frac{Z e^2}{\bar{s}} \right)^2 \frac{2 \bar{\boldsymbol{v}}_r \cdot \boldsymbol{v}_i}{m \bar{v}_r^4} \, . \tag{3.151}$$

This is (up to regularization) just the stretched helix case obtained before by elementary considerations in (2.32). There remains in (3.129) the contribution $\langle \Delta E_i^{(2)} \rangle_{\mathrm{I2}}$ to the energy loss, which involves the function

$$\begin{aligned} g_2(\boldsymbol{k}, \boldsymbol{k}') = \boldsymbol{k} \cdot [\boldsymbol{k}' \times \boldsymbol{b}] &= b_\|^{(r)} \boldsymbol{k}_\perp^{(r)} \cdot [\boldsymbol{k}_\perp'^{(r)} \times \bar{\boldsymbol{n}}_r] + k_\|'^{(r)} \boldsymbol{k}_\perp'^{(r)} \cdot [\bar{\boldsymbol{n}}_r \times \boldsymbol{b}_\perp^{(r)}] \\ &= b_\|^{(r)} k_\perp^{(r)} k_\perp'^{(r)} \sin(\vartheta - \vartheta') - b_\perp^{(r)} k_\perp^{(r)} k_\|'^{(r)} \sin \vartheta + b_\perp^{(r)} k_\|^{(r)} k_\perp'^{(r)} \sin \vartheta' \, , \end{aligned} \tag{3.152}$$

see Fig. 3.6. Because of the sine functions the integrations with respect to the azimuthal angles ϑ and ϑ' yield zero.

These results are obtained in the limit $a \to 0$ where the electrons move along their guiding center trajectories. Moreover, for $\Omega_e \to \infty$ also the pitch $\delta \to 0$ and these trajectories are rectilinear along the lines of the magnetic field and the energy transfer is given by $\langle \Delta E_i^{(2)} \rangle_{\mathrm{I0}}$. For a finite Ω_e corresponding to a finite pitch the contribution $\langle \Delta E_i^{(2)} \rangle_{\mathrm{I1}}$ describes the perturbation of the guiding center trajectory.

The quadratic term in the expansion of the Bessel function $J_0(aK(\tau)) = 1 - (aK(\tau))^2/4 + \ldots$ accounts for the finite cyclotron motion. Using the same techniques as before a tedious but straightforward calculation yields

$$\left\langle \Delta E_i^{(2)} \right\rangle_{\mathrm{II}} = \left(\frac{Ze^2 a}{\bar{s}^2} \right)^2 \frac{1}{2m\bar{v}_r^2} \left(Q_0 + Q_1 + Q_2 \right) . \tag{3.153}$$

Here we restrict ourselves to give the result for the screened potential (3.2), (3.6):

$$\begin{aligned} Q_0 = &\frac{v_{e\|}\bar{v}_{r\|}v_{i\perp}^2}{\bar{v}_r^4} \left(\frac{v_{i\perp}^2}{8\bar{v}_r^2} - \frac{\bar{v}_{r\|}^2}{\bar{v}_r^2} \right) \left[\rho^2 K_0(\rho) \right]^2 + \frac{v_{i\perp}^2}{\bar{v}_r^2} \left[\frac{v_{e\|}v_{i\|} - v_i^2}{\bar{v}_r^2} \right. \\ &\times \left. \left(1 - \frac{3v_{i\perp}^2}{4\bar{v}_r^2} \right) + \frac{v_{e\|}\bar{v}_{r\|}}{\bar{v}_r^2} \left(1 - \frac{3v_{i\perp}^2}{2\bar{v}_r^2} \right) \right] \left[\rho^2 K_1(\rho) \right]^2 + \frac{3v_{e\|}\bar{v}_{r\|}v_{i\perp}^4}{8\bar{v}_r^6} \left[\rho^2 K_2(\rho) \right]^2 , \end{aligned} \tag{3.154}$$

$$\begin{aligned} Q_1 = &\left\{ \frac{v_{e\|}\bar{v}_{r\|}v_{i\perp}^2}{\bar{v}_r^4} \left(1 - \frac{17v_{i\perp}^2}{8\bar{v}_r^2} \right) + \frac{\bar{v}_{r\|}^2 \left(v_{e\|}v_{i\|} - v_i^2 \right)}{\bar{v}_r^4} \left(1 + \frac{v_{i\perp}^2}{2\bar{v}_r^2} \right) \right. \\ &+ \frac{1}{(\kappa_1\delta)^2} \frac{v_{i\perp}^2}{\bar{v}_r^2} \left[\frac{v_{e\|}\bar{v}_{r\|}}{\bar{v}_r^2} \left(5 - \frac{12v_{i\perp}^2}{\bar{v}_r^2} \right) + \frac{v_{e\|}v_{i\|} - v_i^2}{\bar{v}_r^2} \left(10 - \frac{11v_{i\perp}^2}{\bar{v}_r^2} \right) \right] \\ &+ \left. \frac{2}{(\kappa_1\delta)^4} \frac{v_{i\perp}^2}{\bar{v}_r^2} \left[\frac{v_{e\|}\bar{v}_{r\|}}{\bar{v}_r^2} \left(1 - \frac{2v_{i\perp}^2}{\bar{v}_r^2} \right) + \frac{v_{e\|}v_{i\|} - v_i^2}{\bar{v}_r^2} \left(7 - \frac{8v_{i\perp}^2}{\bar{v}_r^2} \right) \right] \right\} \left[\rho_1^2 K_0(\rho_1) \right]^2 \\ &+ \left\{ -\frac{3v_{e\|}\bar{v}_{r\|}^3 v_{i\perp}^2}{\bar{v}_r^6} + \frac{v_{e\|}v_{i\|} - v_i^2}{\bar{v}_r^2} \left(\frac{3v_{i\perp}^4}{4\bar{v}_r^4} + \frac{2\bar{v}_{r\|}^2}{\bar{v}_r^2} \right) \right. \\ &+ \left. \frac{1}{(\kappa_1\delta)^2} \left[\frac{v_{e\|}\bar{v}_{r\|}v_{i\perp}^2}{\bar{v}_r^4} \left(\frac{13v_{i\perp}^2}{\bar{v}_r^2} - 7 \right) + \frac{v_{e\|}v_{i\|} - v_i^2}{\bar{v}_r^2} \left(\frac{16v_{i\perp}^4}{\bar{v}_r^4} - \frac{23v_{i\perp}^2}{\bar{v}_r^2} + 6 \right) \right] \right\} \left[\rho_1^2 K_1(\rho_1) \right]^2 \\ &+ \left[\frac{v_{e\|}\bar{v}_{r\|}v_{i\perp}^2}{2\bar{v}_r^4} \left(4 - \frac{7v_{i\perp}^2}{4\bar{v}_r^2} \right) - \frac{v_{e\|}v_{i\|} - v_i^2}{\bar{v}_r^2} \left(1 - \frac{3v_{i\perp}^2}{2\bar{v}_r^2} + \frac{v_{i\perp}^4}{4\bar{v}_r^4} \right) \right] \left[\rho_1^2 K_2(\rho_1) \right]^2 \\ &- \frac{1}{(\kappa_1\delta)^2} \frac{v_{e\|}v_{i\|} - v_i^2}{\bar{v}_r^2} \left[\frac{v_{i\perp}^2}{\bar{v}_r^2} \left(2 - \frac{3v_{i\perp}^2}{2\bar{v}_r^2} \right) + \frac{2}{(\kappa_1\delta)^2} \left(2 - \frac{3v_{i\perp}^2}{\bar{v}_r^2} + \frac{2v_{i\perp}^4}{\bar{v}_r^4} \right) \right. \\ &+ \left. \frac{4}{(\kappa_1\delta)^4} \frac{\bar{v}_{r\|}^2 v_{i\perp}^2}{\bar{v}_r^4} \right] \rho_1^5 K_0(\rho_1) K_1(\rho_1) , \end{aligned} \tag{3.155}$$

$$
\begin{aligned}
Q_2 &= \frac{v_{i\perp}^4}{\bar{v}_r^4}\left(1+\frac{8}{(\kappa_2\delta)^2}\right)\left[\frac{v_{e\parallel}\bar{v}_{r\parallel}}{\bar{v}_r^2}+\frac{v_{e\parallel}v_{i\parallel}-v_i^2}{2\bar{v}_r^2}\left(1+\frac{8}{(\kappa_2\delta)^2}\right)\right]\left[\rho_2^2K_0(\rho_2)\right]^2 \\
&+\frac{2v_{i\perp}^2}{\bar{v}_r^2}\left\{\frac{v_{e\parallel}\bar{v}_{r\parallel}}{\bar{v}_r^2}\left(1-\frac{3v_{i\perp}^2}{4\bar{v}_r^2}\right)+\frac{4}{(\kappa_2\delta)^2}\right. \\
&\times\left.\left[\frac{2\left(v_{e\parallel}v_{i\parallel}-v_i^2\right)}{\bar{v}_r^2}\left(1+\frac{\bar{v}_{r\parallel}^2}{\bar{v}_r^2}\right)-\frac{v_{e\parallel}\bar{v}_{r\parallel}v_{i\perp}^2}{\bar{v}_r^4}\right]\right\}\left[\rho_2^2K_1(\rho_2)\right]^2 \qquad (3.156)\\
&+\left[\frac{v_{e\parallel}v_{i\parallel}-v_i^2}{\bar{v}_r^2}\left(\frac{2\bar{v}_{r\parallel}^2}{\bar{v}_r^2}+\frac{v_{i\perp}^4}{4\bar{v}_r^4}\right)+\frac{\bar{v}_{r\parallel}v_{e\parallel}v_{i\perp}^2}{\bar{v}_r^4}\left(\frac{v_{i\perp}^2}{2\bar{v}_r^2}-2\right)\right]\left[\rho_2^2K_2(\rho_2)\right]^2 .
\end{aligned}
$$

Here $\kappa_2^2 = 4\delta^{-2} + \lambda^{-2}$, $\rho_2 = \kappa_2\bar{s}$. The cyclotron motion and the drift of the guiding center of the electron are coupled to each other. Therefore the perturbation of the cyclotron motion causes an additional perturbation of the guiding center motion. This effect is given by the first term Q_0 in (3.153) which depends on magnetic field through the cyclotron radius a in the prefactor, while the arguments of the modified Bessel functions in (3.154) do not depend on magnetic field. In the other terms Q_1 and Q_2 the arguments of the Bessel functions ρ_1 and ρ_2 correspond to the first and second cyclotron harmonic perturbations, respectively.

For $v_{i\perp} = 0$ we have $Q_0 = 0$ and

$$
\begin{aligned}
Q_1 &= \frac{v_{i\parallel}}{\bar{v}_{r\parallel}}\left\{\rho_1^4\left[K_0^2(\rho_1)-K_2^2(\rho_1)\right]+\left[2+\frac{6}{(\kappa_1\delta)^2}\right]\rho_1^4K_1^2(\rho_1)\right. \qquad (3.157)\\
&-\left.\frac{4}{(\kappa_1\delta)^4}\rho_1^5K_0(\rho_1)K_1(\rho_1)\right\} ,
\end{aligned}
$$

$$
Q_2 = \frac{2v_{i\parallel}}{\bar{v}_{r\parallel}}\left[\rho_2^2K_2(\rho_2)\right]^2 . \qquad (3.158)
$$

With the help of the recursion relations of the Bessel functions it is easy to see that the resulting energy transfer $\langle\Delta E_{i\parallel}\rangle_{\mathrm{II}}$ agrees with the corresponding $O(a^2/\delta^2)$–term of the energy transfer (3.118), where the limit of parallel ion motion $v_{i\perp} \to 0$ was taken before the limit $a \ll \delta$.

3.9 Binary Collisions in a Weak Field

We return to the case of a weak magnetic field, where the cyclotron radius a is larger than other characteristic length scales like the screening length λ while the cyclotron frequency Ω_e is small compared to the plasma frequency. Very high cyclotron harmonics with $n \gg 1$ will now contribute to the motion of the particles as described by the expansion (3.91). We start therefore before, on the level of (3.87)-(3.90), and expand up to terms $O(\Omega_e^2)$. For that purpose we rewrite the helical motion (3.62) using $\boldsymbol{v}_{e\perp} = v_{e\perp}\boldsymbol{u}$

$$
\begin{aligned}
\boldsymbol{r}_0(t) &= \bar{\boldsymbol{R}}_0 + \bar{\boldsymbol{v}}_r t + a\left[\boldsymbol{u}\sin(\Omega_e t) - (\boldsymbol{b}\times\boldsymbol{u})\cos(\Omega_e t)\right] \\
&= \boldsymbol{R}_0 + \bar{\boldsymbol{v}}_r t + \frac{1}{\Omega_e}\left[\boldsymbol{v}_{e\perp}\sin(\Omega_e t) + (\boldsymbol{b}\times\boldsymbol{v}_{e\perp})(1-\cos(\Omega_e t))\right] \qquad (3.159)\\
&\simeq \boldsymbol{R}_0 + \boldsymbol{v}_r t + \frac{\Omega_e t^2}{2}(\boldsymbol{b}\times\boldsymbol{v}_{e\perp}) - \frac{\Omega_e^2 t^3}{6}\boldsymbol{v}_{e\perp} = \tilde{\boldsymbol{r}}_0(t) + \tilde{\boldsymbol{r}}_1(t) + \tilde{\boldsymbol{r}}_2(t)\,.
\end{aligned}
$$

Here $\tilde{\boldsymbol{r}}_0(t) = \boldsymbol{R}_0 + \boldsymbol{v}_r t$ with $\boldsymbol{R}_0 = \bar{\boldsymbol{R}}_0 - (\boldsymbol{b}\times\boldsymbol{v}_{e\perp})$ and $\boldsymbol{v}_r = \bar{\boldsymbol{v}}_r + \boldsymbol{v}_{e\perp}$ describes the uniform motion of the field free case and $\tilde{\boldsymbol{r}}_1(t) = \frac{\Omega_e t^2}{2}(\boldsymbol{b}\times\boldsymbol{v}_{e\perp})$ and $\tilde{\boldsymbol{r}}_2(t) = -\frac{\Omega_e^2 t^3}{6}\boldsymbol{v}_{e\perp}$ are the first and second order trajectory corrections, respectively. The second–order energy transfer (3.87) involves the multiple time integrals (3.90) and these in turn functions like $\exp[i\boldsymbol{k}\cdot\boldsymbol{r}_0(t)]$. Their expansions

$$
\mathrm{e}^{\mathrm{i}\boldsymbol{k}\cdot\boldsymbol{r}_0(t)} \simeq \mathrm{e}^{\mathrm{i}\boldsymbol{k}\cdot\tilde{\boldsymbol{r}}_0(t)}\left[1 + \mathrm{i}\boldsymbol{k}\cdot\tilde{\boldsymbol{r}}_1(t) + \mathrm{i}\boldsymbol{k}\cdot\tilde{\boldsymbol{r}}_2(t) - \frac{(\boldsymbol{k}\cdot\tilde{\boldsymbol{r}}_1(t))^2}{2}\right] \qquad (3.160)
$$

are inserted into $\psi_0(\boldsymbol{k},\boldsymbol{k}')$ and $\psi_\pm(\boldsymbol{k},\boldsymbol{k}')$. This yields up to second order

$$
\begin{aligned}
\psi_0(\boldsymbol{k},\boldsymbol{k}') &= \int_{-\infty}^{\infty}\mathrm{d}t\,\mathrm{e}^{\mathrm{i}\boldsymbol{k}\cdot\boldsymbol{r}_0(t)}\int_{-\infty}^{t}\mathrm{d}\tau(t-\tau)\mathrm{e}^{\mathrm{i}\boldsymbol{k}'\cdot\boldsymbol{r}_0(\tau)} \qquad (3.161)\\
&\simeq \int_{-\infty}^{\infty}\mathrm{d}t\,\mathrm{e}^{\mathrm{i}\boldsymbol{k}\cdot\tilde{\boldsymbol{r}}_0(t)}\int_{-\infty}^{t}\mathrm{d}\tau(t-\tau)\mathrm{e}^{\mathrm{i}\boldsymbol{k}'\cdot\tilde{\boldsymbol{r}}_0(t')}\Big[1 + \mathrm{i}\boldsymbol{k}\cdot\tilde{\boldsymbol{r}}_1(t) + \mathrm{i}\boldsymbol{k}'\cdot\tilde{\boldsymbol{r}}_1(\tau)\\
&\quad + \mathrm{i}\boldsymbol{k}\cdot\tilde{\boldsymbol{r}}_2(t) + \mathrm{i}\boldsymbol{k}'\cdot\tilde{\boldsymbol{r}}_2(\tau) - \frac{1}{2}\left(\boldsymbol{k}\cdot\tilde{\boldsymbol{r}}_1(t) + \boldsymbol{k}'\cdot\tilde{\boldsymbol{r}}_1(\tau)\right)^2\Big]\\
&= \psi_0^{(0)}(\boldsymbol{k},\boldsymbol{k}') + \psi_0^{(1)}(\boldsymbol{k},\boldsymbol{k}') + \psi_0^{(2)}(\boldsymbol{k},\boldsymbol{k}')\,,
\end{aligned}
$$

$$
\begin{aligned}
\psi_\pm(\boldsymbol{k},\boldsymbol{k}') &= \frac{1}{\pm 2\mathrm{i}\Omega_e}\int_{-\infty}^{\infty}\mathrm{d}t\,\mathrm{e}^{\mathrm{i}\boldsymbol{k}\cdot\boldsymbol{r}_0(t)}\int_{-\infty}^{t}\mathrm{d}\tau\left(\mathrm{e}^{\pm\mathrm{i}\Omega_e(t-\tau)} - 1\right)\mathrm{e}^{\mathrm{i}\boldsymbol{k}'\cdot\boldsymbol{r}_0(\tau)} \qquad (3.162)\\
&\simeq \frac{1}{\pm 2\mathrm{i}\Omega_e}\int_{-\infty}^{\infty}\mathrm{d}t\,\mathrm{e}^{\mathrm{i}\boldsymbol{k}\cdot\tilde{\boldsymbol{r}}_0(t)}\int_{\infty}^{t}\mathrm{d}\tau\mathrm{e}^{\mathrm{i}\boldsymbol{k}\cdot\tilde{\boldsymbol{r}}_0(\tau)}\Big\{\pm\mathrm{i}\Omega_e(t-\tau)\Big[1 + \mathrm{i}\boldsymbol{k}\cdot\tilde{\boldsymbol{r}}_1(t)\\
&\quad + \mathrm{i}\boldsymbol{k}'\cdot\tilde{\boldsymbol{r}}_1(\tau) + \mathrm{i}\boldsymbol{k}\cdot\tilde{\boldsymbol{r}}_2(t) + \mathrm{i}\boldsymbol{k}'\cdot\tilde{\boldsymbol{r}}_2(\tau) - \frac{1}{2}\left(\boldsymbol{k}\cdot\tilde{\boldsymbol{r}}_1(t) + \boldsymbol{k}'\cdot\tilde{\boldsymbol{r}}_1(\tau)\right)^2\Big]\\
&\quad - \Omega_e^2(t-\tau)^2\left[1 + \mathrm{i}\boldsymbol{k}\cdot\tilde{\boldsymbol{r}}_1(t) + \mathrm{i}\boldsymbol{k}'\cdot\tilde{\boldsymbol{r}}_1(\tau)\right] \mp \mathrm{i}\Omega_e^3(t-\tau)^3\Big\}\\
&= \psi_\pm^{(0)}(\boldsymbol{k},\boldsymbol{k}') + \psi_\pm^{(1)}(\boldsymbol{k},\boldsymbol{k}') + \psi_\pm^{(2)}(\boldsymbol{k},\boldsymbol{k}')\,.
\end{aligned}
$$

Here

$$
\psi_0^{(0)}(\boldsymbol{k},\boldsymbol{k}') = 2\psi_\pm^{(0)}(\boldsymbol{k},\boldsymbol{k}') = \int_{-\infty}^{\infty}\mathrm{d}t\,\mathrm{e}^{\mathrm{i}\boldsymbol{k}\cdot\tilde{\boldsymbol{r}}_0(t)}\int_{-\infty}^{t}\mathrm{d}\tau\mathrm{e}^{\mathrm{i}\boldsymbol{k}'\cdot\tilde{\boldsymbol{r}}_0(\tau)}(t-\tau) \qquad (3.163)
$$

are the zero–order terms

$$
\begin{aligned}
\psi_0^{(1)}(\boldsymbol{k},\boldsymbol{k}') &= \mathrm{i}\int_{-\infty}^{\infty}\mathrm{d}t\,\mathrm{e}^{\mathrm{i}\boldsymbol{k}\cdot\tilde{\boldsymbol{r}}_0(t)}\int_{-\infty}^{t}\mathrm{d}\tau\mathrm{e}^{\mathrm{i}\boldsymbol{k}'\cdot\tilde{\boldsymbol{r}}_0(\tau)}(t-\tau)\left[\boldsymbol{k}\cdot\tilde{\boldsymbol{r}}_1(t) + \boldsymbol{k}'\cdot\tilde{\boldsymbol{r}}_1(\tau)\right]\,,\\
\psi_\pm^{(1)}(\boldsymbol{k},\boldsymbol{k}') &= \frac{1}{2}\psi_0^{(1)}(\boldsymbol{k},\boldsymbol{k}') \pm \frac{1}{2}\Theta_1(\boldsymbol{k},\boldsymbol{k}') \qquad (3.164)
\end{aligned}
$$

with

$$\Theta_1(\boldsymbol{k},\boldsymbol{k}') = \frac{\mathrm{i}\Omega_e}{2}\int_{-\infty}^{\infty}\mathrm{d}t\,\mathrm{e}^{\mathrm{i}\boldsymbol{k}\cdot\tilde{\boldsymbol{r}}_0(t)}\int_{-\infty}^{t}\mathrm{d}\tau\mathrm{e}^{\mathrm{i}\boldsymbol{k}'\cdot\tilde{\boldsymbol{r}}_0(\tau)}(t-\tau)^2 \tag{3.165}$$

are the first–order terms and

$$\psi_0^{(2)}(\boldsymbol{k},\boldsymbol{k}') = \int_{-\infty}^{\infty}\mathrm{d}t\,\mathrm{e}^{\mathrm{i}\boldsymbol{k}\cdot\tilde{\boldsymbol{r}}_0(t)}\int_{-\infty}^{t}\mathrm{d}\tau\mathrm{e}^{\mathrm{i}\boldsymbol{k}'\cdot\tilde{\boldsymbol{r}}_0(\tau)}(t-\tau)\left[\mathrm{i}\boldsymbol{k}\cdot\tilde{\boldsymbol{r}}_2(t)+\mathrm{i}\boldsymbol{k}'\cdot\tilde{\boldsymbol{r}}_2(\tau)\right.$$
$$\left. -\frac{1}{2}\left(\boldsymbol{k}\cdot\tilde{\boldsymbol{r}}_1(t)+\boldsymbol{k}'\cdot\tilde{\boldsymbol{r}}_1(\tau)\right)^2\right], \tag{3.166}$$

$$\psi_\pm^{(2)}(\boldsymbol{k},\boldsymbol{k}') = \frac{1}{2}\,\psi_0^{(2)}(\boldsymbol{k},\boldsymbol{k}') \mp \frac{1}{2}\,\Theta_{21}(\boldsymbol{k},\boldsymbol{k}') - \frac{1}{2}\,\Theta_{22}(\boldsymbol{k},\boldsymbol{k}')$$

with

$$\Theta_{21}(\boldsymbol{k},\boldsymbol{k}') = \frac{\Omega_e}{2}\int_{-\infty}^{\infty}\mathrm{d}t\mathrm{e}^{\mathrm{i}\boldsymbol{k}\cdot\tilde{\boldsymbol{r}}_0(t)}\int_{-\infty}^{t}\mathrm{d}\tau\mathrm{e}^{\mathrm{i}\boldsymbol{k}'\cdot\tilde{\boldsymbol{r}}_0(\tau)}(t-\tau)^2 \tag{3.167}$$
$$\times\left[\boldsymbol{k}\cdot\tilde{\boldsymbol{r}}_1(t)+\boldsymbol{k}'\cdot\tilde{\boldsymbol{r}}_1(\tau)\right],$$

$$\Theta_{22}(\boldsymbol{k},\boldsymbol{k}') = \frac{\Omega_e^2}{6}\int_{-\infty}^{\infty}\mathrm{d}t\mathrm{e}^{\mathrm{i}\boldsymbol{k}\cdot\tilde{\boldsymbol{r}}_0(t)}\int_{-\infty}^{t}\mathrm{d}\tau\mathrm{e}^{\mathrm{i}\boldsymbol{k}'\cdot\tilde{\boldsymbol{r}}_0(\tau)}(t-\tau)^3 \tag{3.168}$$

are the second–order terms. It should be noted that the first and second order corrections are related among each other by

$$\psi_+^{(1)}(\boldsymbol{k},\boldsymbol{k}') + \psi_-^{(1)}(\boldsymbol{k},\boldsymbol{k}') = \psi_0^{(1)}(\boldsymbol{k},\boldsymbol{k}') \tag{3.169}$$

and

$$\psi_+^{(2)}(\boldsymbol{k},\boldsymbol{k}') + \psi_-^{(2)}(\boldsymbol{k},\boldsymbol{k}') = \psi_0^{(2)}(\boldsymbol{k},\boldsymbol{k}') - \Theta_{22}(\boldsymbol{k},\boldsymbol{k}')\,. \tag{3.170}$$

The energy transfer can be represented as a sum of three terms, namely field free, first and second order with respect to the magnetic field. Inserting the zero–order term

$$g_0(\boldsymbol{k},\boldsymbol{k}')\psi_0^{(0)}(\boldsymbol{k},\boldsymbol{k}') + g_-(\boldsymbol{k},\boldsymbol{k}')\psi_+(\boldsymbol{k},\boldsymbol{k}') + g_+(\boldsymbol{k},\boldsymbol{k}')\psi_-(\boldsymbol{k},\boldsymbol{k}') \tag{3.171}$$
$$= -(\boldsymbol{k}\cdot\boldsymbol{k}')\psi_0^{(0)}(\boldsymbol{k},\boldsymbol{k}')$$

into (3.87) yields

$$\left(\Delta E_i^{(2)}\right)_0 = \frac{\mathrm{i}}{m}\int\mathrm{d}\boldsymbol{k}\int\mathrm{d}\boldsymbol{k}'U(\boldsymbol{k})U(\boldsymbol{k}')(\boldsymbol{k}\cdot\boldsymbol{v}_i)(\boldsymbol{k}\cdot\boldsymbol{k}')\mathrm{e}^{\mathrm{i}(\boldsymbol{k}+\boldsymbol{k}')\cdot\boldsymbol{R}_0} \tag{3.172}$$
$$\times\int_{-\infty}^{\infty}\mathrm{d}t\,\mathrm{e}^{\mathrm{i}(\boldsymbol{k}+\boldsymbol{k}')\cdot\boldsymbol{v}_r t}\int_{-\infty}^{t}\mathrm{d}\tau(t-\tau)\,\mathrm{e}^{\mathrm{i}\boldsymbol{k}'\cdot\boldsymbol{v}_r\tau}\,,$$

which agrees, of course, with (3.43) and leads to the results (3.53), (3.59) in the absence of a magnetic field.

For the first–order correction we insert

$$g_0(\boldsymbol{k},\boldsymbol{k}')\psi_0^{(1)}(\boldsymbol{k},\boldsymbol{k}') + g_-(\boldsymbol{k},\boldsymbol{k}')\psi_+^{(1)}(\boldsymbol{k},\boldsymbol{k}') + g_+(\boldsymbol{k},\boldsymbol{k}')\psi_-^{(1)}(\boldsymbol{k},\boldsymbol{k}')$$
$$= -(\boldsymbol{k}\cdot\boldsymbol{k}')\psi_0^{(1)}(\boldsymbol{k},\boldsymbol{k}') - \mathrm{i}\boldsymbol{k}\cdot(\boldsymbol{k}'\times\boldsymbol{b})\Theta_1(\boldsymbol{k},\boldsymbol{k}') \tag{3.173}$$

into (3.87)

$$\left(\Delta E_i^{(2)}\right)_1 = \frac{1}{m}\int \mathrm{d}\boldsymbol{k}\int \mathrm{d}\boldsymbol{k}' U(\boldsymbol{k})U(\boldsymbol{k}')(\boldsymbol{k}\cdot\boldsymbol{v}_i)$$
$$\times\left[\mathrm{i}(\boldsymbol{k}\cdot\boldsymbol{k}')\psi_0^{(1)}(\boldsymbol{k}\cdot\boldsymbol{k}') - (\boldsymbol{k}\cdot(\boldsymbol{k}'\times\boldsymbol{b}))\,\Theta_1(\boldsymbol{k},\boldsymbol{k}')\right]. \tag{3.174}$$

Here the explicit expression for $\tilde{\boldsymbol{r}}_1(t)$ has to be inserted into (3.164) in order to do the time integrals. It turns out to be advantageous to do this in a systematic manner by defining the operators

$$\hat{R}_{\mu,\nu}(\omega,\omega') = \frac{\mathrm{i}^{\nu-\mu+1}}{2\pi}\int_{-\infty}^{\infty}\mathrm{d}t\, t^{\nu}\mathrm{e}^{\mathrm{i}(\omega+\omega')t}\int_0^{\infty}\mathrm{d}t'\, t'^{\mu}\mathrm{e}^{-\mathrm{i}\omega' t'}$$
$$= \left[\frac{\partial^{\mu}}{\partial\omega'^{\mu}}\frac{1}{\omega'-\mathrm{i}0}\right]\frac{\partial^{\nu}}{\partial\omega^{\nu}}\delta(\omega+\omega'). \tag{3.175}$$

Their action on a test function $\Phi(\omega,\omega')$ is obtained by repeated partial integrations

$$\int_{-\infty}^{\infty}\mathrm{d}\omega'\int_{-\infty}^{\infty}\mathrm{d}\omega\,\hat{R}_{\mu,\nu}(\omega,\omega')\Phi(\omega,\omega') = (-1)^{\nu+\mu}\int_{-\infty}^{\infty}\frac{\mathrm{d}\omega'}{\omega'-\mathrm{i}0}\frac{\partial^{\mu}}{\partial\omega'^{\mu}}\left[\frac{\partial^{\nu}}{\partial\omega^{\nu}}\Phi(\omega,\omega')\right]_{\omega=-\omega'}. \tag{3.176}$$

We note in passing that (3.44) defines and (3.49) involves $\hat{R}_{1,0}$. Then

$$\psi_0^{(1)}(\boldsymbol{k},\boldsymbol{k}') = -\mathrm{i}\pi\Omega_e\mathrm{e}^{\mathrm{i}(\boldsymbol{k}+\boldsymbol{k}')\cdot\boldsymbol{R}_0}\left[((\boldsymbol{k}+\boldsymbol{k}')\cdot(\boldsymbol{b}\times\boldsymbol{v}_{e\perp}))\hat{R}_{1,2}(\omega,\omega')\right. \tag{3.177}$$
$$\left.+\,(\boldsymbol{k}'\cdot(\boldsymbol{b}\times\boldsymbol{v}_{e\perp}))\left(2\hat{R}_{2,1}(\omega,\omega')+\hat{R}_{3,0}(\omega,\omega')\right)\right]$$

and

$$\Theta_1(\boldsymbol{k},\boldsymbol{k}') = -\pi\Omega_e\mathrm{e}^{\mathrm{i}(\boldsymbol{k}+\boldsymbol{k}')\cdot\boldsymbol{R}_0}\hat{R}_{2,0}(\omega,\omega') \tag{3.178}$$

with $\omega = \boldsymbol{k}\cdot\boldsymbol{v}_r$, $\omega' = \boldsymbol{k}'\cdot\boldsymbol{v}_r$. The first–order energy transfer (3.174) is then

$$\left(\Delta E_i^{(2)}\right)_1 = \frac{\pi\Omega_e}{m}\int \mathrm{d}\boldsymbol{k}\int \mathrm{d}\boldsymbol{k}' U(\boldsymbol{k})U(\boldsymbol{k}')(\boldsymbol{k}\cdot\boldsymbol{v}_i)\,\mathrm{e}^{\mathrm{i}(\boldsymbol{k}+\boldsymbol{k}')\cdot\boldsymbol{R}_0} \tag{3.179}$$
$$\times\left[(\gamma+\gamma')\hat{R}_{1,2}+\gamma'\left(2\hat{R}_{2,1}+\hat{R}_{3,0}\right)+g_2R_{2,0}\right],$$

where all operators are understood as $\hat{R}_{\mu,\nu}(\omega,\omega')$ and

$$\gamma = (\boldsymbol{k}\cdot\boldsymbol{k}')(\boldsymbol{k}\cdot(\boldsymbol{b}\times\boldsymbol{v}_{e\perp})),\ \gamma' = (\boldsymbol{k}\cdot\boldsymbol{k}')(\boldsymbol{k}'\cdot(\boldsymbol{b}\times\boldsymbol{v}_{e\perp})),\ g_2 = \boldsymbol{k}\cdot(\boldsymbol{k}'\times\boldsymbol{b}). \tag{3.180}$$

The term involving $g_2(\boldsymbol{k},\boldsymbol{k}')$ vanishes upon the angular integration, see the discussion in connection with (3.152). The other terms are linear in the vector $\boldsymbol{v}_{e\perp}$ and will give no contribution when folding with a velocity distribution which is even in $\boldsymbol{v}_{e\perp}$. Then the first non–vanishing correction to the field free case is $\mathrm{O}(\Omega_e^2)$.

Proceeding as before the second order term is

$$\begin{aligned}
\left(\Delta E_i^{(2)}\right)_2 &= \frac{\mathrm{i}}{m}\int \mathrm{d}\boldsymbol{k}\int \mathrm{d}\boldsymbol{k}' U(\boldsymbol{k})U(\boldsymbol{k}')\,(\boldsymbol{k}\cdot\boldsymbol{v}_i) \\
&\quad\times\left[(\boldsymbol{k}\cdot\boldsymbol{k}')\psi_0^{(2)}(\boldsymbol{k},\boldsymbol{k}') - \mathrm{i}g_2(\boldsymbol{k},\boldsymbol{k}')\Theta_{21}(\boldsymbol{k},\boldsymbol{k}') + g_1(\boldsymbol{k},\boldsymbol{k}')\Theta_{22}(\boldsymbol{k},\boldsymbol{k}')\right] \\
&= \frac{\pi \mathrm{i}\Omega_e^2}{m}\int \mathrm{d}\boldsymbol{k}\int \mathrm{d}\boldsymbol{k}' U(\boldsymbol{k})U(\boldsymbol{k}')\,(\boldsymbol{k}\cdot\boldsymbol{v}_i)\,\mathrm{e}^{\mathrm{i}(\boldsymbol{k}+\boldsymbol{k}')\cdot\boldsymbol{R}_0} \\
&\quad\times\left\{-\frac{2\alpha_3+\alpha_1+\alpha_1'}{4}\hat{R}_{1,4} + \frac{\alpha_2+\alpha_2'}{3}\hat{R}_{1,3} - (\alpha_3+\alpha_1')\hat{R}_{2,3}\right. \\
&\quad+\frac{2\alpha_2'-\alpha_4'-\alpha_4}{2}\hat{R}_{2,2} - \frac{g_1}{3}\hat{R}_{3,0} + (\alpha_2'-\alpha_4')\hat{R}_{3,1} \\
&\quad\left.-\frac{\alpha_3+3\alpha_1'}{2}\hat{R}_{3,2} - \alpha_1'\hat{R}_{4,1} + \frac{2\alpha_2'-3\alpha_4'}{6}\hat{R}_{4,0} - \frac{\alpha_1'}{4}\hat{R}_{5,0}\right\}
\end{aligned} \tag{3.181}$$

with

$$\begin{aligned}
\alpha_1 &= (\boldsymbol{k}\cdot\boldsymbol{k}')\,(\boldsymbol{k}\cdot(\boldsymbol{b}\times\boldsymbol{v}_{e\perp}))^2\,, \quad \alpha_1' = (\boldsymbol{k}\cdot\boldsymbol{k}')\,(\boldsymbol{k}'\cdot(\boldsymbol{b}\times\boldsymbol{v}_{e\perp}))^2\,, \\
\alpha_2 &= (\boldsymbol{k}\cdot\boldsymbol{k}')\,(\boldsymbol{k}\cdot\boldsymbol{v}_{e\perp})\,, \quad \alpha_2' = (\boldsymbol{k}\cdot\boldsymbol{k}')\,(\boldsymbol{k}'\cdot\boldsymbol{v}_{e\perp})\,, \\
\alpha_3 &= (\boldsymbol{k}\cdot\boldsymbol{k}')\,(\boldsymbol{k}\cdot(\boldsymbol{b}\times\boldsymbol{v}_{e\perp}))\,(\boldsymbol{k}'\cdot(\boldsymbol{b}\times\boldsymbol{v}_{e\perp}))\,, \\
\alpha_4 &= (\boldsymbol{k}\cdot(\boldsymbol{k}'\times\boldsymbol{b}))\,(\boldsymbol{k}\cdot(\boldsymbol{b}\times\boldsymbol{v}_{e\perp}))\,, \quad \alpha_4' = (\boldsymbol{k}\cdot(\boldsymbol{k}'\times\boldsymbol{b}))\,(\boldsymbol{k}'\cdot(\boldsymbol{b}\times\boldsymbol{v}_{e\perp}))\,.
\end{aligned} \tag{3.182}$$

In the first–order correction (3.179) the integrand must be differentiated up to three times, in the second–order correction even up to five times. This leads to tedious, but straightforward calculations.

3.10 Impact Parameter Integration and Velocity Averaging

In our treatment we distinguished between the impact parameters $\bar{\boldsymbol{s}}$ and the relative velocity $\bar{\boldsymbol{v}}_r$ of the guiding center for magnetized electrons and the impact parameter $\boldsymbol{s}$ and the relative velocity $\boldsymbol{v}_r = \bar{\boldsymbol{v}}_r + \boldsymbol{v}_{e\perp}$ in the field–free case. The energy loss (2.9) involves

$$v_r\int \mathrm{d}s\; s = v_r\int \mathrm{d}\bar{s}\;\bar{s}\;J(s,\bar{s}) \to \bar{v}_r\int \mathrm{d}\bar{s}\;\bar{s} \tag{3.183}$$

as the Jacobian $J(s,\bar{s}) \sim \bar{v}_r/v_r$ cancels the particle fluxes up to an unimportant phase factor. The regularized potential (3.3), (3.7) obviates the introduction of Coulomb logarithms when doing the integration with respect to the impact parameter. We will show this explicitly for the cases of parallel ion motion and infinitly strong field. It is advantageous to do the $\bar{s}$–integration before the $k_\perp$–integration in passing from (3.100) to (3.104). Using

$$\int_0^\infty \mathrm{d}\bar{s}\;\bar{s}J_n(k_\perp\bar{s})J_n(k_\perp'\bar{s}) = \frac{1}{k_\perp}\,\delta(k_\perp - k_\perp') \tag{3.184}$$

one obtains

$$\int_0^\infty \mathrm{d}\bar{s}\,\bar{s}\left\langle \Delta E_{i\|}^{(2)}\right\rangle = \frac{4Z^2e^4 v_{i\|}}{m\bar{v}_{r\|}^3\delta^2}\sum_{n=1}^{\infty} n^2\left\{3\Phi_n(k_\|,a) + k_\|\frac{\partial}{\partial k_\|}\Phi_n(k_\|,a)\right.$$
$$\left. + \frac{\delta^2}{a}\frac{\partial}{\partial a}\Phi_n(k_\|,a)\right\}_{k_\|=\frac{n}{\delta}} \tag{3.185}$$

with

$$\Phi_n(k_\|,a) = (2\pi^2)^2\int_0^\infty \mathrm{d}k_\perp k_\perp u_R^2(k_\perp,k_\|)J_n^2(k_\perp a)\,. \tag{3.186}$$

For the convergence of the series (3.185) as well as for the limiting behavior at small relative velocities $\bar{v}_{r\|}\to 0$ we study

$$\lim_{n/\delta\to\infty} u_R\left(k_\perp, k_\| = \frac{n}{\delta}\right) = \frac{1}{2\pi^2}\frac{\lambda'^{-2}-\lambda^{-2}}{(k_\perp^2+n^2/\delta^2)^2} = \mathrm{O}\left(\frac{\delta^4}{n^4}\right)\,. \tag{3.187}$$

From (3.102) and (3.109), we find in this limit with $\kappa = n/\delta$

$$\int_0^\infty \frac{\mathrm{d}k_\perp k_\perp}{k_\perp^2+\kappa^2}J_n^2(k_\perp a) \to \frac{1}{2}\frac{1}{\sqrt{n^2+\kappa^2a^2}} \to \frac{1}{2n}\frac{\delta}{a}\,. \tag{3.188}$$

For Φ_n this has to be differentiated three times with respect to κ. There results

$$\Phi_n(k_\|,a) \to \frac{5}{32k_\|^7 a}\left(\lambda'^{-2}-\lambda^{-2}\right)^2 = \frac{5\delta^7}{32n^7 a}\left(\lambda'^{-2}-\lambda^{-2}\right)^2$$
$$= \frac{5}{32n^7 a}\left(\lambda'^{-2}-\lambda^{-2}\right)^2\left(\frac{|\bar{v}_{r\|}|}{\Omega_e}\right)^7 \tag{3.189}$$

and this is sufficient to guarantee a regular behavior of the integrated energy transfer (3.185) at small relative velocities $\bar{v}_{r\|}\to 0$

$$\int_0^\infty \mathrm{d}\bar{s}\bar{s}\,\left\langle \Delta E_{i\|}^{(2)}\right\rangle = \mathrm{O}(\bar{v}_{r\|}^2) \tag{3.190}$$

for the regularized potential (3.3). This should be compared to the cases of the screened (3.2) and the Coulomb potential (3.1) in which a self–cutting Coulomb logarithm must be kept under the integral for the velocity averaging, see equation (2.37).

Similar calculations yield the stopping power $\mathrm{d}\mathcal{E}_i/\mathrm{d}l$ for an infinite magnetic field and an arbitrary direction of the motion of the ion. The transforms (3.140) are inserted into the energy transfer (3.138). The integrated energy transfer involves integrals like

$$\int_0^\infty \mathrm{d}\bar{s}\bar{s}\,T_{0,3}(0,\bar{s})T_{0,1}(0,\bar{s}) = \frac{Z^2e^4}{(2\pi^2)^2}\int_0^\infty \mathrm{d}\rho\rho\left[K_0(\varkappa\rho)-K_0(\rho)\right]\left[K_0(\rho)-\varkappa^2K_0(\varkappa\rho)\right]$$
$$= \frac{Z^2e^4}{(2\pi^2)^2}\mathcal{U}_0(\varkappa)\,, \tag{3.191}$$

where $\varkappa = \lambda/\lambda' = 1 + \lambda/\lambda\!\!\!^{-}$ and

$$\mathcal{U}_0 = \mathcal{U}_0(\varkappa) = \frac{\varkappa^2 + 1}{\varkappa^2 - 1} \ln \varkappa - 1 \tag{3.192}$$

takes the role of the modified Coulomb logarithm (2.11), see also Sect. 4.7. Likewise

$$\int_0^\infty \mathrm{d}\bar{s}\bar{s}\, T_{1,2}^2(0, \bar{s}) = \frac{Z^2 \not{e}^4}{(2\pi^2)^2} \mathcal{U}_0(\varkappa) \,, \tag{3.193}$$

so the integral energy transfer is

$$2\pi \int_0^\infty \mathrm{d}\bar{s}\bar{s} \left\langle \Delta E_i^{(2)} \right\rangle_{\mathrm{I0}} = \frac{2\pi Z^2 \not{e}^4}{m} \frac{v_{i\perp}^2}{\bar{v}_r^6} \mathcal{U}_0(\varkappa) \left[v_{e\|}(v_{e\|} - v_{i\|}) + v_{e\|}^2 - v_i^2 \right] . \tag{3.194}$$

Now the energy loss (2.9)

$$\frac{\mathrm{d}E_i}{\mathrm{d}l} = \frac{2\pi n_e \bar{v}_r}{v_i} \int_0^\infty \mathrm{d}\bar{s}\bar{s} \left\langle \Delta E_i^{(2)} \right\rangle_{\mathrm{I0}} \tag{3.195}$$

is averaged with respect to the anisotropic electron velocity distributions (2.28). There results a one–dimensional integral for the stopping power

$$\begin{aligned} \frac{\mathrm{d}\mathcal{E}_i}{\mathrm{d}l} &= \int \mathrm{d}\boldsymbol{v}_e f_0(\boldsymbol{v}_e) \frac{\mathrm{d}E_i}{\mathrm{d}l} = \frac{(2\pi)^{1/2} n_e v_i}{m v_{\mathrm{th}\|}} Z^2 \not{e}^4 \mathcal{U}_0(\varkappa) \sin^2\alpha \\ &\times \int_{-\infty}^\infty \mathrm{d}v_{e\|} \mathrm{e}^{-v_{e\|}^2/2v_{\mathrm{th}\|}^2} \frac{2v_{e\|}^2 - v_{e\|} v_i \cos\alpha - v_i^2}{(v_{e\|}^2 - 2v_{e\|} v_i \cos\alpha + v_i^2)^{5/2}} \end{aligned} \tag{3.196}$$

with $\sin\alpha = v_{i\perp}/v_i$. This can be further simplified by introducing the integration variable $y = v_{e\|}/v_{\mathrm{th}\|}$ and a partial integration

$$-\frac{\mathrm{d}\mathcal{E}_i}{\mathrm{d}l} = \frac{(2\pi)^{1/2} n_e v_i}{m v_{\mathrm{th}\|}^3} Z^2 \not{e}^4 \mathcal{U}_0(\varkappa) \sin^2\alpha \int_{-\infty}^\infty \mathrm{d}y \, \frac{y^2 \mathrm{e}^{-y^2/2}}{(y^2 - 2xy\cos\alpha + x^2)^{3/2}} \tag{3.197}$$

with $x = v_i/v_{\mathrm{th}\|}$. A further partial integration shows that the integral

$$\begin{aligned} \mathfrak{I}(\alpha, x) &= \int_{-\infty}^\infty \mathrm{d}y \frac{y^2 \mathrm{e}^{-y^2/2}}{(y^2 - 2xy\cos\alpha + x^2)^{3/2}} \\ &= \frac{1}{x^2 \sin^2\alpha} \int_{-\infty}^\infty \mathrm{d}y \, \mathrm{e}^{-y^2/2} \frac{(y - x\cos\alpha)(y^3 - 2y)}{(y^2 - 2xy\cos\alpha + x^2)^{1/2}} \end{aligned} \tag{3.198}$$

is singular in the limit $\alpha \to 0$

$$\lim_{\alpha \to 0} \mathfrak{I}(\alpha, x) = \frac{2}{\sin^2\alpha} \, \mathrm{e}^{-x^2/2} \,. \tag{3.199}$$

So the stopping power (3.197) remains finite for ions moving parallel to the field, while the energy loss in collisions with monoenergetic electrons (3.195) with

(3.194) vanishes. The energy transfer (3.194) diverges too strongly for small relative velocities $\bar{v}_r \to 0$. This indicates a failure of the perturbation treatment. The regularization in the potential (3.3) is sufficient to guarantee the existence of the integrated energy transfer (3.194) but there remains the problem of treating hard collisions. For a perturbation treatment the change in velocity must be small compared to $\bar{v}_r$ and this condition is increasingly difficult to fulfill in the regime $\bar{v}_r \to 0$. This suggests a physically reasonable procedure: The potential must be softened near the origin. In fact the parameter $\lambda\!\!\!\bar{}$ which describes the effects of quantum diffraction should be related to the de Broglie wavelength which is inversely proportional to $\bar{v}_r$. Then the function $\mathcal{U}_0(\varkappa)$ (3.192) must remain within the velocity averaging integral. As this function is self–cutting

$$\mathcal{U}_0(\varkappa) \sim (\varkappa - 1)^2 \sim (\lambda/\lambda\!\!\!\bar{})^2 \sim \bar{v}_r^2 \tag{3.200}$$

for $\bar{v}_r \to 0$ the integrand becomes less singular by two orders. Then the integral $\mathfrak{I}(\alpha, x)$ diverges only logarithmically in the limit $\alpha \to 0$, and the stopping power vanishes as $\propto \alpha^2 \ln \alpha$.

A similar argument applies to the limit of small ion velocities $x = v_i/v_{\text{th}\|} \to 0$. Performing a multipole expansion of the integrand in (3.198) the leading behavior is

$$\lim_{x\to 0} \mathfrak{I}(\alpha, x) = \ln \frac{8}{x^2 \sin^2 \alpha} - \gamma + 2\left(\frac{\cos^2 \alpha}{\sin^2 \alpha} - 1\right). \tag{3.201}$$

Here $\gamma = 0.5772$ is the Euler's constant. On the other hand if (3.200) is included in the velocity averaging this singular behavior is changed to $\propto x^2 \ln(1/x^2)$.

Before closing this section we would like to check the range of validity of the impact parameter integrated second–order energy transfer in the case of an infinitely strong magnetic field, i.e. expression (3.194), by comparing it with the numerical CTMC calculations of Sect. 2.3. In the present limit of vanishing electron gyration $a \to 0$ the numerical simulations are done for purely one dimensional electron motion, where $\boldsymbol{v}(t) = \bar{v}_{r\|}\boldsymbol{b} - \boldsymbol{v}_{i\perp} + \delta v_\|(t)\boldsymbol{b}$ and $\Delta\boldsymbol{v} = \boldsymbol{v}(t \to \infty) - \boldsymbol{v}(t \to -\infty) = \boldsymbol{b}\delta v_\|(t \to \infty) = \boldsymbol{b}\Delta v_\|$. Note that under this constraint of electron motion along the magnetic field lines neither the energy nor the quantity K (3.18) are conserved. The related equations of motion which have to be solved numerically are, after scaling as introduced in the beginning of Sect. 3.7, given by

$$\frac{d\tilde{v}_\|}{d\tilde{t}} = \boldsymbol{b} \cdot \frac{\partial}{\partial \tilde{\boldsymbol{r}}}\left[\frac{1}{\tilde{r}}\left(1 - e^{-\tilde{r}(\lambda/\lambda\!\!\!\bar{})}\right)e^{-\tilde{r}}\right], \quad \frac{d\tilde{\boldsymbol{r}}}{d\tilde{t}} = \tilde{\boldsymbol{v}} = \tilde{v}_\|(t)\boldsymbol{b} - \tilde{\boldsymbol{v}}_{i\perp} \tag{3.202}$$

when using the regularized screened interaction U_R (3.3). It has been tested numerically that the general case with finite magnetic field governed by the equations of motion (3.3) and (3.122), respectively, does indeed converge for increasingly strong magnetic field towards the dynamics according to equations (3.202).

Scaling the perturbative expression (3.194) using λ as unit of length and v_s (3.121) as unit of velocity and rewriting it in terms of $\bar{v}_{r\|} = v_{e\|} - v_{i\|}$, i.e. the initial relative velocity along $\boldsymbol{b}$, and the components of the ion velocity $v_{i\|}, v_{i\perp}$ yields

$$\sigma_{\mathrm{I0}}^{(2)} := -\int_0^\infty \frac{\mathrm{d}\bar{s}\bar{s}}{\lambda^2}\frac{\left\langle \Delta E_i^{(2)}\right\rangle_{\mathrm{I0}}}{mv_s^2} = \frac{v_s^2 v_{i\perp}^2}{(\bar{v}_{r\|}^2 + v_{i\perp}^2)^3}\left(v_{i\perp}^2 - 2\bar{v}_{r\|}^2 - 3v_{i\|}\bar{v}_{r\|}\right)\mathcal{U}_0(\varkappa) \quad (3.203)$$

with $\mathcal{U}_0(\varkappa)$ as defined in (3.192). The quantity σ introduced here can be interpreted as a dimensionless generalized cross section for the energy loss, cf. equations (3.195) and (2.9).

Applying on the other hand the energy transfer in the form (3.32) to the present case with $\boldsymbol{v}_{e0} = (\bar{v}_{r\|} + v_{i\|})\boldsymbol{b}$ and $\delta\boldsymbol{v}(t \to \infty) = \Delta v_\| \boldsymbol{b}$, we arrive at the definition of the generalized cross section in terms of $\Delta v_\|$

$$\sigma = -\int \frac{\mathrm{d}^2\bar{\mathbf{s}}}{2\pi\lambda^2}\frac{\Delta E_i}{mv_s^2} = \int \frac{\mathrm{d}^2\bar{\mathbf{s}}}{2\pi\lambda^2}\left[\frac{v_{i\|}}{v_s}\frac{\Delta v_\|}{v_s} + \frac{\bar{v}_{r\|}}{v_s}\frac{\Delta v_\|}{v_s} + \frac{1}{2}\left(\frac{\Delta v_\|}{v_s}\right)^2\right]. \quad (3.204)$$

It is now advantageous to split (3.204) into two terms, one related to the average of the velocity transfer $\Delta v_\|$, named $\sigma_\|$, and a remaining part σ_r which is independent of the parallel ion motion $v_{i\|}$, defining $\sigma = -(v_{i\|}/v_s)\sigma_\| + \sigma_r$. Inserting this into (3.204) results in

$$\sigma_\|(\eta,\xi) = -\int \frac{\mathrm{d}^2\bar{\mathbf{s}}}{2\pi\lambda^2}\frac{\Delta v_\|}{v_s}, \quad (3.205)$$

$$\sigma_r(\eta,\xi) = \int \frac{\mathrm{d}^2\bar{\mathbf{s}}}{2\pi\lambda^2}\left[\frac{\bar{v}_{r\|}}{v_s}\frac{\Delta v_\|}{v_s} + \frac{1}{2}\left(\frac{\Delta v_\|}{v_s}\right)^2\right] \quad (3.206)$$

which are functions of the scaled transverse ion velocity $\eta = v_{i\perp}/v_s$ and the initial parallel relative velocity $\xi = \bar{v}_{r\|}/v_s$. These expressions allow to discuss the average energy transfer (3.203) in terms of the scaled velocities η, ξ independently of the actual value of $v_{i\|}$. The corresponding expressions for $\sigma_\|, \sigma_r$ within the second order perturbation treatment can be easily derived by matching (3.203) with (3.204). This yields

$$\sigma_{\|\,\mathrm{I0}}^{(2)}(\eta,\xi) = 3\frac{\eta^2\xi}{(\eta^2+\xi^2)^3}\mathcal{U}_0(\varkappa), \quad (3.207)$$

$$\sigma_{r\,\mathrm{I0}}^{(2)}(\eta,\xi) = \frac{\eta^2(\eta^2-2\xi^2)}{(\eta^2+\xi^2)^3}\mathcal{U}_0(\varkappa). \quad (3.208)$$

Some examples comparing (3.205) and (3.206) determined by the CTMC method from about 10^7 calculated trajectories for each given η, ξ with the perturbative treatment (3.207) and (3.208) are presented in Figs. 3.7 and 3.8. Shown are $\sigma_\|$ (left column) and σ_r (right column) as functions of $\xi = \bar{v}_{r\|}/v_s$ for increasing transverse ion velocity $\eta = v_{i\perp}/v_s$ from $\eta = 1$ (Fig. 3.7 top) to $\eta = 18$ (Fig. 3.8 bottom). The CTMC results are indicated by open squares for a regularized screened potential U_R (3.3) with $\lambda\!\!\!^-/\lambda = 0.05$ and by filled triangles for $\lambda\!\!\!^-/\lambda = 0.005$. The corresponding second order predictions (3.207), (3.208) are given by the solid curves. Note the logarithmic scale for $\sigma_\|, \sigma_r$ and the use of the absolute value $|\sigma_r|$ for $\eta = 1$ and $\eta = 2$ (Fig. 3.7 top and center).

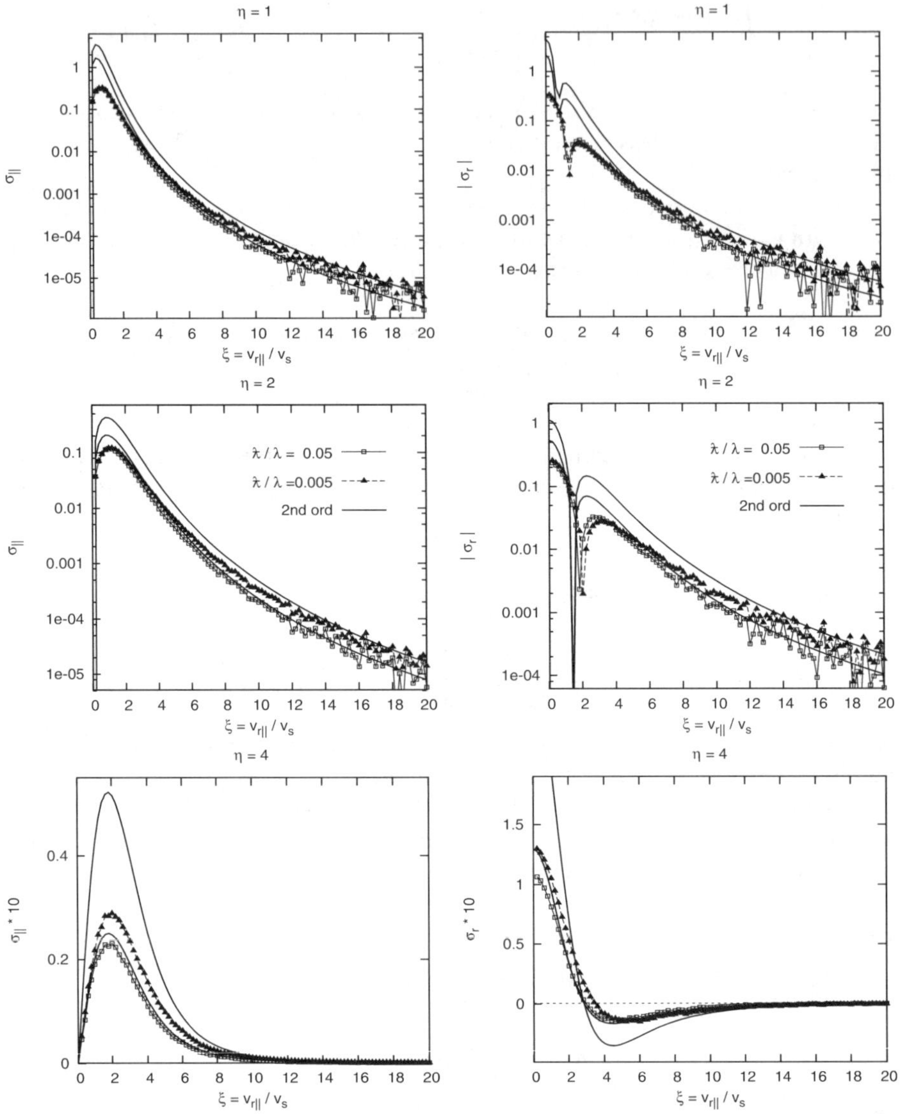

Fig. 3.7. Generalized cross sections $\sigma_{\|}$ (3.205) and σ_r (3.206) as function of the initial parallel relative velocity $\xi = \bar{v}_{r\|}/v_s$ scaled in units of v_s (3.121) for different transverse ion velocities $\eta = v_{i\perp}/v_s$ and regularized screened potentials (3.3) with $\lambda\!\!\!^{-}/\lambda = 0.05$ and 0.005. Compared are the results of CTMC simulations (symbols) with the second order perturbative treatment (solid curves) given by expressions (3.207) and (3.208), respectively. Note the logarithmic scale for $\sigma_{\|,r}$ and the use of the absolute value $|\sigma_r|$ in the upper and central panels.

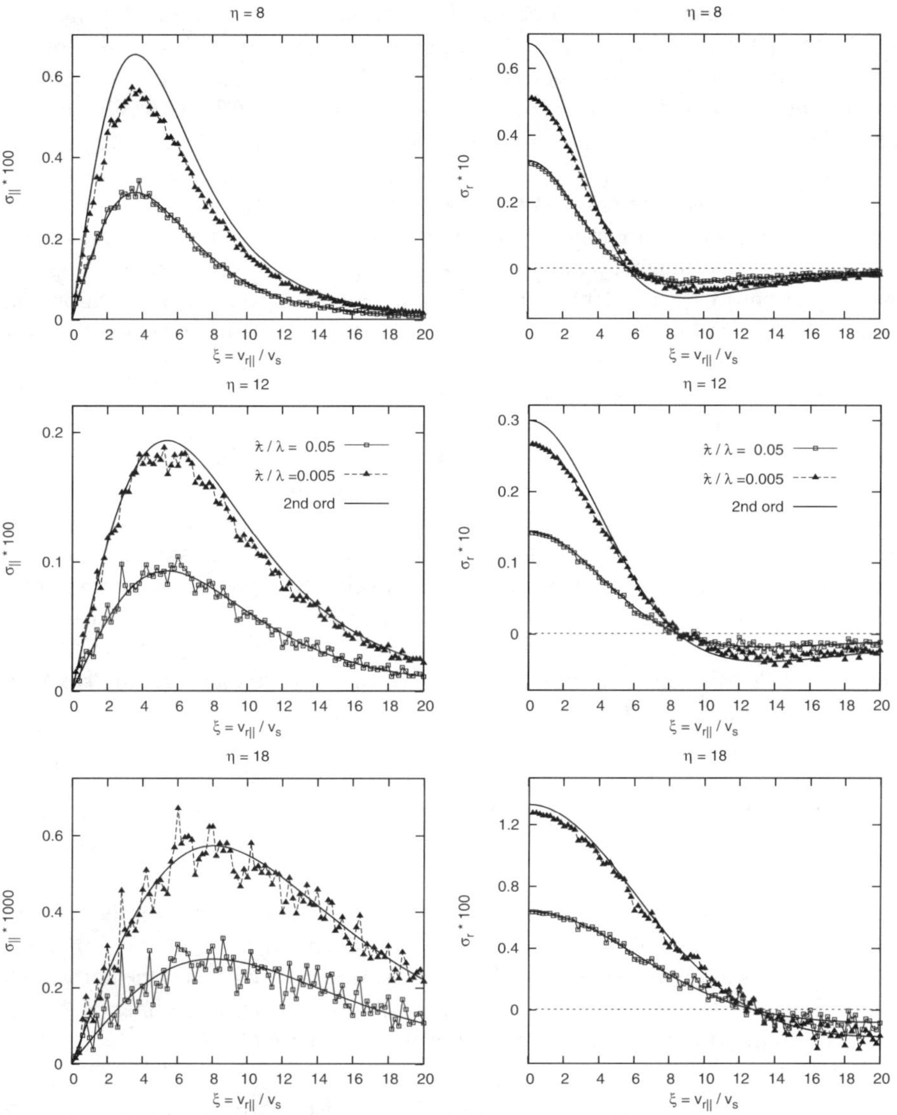

Fig. 3.8. Same as in Fig. 3.7 for larger transverse ion velocities $\eta = v_{i\perp}/v_s$.

Despite the unavoidable numerical fluctuations, which considerably increase when the energy transfer, i.e. $|\sigma_{\parallel}|$, $|\sigma_r|$, decreases, we can clearly observe the following behavior: The larger the velocities η, ξ, and thus, the smaller the cross sections $\sigma_{\parallel}, \sigma_r$, the better is the agreement between the numerical CTMC results and the second order perturbative treatment. With respect to the 'softness' of the regularized potential the agreement improves, keeping all other parameters fixed, for increasing $\tilde{\lambda}/\lambda$, that is, a 'weaker' interaction U_R. This is consistent with the observations made in Sect. 3.7 for the energy transfer as function of the impact parameter $\bar{s}$ and $\eta = 0$, but arbitrary finite magnetic field. In view of the previous discussion in connection with equations (3.196)-(3.199) and (3.201) concerning the velocity integration with respect to $\bar{v}_{r\parallel}$ and the behavior of the stopping power in the limit of small $v_{i\perp}$ and $v_{i\parallel}$, we like to point out the large deviations found for small $\bar{v}_{r\parallel}$ and $v_{i\perp}$. At small ξ the second order treatment considerably overestimates the energy transfer at decreasing η. For $\eta = 1$ and small ξ the second order $\sigma_{\parallel}, \sigma_r$ are already about an order of magnitude larger than the CTMC results (see the top panel of Fig. 3.7). And this discrepancy further strongly increases for smaller η, that is $v_{i\perp}/v_s \ll 1$ (not shown). In this regime the second order perturbative treatment is clearly invalid and a non-perturbative description is required. Any attempts to eliminate this overshooting of the perturbative treatment by a velocity dependent softening of the interaction should be considered with caution. While some straightforward assumptions for such a velocity dependent interaction, that is $\tilde{\lambda} = \tilde{\lambda}(\eta, \xi)$ like in (3.200), might cure the discrepancies in the velocity averaged results on a qualitative level, it is not guaranteed that they will do so quantitatively.

3.11 Velocity Diffusion (Straggling) of Charged Particles in a Magnetic Field

An alternative approach to stopping and velocity diffusion starts from the Fokker–Planck equation, see for example [72, 112]. The velocity distribution $f(\boldsymbol{v}_i, t)$ of ions at time t varies through collisions with electrons. If $w(\boldsymbol{v}_i - \Delta\boldsymbol{v}_i, \Delta\boldsymbol{v}_i)$ is the normalized probability that the ion velocity changes from $\boldsymbol{v}_i - \Delta\boldsymbol{v}_i$ to $\boldsymbol{v}_i$ in a time interval Δt then the distribution at time $t + \Delta t$ is obtained from that at time t by

$$f(\boldsymbol{v}_i, t + \Delta t) = \int \mathrm{d}(\Delta\boldsymbol{v}_i) f(\boldsymbol{v}_i - \Delta\boldsymbol{v}_i, t)\, w(\boldsymbol{v}_i - \Delta\boldsymbol{v}_i, \Delta\boldsymbol{v}_i)\,. \tag{3.209}$$

Excluding hard collisions one may expand both sides as $\Delta\boldsymbol{v}_i$ remains small for small time intervals

$$f(\boldsymbol{v}_i, t) + \Delta t\frac{\partial f}{\partial t} = \int \mathrm{d}(\Delta\boldsymbol{v}_i)\left[1 - \Delta\boldsymbol{v}_i \cdot \frac{\partial}{\partial \boldsymbol{v}_i} + \frac{1}{2}(\Delta\boldsymbol{v}_i\Delta\boldsymbol{v}_i) : \frac{\partial^2}{\partial\boldsymbol{v}_i\partial\boldsymbol{v}_i}\right] f(\boldsymbol{v}_i, t) w(\boldsymbol{v}_i, \Delta\boldsymbol{v}_i)\,. \tag{3.210}$$

The average change in the velocity moments is

$$\langle \Delta \boldsymbol{v}_i \rangle = \int \mathrm{d}(\Delta \boldsymbol{v}_i) \Delta \boldsymbol{v}_i \, w(\boldsymbol{v}_i, \Delta \boldsymbol{v}_i) \,, \tag{3.211}$$

$$\langle \Delta \boldsymbol{v}_i \Delta \boldsymbol{v}_i \rangle = \int \mathrm{d}(\Delta \boldsymbol{v}_i) \Delta \boldsymbol{v}_i \Delta \boldsymbol{v}_i \, w(\boldsymbol{v}_i, \Delta \boldsymbol{v}_i) \,.$$

Insertion yields the Fokker–Planck equation

$$\frac{\partial f}{\partial t} = -\frac{\partial}{\partial \boldsymbol{v}_i} \left(\frac{\langle \Delta \boldsymbol{v}_i \rangle}{\Delta t} f \right) + \frac{1}{2} \frac{\partial^2}{\partial \boldsymbol{v}_i \partial \boldsymbol{v}_i} : \left(\frac{\langle \Delta \boldsymbol{v}_i \Delta \boldsymbol{v}_i \rangle}{\Delta t} f \right) . \tag{3.212}$$

Here the friction coefficient

$$\frac{\langle \Delta \boldsymbol{v}_i \rangle}{\Delta t} = -\frac{\boldsymbol{F}}{M} \tag{3.213}$$

is related to the force $-\boldsymbol{F}$ on the ion, see Sect. 3.2, and

$$\mathsf{D}(\boldsymbol{v}_i) = \frac{\langle \Delta \boldsymbol{v}_i \Delta \boldsymbol{v}_i \rangle}{\Delta t} \tag{3.214}$$

is the (velocity) diffusion tensor. While the friction coefficient describes the stopping of the ion as discussed in previous Sections, velocity diffusion is related to straggling. For the leading behavior of $\mathsf{D}(\boldsymbol{v}_i)$ the first–order velocity transfer is required

$$\Delta \boldsymbol{v}_i = -\frac{1}{m} \int_{-\infty}^{\infty} \mathrm{d}t \boldsymbol{F}_0(\boldsymbol{r}_0(t)) = \frac{\mathrm{i}}{m} \int \mathrm{d}\boldsymbol{k} U(\boldsymbol{k}) \boldsymbol{k} \exp\left[\mathrm{i}\boldsymbol{k} \cdot \boldsymbol{r}_0(t)\right] . \tag{3.215}$$

Without magnetic field the zero–order motion $\boldsymbol{r}_0(t)$ is uniform (3.39) and

$$\Delta \boldsymbol{v}_i = -\frac{2\pi \mathrm{i}}{m} \int \mathrm{d}\boldsymbol{k} U(\boldsymbol{k}) \boldsymbol{k} \, \mathrm{e}^{\mathrm{i}\boldsymbol{k}\cdot\boldsymbol{s}} \delta(\boldsymbol{k} \cdot \boldsymbol{v}_r) \,, \tag{3.216}$$

where $\boldsymbol{v}_r$ is the relative velocity and $\boldsymbol{s} \perp \boldsymbol{v}_r$ the impact parameter, see Fig. 2.3. Then

$$\Delta \boldsymbol{v}_i \Delta \boldsymbol{v}_i = -\frac{4\pi^2}{m^2} \int \mathrm{d}\boldsymbol{k} \int \mathrm{d}\boldsymbol{k}' U(\boldsymbol{k}) U(\boldsymbol{k}') (\boldsymbol{k}\boldsymbol{k}') \delta(\boldsymbol{k} \cdot \boldsymbol{v}_r) \delta(\boldsymbol{k}' \cdot \boldsymbol{v}_r) \, \mathrm{e}^{\mathrm{i}(\boldsymbol{k}+\boldsymbol{k}')\cdot\boldsymbol{s}} \,. \tag{3.217}$$

Introducing cylindrical coordinates about $\boldsymbol{v}_r$ and averaging with respect to $\hat{\boldsymbol{s}}$ yields with the same techniques as used between (3.43) and (3.53)

$$\langle \Delta \boldsymbol{v}_i \Delta \boldsymbol{v}_i \rangle_{\hat{s}} = \frac{(2\pi)^4}{2m^2 v_r^2} T_{1,2}^2(0,s) \begin{pmatrix} 1 & 0 & 0 \\ 0 & 1 & 0 \\ 0 & 0 & 0 \end{pmatrix} , \tag{3.218}$$

where the transform $T_{1,2}^2(0,s)$ has been defined in (3.54), and explicit expressions for the regularized and the screened potential have been given in (3.57) and (3.56), respectively. For an arbitrary direction of relative motion the replacement

$$\begin{pmatrix} 1 & 0 & 0 \\ 0 & 1 & 0 \\ 0 & 0 & 0 \end{pmatrix} \longrightarrow \mathsf{1} - \frac{\boldsymbol{v}_r \boldsymbol{v}_r}{v_r^2} \tag{3.219}$$

must be made. In a ring between s and $s + \mathrm{d}s$ there are $2\pi n_e v_r \Delta t$ scattering events during a time interval Δt. Thus

$$\frac{\overline{\Delta \boldsymbol{v}_i \Delta \boldsymbol{v}_i}}{\Delta t} = \frac{(2\pi)^5 n_e}{2m^2 v_r} \int \mathrm{d}s s\, T_{1,2}^2(0, s) \left(1 - \frac{\boldsymbol{v}_r \boldsymbol{v}_r}{v_r^2}\right) \tag{3.220}$$

for a general interaction. For the regularized interaction (3.3) the impact parameter integration yields

$$\frac{\overline{\Delta \boldsymbol{v}_i \Delta \boldsymbol{v}_i}}{\Delta t} = \frac{4\pi n_e Z^2 e^4}{m^2 v_r} \mathcal{U}_0(\varkappa) \left(1 - \frac{\boldsymbol{v}_r \boldsymbol{v}_r}{v_r^2}\right) \tag{3.221}$$

with the cut–off function $\mathcal{U}_0$ defined in (3.192). In the special case of the Coulomb interaction this is replaced by [72]

$$\frac{\overline{\Delta \boldsymbol{v}_i \Delta \boldsymbol{v}_i}}{\Delta t} = \frac{4\pi n_e Z^2 e^4}{m^2 v_r} L(s_{\max}, s_{\min}) \left(1 - \frac{\boldsymbol{v}_r \boldsymbol{v}_r}{v_r^2}\right). \tag{3.222}$$

In view of the discussion in Sect. 2.1 the conventional Coulomb logarithm (2.23) should be replaced by the modified logarithm (2.11). The latter is self–cutting for small relative velocities $v_r \to 0$, where the perturbation treatment becomes doubtful. Without magnetic field this region of velocity space has sufficiently small weight so average values of the Coulomb logarithm may be taken out of the three–dimensional velocity integrals. Then

$$\mathsf{D}(\boldsymbol{v}_i) = \frac{4\pi n_e Z^2 e^4}{m^2} \Lambda_{\mathrm{th}} \frac{\partial^2}{\partial \boldsymbol{v}_i \partial \boldsymbol{v}_i} \left.\frac{\partial^2 H(\eta, \boldsymbol{v}_i)}{\partial \eta^2}\right|_{\eta=0} = \frac{4\pi n_e Z^2 e^4}{m^2} \Lambda_{\mathrm{th}}\, \mathsf{d}(\boldsymbol{v}_i) \tag{3.223}$$

with the pseudopotential [112]

$$H(\eta, \boldsymbol{v}_i) = \int \mathrm{d}\boldsymbol{v}_e f_0(\boldsymbol{v}_e) \frac{\mathrm{e}^{-\eta|\boldsymbol{v}_e - \boldsymbol{v}_i|}}{|\boldsymbol{v}_e - \boldsymbol{v}_i|} \tag{3.224}$$

and a convergence enforcing parameter η. Using cylindrical coordinates we obtain for a distribution like (2.28)

$$\begin{aligned} H(\eta, \boldsymbol{v}_i) = 2\pi &\int_0^\infty \frac{\mathrm{d}k_\perp k_\perp J_0(k_\perp v_{i\perp})}{\sqrt{k_\perp^2 + \eta^2}} \int_{-\infty}^\infty \mathrm{d}v_{e\|} \mathrm{e}^{-|v_{e\|} - v_{i\|}| \sqrt{k_\perp^2 + \eta^2}} \\ &\times \int_0^\infty \mathrm{d}v_{e\perp} v_{e\perp} f_0(v_{e\|}, v_{e\perp}) J_0(k_\perp v_{e\perp}) \end{aligned} \tag{3.225}$$

and

$$\begin{aligned} \left.\frac{\partial^2 H(\eta, \boldsymbol{v}_i)}{\partial \eta^2}\right|_{\eta=0} = \frac{v_{\mathrm{th}\perp}}{\sqrt{2}} &\left\{ \mathrm{e}^{-x_\|^2} \int_0^\infty \mathrm{d}y \mathrm{e}^{-y^2} F_+(x_\|, \tau_1 y) J_0(2x_\perp y) + x_\perp \mathrm{e}^{-x_\|^2} \right. \\ &\left. \times \int_0^\infty \mathrm{d}y \mathrm{e}^{-y^2} F_+(x_\|, \tau_1 y) \frac{J_1(2x_\perp y)}{y} + 2\tau_1 \left[x_\| \operatorname{erf}(x_\|) + \frac{1}{\sqrt{\pi}} \mathrm{e}^{-x_\|^2} \right] \right\}. \end{aligned} \tag{3.226}$$

Here

$$x_{\|} = \frac{v_{i\|}}{v_{\mathrm{th}\|}\sqrt{2}}, \quad x_{\perp} = \frac{v_{i\perp}}{v_{\mathrm{th}\perp}\sqrt{2}}, \quad \tau_1 = \frac{v_{\mathrm{th}\|}}{v_{\mathrm{th}\perp}} \tag{3.227}$$

and

$$F_{\pm}(a, x) = \mathrm{e}^{(x+a)^2}\,\mathrm{erfc}(x+a) \pm \mathrm{e}^{(x-a)^2}\,\mathrm{erfc}(x-a) \tag{3.228}$$

with the complementary error functions erfc(z). Note that the parameter τ_1 is related to the temperature anisotropy τ introduced in Sect. 2.4 as $\tau_1 = \tau^{-1/2}$. The derivatives of these potentials required in (3.223) for the diffusion tensor can be evalulated with the help of the relation

$$\frac{\partial}{\partial a}\left[\mathrm{e}^{-a^2} F_{+}(a, x)\right] = 2x\,\mathrm{e}^{-a^2} F_{-}(a, x)\,. \tag{3.229}$$

For small and large ion velocities the diffusion tensor can be calculated without specifying the distribution function. At $\boldsymbol{v}_i \to 0$ the tensor d is diagonal

$$d_{\alpha\beta} \simeq 4\pi\delta_{\alpha\beta} \int_0^{\infty} \mathrm{d}v_{e\|} \int_0^{\infty} \frac{\mathrm{d}v_{e\perp} v_{e\perp}}{v_e} f_0(v_{e\|}, v_{e\perp}) \begin{cases} 1 - \frac{v_{e\perp}^2}{2v_e^2} & \alpha, \beta = x, y \\ \frac{v_{e\perp}^2}{v_e^2} & \alpha, \beta = z \end{cases} . \tag{3.230}$$

For the anisotropic Maxwell distribution (2.28)

$$d_{xx} = d_{yy} = \frac{4}{3\sqrt{2\pi}\,v_{\mathrm{th}\perp}} \chi_{\perp}(\tau_1)\,, \quad d_{zz} = \frac{4}{3\sqrt{2\pi}\,v_{\mathrm{th}\perp}} \chi_{\|}(\tau_1) \tag{3.231}$$

with $\chi_{\perp}(1) = \chi_{\|}(1) = 1$,

$$\begin{aligned} \chi_{\perp}(\tau) &= \frac{3\tau}{4(1-\tau^2)}\left(1 + \frac{1-2\tau^2}{\tau\sqrt{1-\tau^2}} \arccos\tau\right), \quad 0 \le \tau < 1 \\ \chi_{\perp}(\tau) &= \frac{3\tau}{4(\tau^2-1)}\left[\frac{2\tau^2-1}{\tau\sqrt{\tau^2-1}} \ln\left(\tau + \sqrt{\tau^2-1}\right) - 1\right], \quad \tau > 1 \end{aligned} \tag{3.232}$$

and

$$\begin{aligned} \chi_{\|}(\tau) &= \frac{3\tau}{2(1-\tau^2)}\left(\frac{1}{\tau\sqrt{1-\tau^2}} \arccos\tau - 1\right), \quad 0 \le \tau < 1 \\ \chi_{\|}(\tau) &= \frac{3\tau}{2(\tau^2-1)}\left[1 - \frac{1}{\tau\sqrt{\tau^2-1}} \ln\left(\tau + \sqrt{\tau^2-1}\right)\right], \quad \tau > 1 \end{aligned} \tag{3.233}$$

At large ion velocities $\boldsymbol{v}_i \to \infty$ and for $\alpha, \beta = x, y$

$$\begin{aligned} d_{\alpha\beta}(\boldsymbol{v}_i) \simeq{}& \frac{1}{v_i}\left(\delta_{\alpha\beta} - \frac{v_{i,\alpha} v_{i,\beta}}{v_i^2}\right)\left[1 + \frac{1}{2}\left(\frac{3v_{i\|}^2}{v_i^2} - 1\right) c_{21} - \frac{1}{4}\left(\frac{3v_{i\|}^2}{v_i^2} + 1\right) c_{03}\right] \\ &+ \frac{v_{i,\alpha} v_{i,\beta}}{v_i^3}\left[\left(1 - \frac{6v_{i\|}^2}{v_i^2}\right) c_{21} + \frac{1}{2}\left(1 + \frac{6v_{i\|}^2}{v_i^2}\right) c_{03}\right], \end{aligned} \tag{3.234}$$

while for $\alpha = x, y$

$$d_{\alpha z} \simeq \frac{v_{i,\alpha} v_{i\|}}{v_i^3} \left[-1 + \frac{3}{2} \left(3 - \frac{5v_{i\|}^2}{v_i^2} \right) c_{21} + \frac{3}{4} \left(\frac{5v_{i\|}^2}{v_i^2} - 1 \right) c_{03} \right] \tag{3.235}$$

and

$$\begin{aligned} d_{zz} &= \frac{v_{i\perp}^2}{v_i^3} + \frac{1}{2v_i} \left[3 \left(\frac{2v_{i\|}^2}{v_i^2} - 1 \right) + \frac{3v_{i\|}^2}{v_i^2} \left(4 - \frac{5v_{i\|}^2}{v_i^2} \right) \right] c_{21} \\ &+ \frac{1}{4v_i} \left[1 + \frac{3v_{i\|}^2}{v_i^2} \left(\frac{5v_{i\|}^2}{v_i^2} - 4 \right) \right] c_{03} , \end{aligned} \tag{3.236}$$

where $v_{i,\alpha}$ is the α-th component of the ion velocity $\boldsymbol{v}_i$ and

$$c_{\mu\nu} = \frac{4\pi}{v_i^2} \int_0^\infty \mathrm{d}v_{e\|} v_{e\|}^\mu \int_0^\infty \mathrm{d}v_{e\perp} v_{e\perp}^\nu f_0(v_{e\|}, v_{e\perp}) . \tag{3.237}$$

For the anisotropic Maxwell distribution (2.28)

$$c_{03} = \frac{2v_{\text{th}\perp}^2}{v_i^2} , \qquad c_{21} = \frac{v_{\text{th}\|}^2}{v_i^2} . \tag{3.238}$$

With magnetic field the zero–order motion is helical (3.62). Using the methods of Sect. 3.5 the first order velocity transfer is

$$\Delta \boldsymbol{v}_i = \frac{2\pi \mathrm{i}}{m} \int \mathrm{d}\boldsymbol{k} U(\boldsymbol{k}) \boldsymbol{k} \mathrm{e}^{\mathrm{i}\boldsymbol{k}\cdot\bar{\boldsymbol{R}}_0} \sum_{n=-\infty}^{\infty} J_n(k_\perp a) \mathrm{e}^{\mathrm{i}n(\varphi-\theta)} \delta(\boldsymbol{k}\cdot\bar{\boldsymbol{v}}_r + n\Omega) \tag{3.239}$$

and after averaging with respect to the phase angle φ and integration with respect to the impact parameter $\bar{s}$ one obtains in the limit $a \to 0$ for a strong magnetic field

$$\frac{\overline{\Delta \boldsymbol{v}_i \cdot \Delta \boldsymbol{v}_i}}{\Delta t} = \frac{(2\pi)^5 n_e}{2m^2 \bar{v}_r} \int \mathrm{d}\bar{s}\, \bar{s} T_{1,2}^2(0, \bar{s}) \left(1 - \frac{\bar{\boldsymbol{v}}_r \bar{\boldsymbol{v}}_r}{\bar{v}_r^2} \right) . \tag{3.240}$$

This differs from the result (3.220) in the field–free case just by the replacement of the kinematic variables of the electrons by those of their guiding center. The average with respect to the electron velocity distribution leads now to one–dimensional integrals. As discussed in Sect. 3.10 the velocity dependent Coulomb logarithm or the cut–off function (3.192) must be kept under the velocity integral so that they suppress the integrand in the region $\bar{v}_r \to 0$ where the perturbation treatment becomes doubtful.

4 Dielectric Theory

4.1 Stopping Power (SP) in Plasmas Without Magnetic field

In this section we analyze expression (2.60) in the case when a projectile ion moves in an anisotropic two-temperature electron plasma (the susceptibility of the ions is neglected and the electronic index e is suppressed) without magnetic field. In the limit of vanishing magnetic field equation (2.49) takes the form $X(t) = Y^2 t^2$, where $Y = [\mu^2 + \tau(1-\mu^2)]^{1/2}$ with $\mu = \cos\beta$. In this limit the plasma dielectric function from (2.48) after changing the integration variable, $t \to t/Y$, now reads

$$\varepsilon(\boldsymbol{k},\omega) = 1 + \frac{1}{k^2\lambda_{\mathrm{D}\|}^2}\left\{1 + \frac{\mathrm{i}\zeta\sqrt{2}}{Y}\int_0^{\infty} \mathrm{d}t\, \mathrm{e}^{\mathrm{i}(\zeta/Y)t\sqrt{2}-t^2} \right. \\ \left. + 2(1-\tau)\frac{\sin^2\beta}{Y^2}\int_0^{\infty} \mathrm{d}t\, t\, \mathrm{e}^{\mathrm{i}(\zeta/Y)t\sqrt{2}-t^2}\right\}, \tag{4.1}$$

where $\zeta = \omega/kv_{\mathrm{th}\|}$. The t-integrals in (4.1) can be evaluated in close form and finally the dielectric function is represented as

$$\varepsilon(\boldsymbol{k},\omega) = 1 + \frac{1}{k^2\lambda_{\mathrm{D}\|}^2}\frac{1}{Y^2}W\left(\frac{\zeta}{Y}\right). \tag{4.2}$$

Here $W(\zeta) = g(\zeta) + \mathrm{i}f(\zeta)$ is the plasma dispersion function [43],

$$g(\zeta) = 1 - \zeta\sqrt{2}\,\mathrm{Di}\left(\frac{\zeta}{\sqrt{2}}\right), \quad f(\zeta) = \sqrt{\frac{\pi}{2}}\,\zeta\,\mathrm{e}^{-\zeta^2/2}, \tag{4.3}$$

where

$$\mathrm{Di}(\zeta) = \mathrm{e}^{-\zeta^2}\int_0^{\zeta}\mathrm{d}t\,\mathrm{e}^{t^2} \tag{4.4}$$

is the Dawson integral [43] which has for large arguments ζ the asymptotic behavior $\mathrm{Di}(\zeta) \simeq 1/(2\zeta) + 1/(4\zeta^3)$. Notice that the real and imaginary parts of plasma dispersion function, $g(\zeta)$ and $f(\zeta)$, are related to the corresponding functions $\mathcal{G}$ and $\mathcal{F}$ for the magnetized plasma (see equation (2.48)) according to $\mathcal{G}|_{B\to 0} = Y^{-2}g(\zeta/Y)$ and $\mathcal{F}|_{B\to 0} = Y^{-2}f(\zeta/Y)$. Thus in the absence of a temperature anisotropy, i.e. $\tau = 1$ and $Y = 1$, equation (4.2) yields the dielectric function of the isotropic plasma, [43].

Substituting (4.2) into (2.60) and performing the k–integration we obtain

$$S_0 = \frac{Z^2 \not{E}^2}{2\pi^2 \lambda_{D\|}^2} \int_0^1 d\mu \int_0^{\pi} d\phi \frac{\cos\Theta}{Y^2} Q_0\left(x \frac{\cos\Theta}{Y}, \xi Y\right), \tag{4.5}$$

where $x = v_i / v_{th\|}$, $\xi = k_{max} \lambda_{D\|}$, $\cos\Theta$ is determined from (2.61) and

$$Q_0(X,\xi) = f(X) \ln \frac{f^2(X) + [\xi^2 + g(X)]^2}{f^2(X) + g^2(X)} + 2g(X) \left[\arctan \frac{g(X)}{f(X)} - \arctan \frac{\xi^2 + g(X)}{f(X)}\right]. \tag{4.6}$$

In the case of an isotropic plasma ($T_\perp = T_\| \equiv T$, and $\tau = 1$)) $Y = 1$ and equation (4.5) after ϕ-integration coincides with the result of e.g. [105]

$$S_0 = \frac{Z^2 \not{E}^2}{2\pi \lambda_D^2 x^2} \int_0^x dy y Q_0(y,\xi), \tag{4.7}$$

where $x = v_i / v_{th}$, $v_{th} = v_{th\|} = v_{th\perp}$, $\lambda_D = v_{th}/\omega_p$, and $\xi = k_{max} \lambda_D$.

When a projectile ion moves slowly through a plasma, the electrons have sufficient time to experience the attractive ion potential. They are accelerated towards the ion, but when they reach its trajectory the ion has already moved forward a little bit. Hence, we expect an increased density of electrons at some place in the wake of the ion. This negative charge density pulls back the positive ion and gives rise to the stopping. This drag force is of particular interest for the electron cooling process. In the limit of small velocities $S \simeq \mathcal{R} v_i$. This looks like the friction law of a viscous fluid, and accordingly $\mathcal{R}$ is called the friction coefficient. However, in this case it should be noted that this law does not depend on the plasma viscosity and is not a consequence of electron–electron collisions which are neglected in the Vlasov equation.

The Taylor expansion of equation (4.5) for small v_i ($v_i \ll \bar{v}_{th}$) yields the friction law

$$S_0 = \frac{Z^2 \not{E}^2}{3\sqrt{2\pi}\, \bar{\lambda}_D^2} \frac{v_i}{\bar{v}_{th}} \psi(\bar{\xi}) \left[I_1(\tau) + I_2(\tau) \sin^2 \alpha\right], \tag{4.8}$$

where $\bar{\xi} = k_{max} \bar{\lambda}_D = \left(1 + v_i^2/\bar{v}_{th}^2\right)/\mathcal{Z} \simeq 1/\mathcal{Z}$. Here $\mathcal{Z}$ is the ion–plasma coupling parameter (2.42) and

$$I_1(\tau) = \frac{3}{\psi(\bar{\xi})} \left(\frac{2\tau + 1}{3}\right)^{3/2} \int_0^1 d\mu \frac{\mu^2 \psi(\xi Y(\mu))}{Y^3(\mu)}, \tag{4.9}$$

$$I_2(\tau) = \frac{3}{2\psi(\bar{\xi})} \left(\frac{2\tau + 1}{3}\right)^{3/2} \int_0^1 d\mu \frac{(1 - 3\mu^2)\psi(\xi Y(\mu))}{Y^3(\mu)}, \tag{4.10}$$

and the function ψ is

$$\psi(\xi) = \ln(1 + \xi^2) - \frac{\xi^2}{1 + \xi^2}. \tag{4.11}$$

In the case of an isotropic plasma ($\tau = 1$) we have $I_1 = 1$ and $I_2 = 0$ and equation (4.8) becomes the usual friction law [105]. For the case of strong temperature anisotropy, when $\tau \ll 1$ ($T_\perp \ll T_\|$) we have $\xi \simeq \sqrt{3}/\mathcal{Z}$ and

$$I_1 \simeq -\frac{\sqrt{3}}{6\psi(\bar{\xi})}\left[\mathrm{Li}_2(1+\xi^2)+\ln(1+\xi^2)\right], \tag{4.12}$$

$$I_2 \simeq \frac{\sqrt{3}}{12\psi(\bar{\xi})}\left[\xi^2+2\ln(1+\xi^2)+3\,\mathrm{Li}_2(1+\xi^2)\right]. \tag{4.13}$$

Here the functions I_1 and I_2 do not depend on τ, and $\mathrm{Li}_2(x)$ is the dilogarithm function [52]. Note that $\mathcal{Z} \ll 1$ and therefore $\bar{\xi} \gg 1$, $\xi \gg 1$ in equations (4.12) and (4.13) where the Coulomb logarithms are the leading terms and

$$I_1 \simeq \frac{\sqrt{3}}{6}\ln\frac{1}{\mathcal{Z}} \ll I_2 \simeq \frac{\sqrt{3}}{8\mathcal{Z}^2}\frac{1}{\ln(1/\mathcal{Z})}. \tag{4.14}$$

The friction coefficient in (4.8) is thus dominated by the second term and increases with increasing α.

In the opposite case, $\tau \gg 1$ ($T_\perp \gg T_\parallel$), the evaluation of equations (4.9) and (4.10) yields

$$I_1 \simeq \frac{\pi\sqrt{6}}{3\psi(\bar{\xi})}\left(\sqrt{1+\frac{3}{2}\bar{\xi}^2}-1-2\ln\frac{1+\sqrt{1+\frac{3}{2}\bar{\xi}^2}}{2}\right), \tag{4.15}$$

$$I_2 \simeq \frac{\pi\sqrt{6}}{6\psi(\bar{\xi})}\left(1+\frac{1}{\sqrt{1+\frac{3}{2}\bar{\xi}^2}}-2\sqrt{1+\frac{3}{2}\bar{\xi}^2}+6\ln\frac{1+\sqrt{1+\frac{3}{2}\bar{\xi}^2}}{2}\right), \tag{4.16}$$

and

$$I_1 \simeq -I_2 \simeq \frac{\pi}{2\mathcal{Z}\ln(1/\mathcal{Z})}. \tag{4.17}$$

Then $I_1 + I_2\sin^2\alpha \simeq I_1\cos^2\alpha$ and the friction coefficient decreases with increasing α in this case.

In Fig. 4.1 the normalized friction coefficient $I(\tau) = I_1(\tau)+I_2(\tau)\sin^2\alpha$ is plotted as a function of temperature anisotropy τ for $\alpha = 0$ (solid line), $\alpha = \pi/6$ (dotted line), $\alpha = \pi/3$ (dashed line), $\alpha = \pi/2$ (dot-dashed line) and for fixed plasma density and average temperature ($\mathcal{Z} = 0.2$). Figure 4.1 shows an enhancement of the friction coefficient when the ion moves along the direction with low temperature.

For arbitrary projectile velocities we evaluated equation (4.5) numerically. In Fig. 4.2 the stopping power is plotted for plasmas with a large temperature anisotropy ($\tau = 10^{-2}$ and $\tau = 10^2$ in left and right panels respectively), $n_e = 10^8\,\mathrm{cm}^{-3}$, $\bar{T} = 0.1\,\mathrm{eV}$ and for four values of α; $\alpha = 0$ (dotted line), $\alpha = \pi/6$ (short-dashed line), $\alpha = \pi/3$ (dashed line), $\alpha = \pi/2$ (dot-dashed line). The solid lines are plotted for an isotropic one-temperature plasma with $T = \bar{T} = 0.1\,\mathrm{eV}$. The general behavior of the stopping power for two anisotropy parameters τ is characterized by an increase by comparison with the isotropic case. At $\alpha \simeq \pi/2$ and $\tau = 10^{-2}$ (left panel, Fig. 4.2) the ion moves in the direction to the electron motion with the lower temperature $T_\perp$ and the maximum of the stopping power is around $v_i \simeq v_{\mathrm{th}\perp}$, whereas the maximum for an ion motion in longitudinal direction is at $v_i \simeq v_{\mathrm{th}\parallel} \gg v_{\mathrm{th}\perp}$.

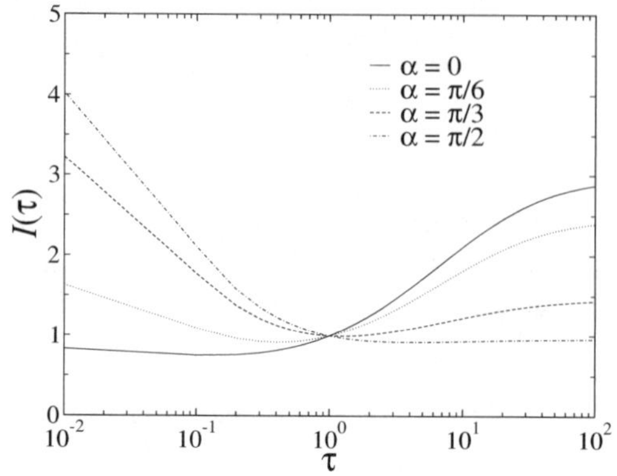

Fig. 4.1. Normalized friction coefficient $I(\tau) = I_1(\tau) + I_2(\tau)\sin^2\alpha$ (see equations (4.8)-(4.10)) in plasma with $\mathcal{Z} = 0.2$ as a function of $\tau = T_\perp/T_\parallel$ for four values of α; $\alpha = 0$ (solid line), $\alpha = \pi/6$ (dotted line), $\alpha = \pi/3$ (dashed line), $\alpha = \pi/2$ (dot-dashed line). After [90].

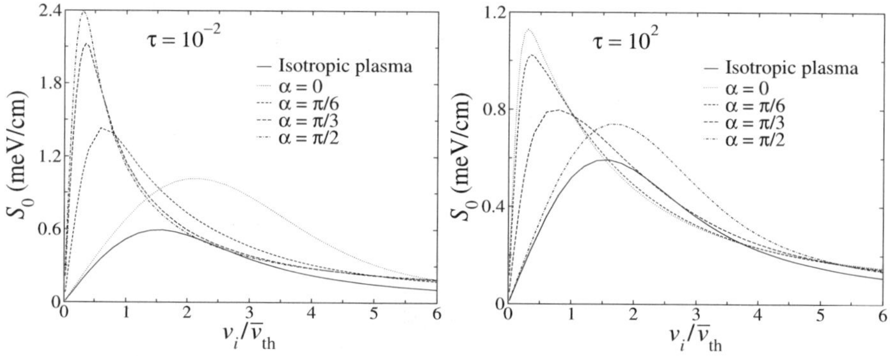

Fig. 4.2. Stopping power (in units of meV/cm) as a function of the projectile velocity v_i (in units of $\bar{v}_{\rm th}$) in a plasma with large temperature anisotropy $\tau = 10^{-2}$ (left) and $\tau = 10^2$ (right) without magnetic field ($\bar{T} = 0.1\,$eV, $n_e = 10^8\,$cm^{-3}) for four values of the angle α, $\alpha = 0$ (dotted lines), $\alpha = \pi/6$ (short-dashed lines), $\alpha = \pi/3$ (dashed lines), $\alpha = \pi/2$ (dot-dashed lines). Solid lines: isotropic plasma with temperature $T = \bar{T} = 0.1\,$eV. After [90].

4.2 Stopping in Plasmas With Weak Magnetic field

For the case when the magnetic field is weak, in the sense that the dimensionless parameter $\eta = \Omega_e/\omega_p$ is much less than unity, the functions $\mathcal{G}$ and $\mathcal{F}$, equations (2.48) and (2.49), which define the dielectric function, can be expanded about the field free values $Y^{-2}g(\zeta/Y)$, $Y^{-2}f(\zeta/Y)$ equations (4.3) and (4.4)

$$\mathcal{G} + \mathrm{i}\mathcal{F} \simeq \frac{1}{Y^2}\left[g\left(\frac{\zeta}{Y}\right) + \mathrm{i}f\left(\frac{\zeta}{Y}\right)\right] + \eta^2 \frac{\sin^2\beta}{Y^4(k\lambda_{\mathrm{D}\parallel})^2}\left[g_1\left(\frac{\zeta}{Y}\right) + \mathrm{i}f_1\left(\frac{\zeta}{Y}\right)\right], \tag{4.18}$$

where

$$g_1(\zeta) + \mathrm{i} f_1(\zeta) = \frac{2}{3}(1-\tau)\int_0^\infty \left(\frac{t^2}{2Y^2}\tau\sin^2\beta - 1\right) \mathrm{e}^{\mathrm{i}\zeta t\sqrt{2}-t^2} t^3 \mathrm{d}t \\ + \frac{\mathrm{i}\zeta\sqrt{2}}{6}\tau \int_0^\infty \mathrm{e}^{\mathrm{i}\zeta t\sqrt{2}-t^2} t^4 \mathrm{d}t , \tag{4.19}$$

$\zeta = \omega/kv_{\mathrm{th}\|} = (v_i/v_{\mathrm{th}\|})\cos\Theta$. The functions $g_1(\zeta)$ and $f_1(\zeta)$ can be evaluated explicitly [52]

$$g_1(\zeta) = -\frac{\tau}{24}\zeta\left[\zeta(5-\zeta^2) + \sqrt{2}(3-6\zeta^2+\zeta^4)\,\mathrm{Di}\left(\frac{\zeta}{\sqrt{2}}\right)\right] \\ + \frac{1-\tau}{6}\left\{\zeta^2 - 2 + \zeta\sqrt{2}(3-\zeta^2)\,\mathrm{Di}\left(\frac{\zeta}{\sqrt{2}}\right)\right. \\ \left. + \frac{\tau\sin^2\beta}{4Y^2}\left[8 - 9\zeta^2 + \zeta^4 - \zeta\sqrt{2}(15-10\zeta^2+\zeta^4)\,\mathrm{Di}\left(\frac{\zeta}{\sqrt{2}}\right)\right]\right\} , \tag{4.20}$$

$$f_1(\zeta) = \frac{1}{6}\sqrt{\frac{\pi}{2}}\zeta \mathrm{e}^{-\zeta^2/2}\left\{\frac{\tau}{4}(3-6\zeta^2+\zeta^4)\right. \\ \left. + (1-\tau)\left[\frac{\tau\sin^2\beta}{4Y^2}(15-10\zeta^2+\zeta^4) + \zeta^2 - 3\right]\right\} . \tag{4.21}$$

Substituting expressions (4.18)-(4.21) into (2.60) leads to

$$S = S_0 + \eta^2 S_1 , \tag{4.22}$$

where S_0 is the stopping power in plasma without magnetic field, equation (4.5) and $\eta^2 S_1$ represents the change due to a weak magnetic field. After some simplifications this becomes

$$S_1 = \sqrt{\frac{\pi}{2}}\frac{Z^2 e^2}{24\pi^2\lambda_{\mathrm{D}\|}^2 (v_i/v_{\mathrm{th}\|})}\int_0^1 \frac{1-\mu^2}{Y^5}\mathrm{d}\mu \int_0^\pi \zeta^2 \mathrm{e}^{-\zeta^2/2Y^2}\frac{\tau\left(7-\zeta^2/Y^2\right)-4Y^2}{f^2(\zeta/Y)+g^2(\zeta/Y)}\mathrm{d}\phi. \tag{4.23}$$

In the isotropic plasma ($\tau = 1$) equation (4.23) coincides with the results by May and Cramer [83] after integration over ϕ. Note that the additional term S_1 does not depend on the cut–off parameter $k_{\max}$. It should be noted that the stopping power (4.23) is valid for sufficiently large velocities in the sense $\eta \lesssim v_i/v_{\mathrm{th}\|}$ as it is intuitively clear from equation (2.49). Obviously in the argument of the cosine function the quantity $\Omega_e/v_{\mathrm{th}\|}$ must be kept small in a limit of vanishing magnetic field. This point will be further discussed in Sect. 4.4 (see also Sect. 5.5).

In the next subsections we evaluate equation (4.23) for small and large projectile velocities.

4.2.1 Small Projectile Velocities

When the projectile ion moves slowly, $v_i < \bar{v}_{\mathrm{th}}$ but $\eta \lesssim v_i/v_{\mathrm{th}\|}$, equation (4.23) leads to the simplified expression

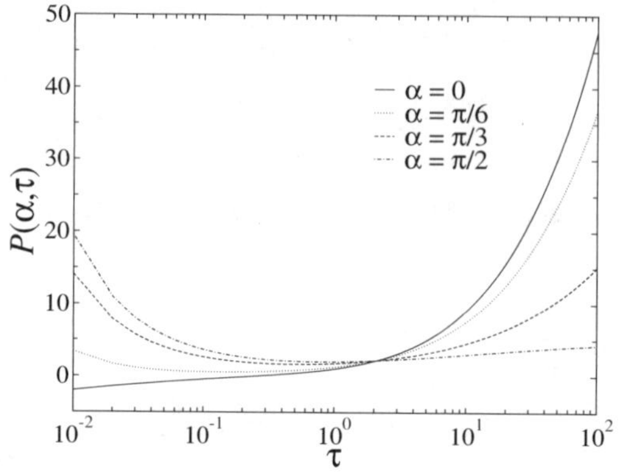

Fig. 4.3. The function $P(\alpha, \tau)$ (see equations (4.24)-(4.27)) as a function of $\tau = T_{\perp}/T_{\|}$ for four values of α, $\alpha = 0$ (solid line), $\alpha = \pi/6$ (dotted line), $\alpha = \pi/3$ (dashed line), $\alpha = \pi/2$ (dot-dashed line). After [90].

$$S_1 \simeq \frac{Z^2 e^2}{60\pi \bar{\lambda}_D^2} \sqrt{\frac{\pi}{2}} \frac{v_i}{\bar{v}_{th}} P(\alpha, \tau), \tag{4.24}$$

with

$$P(\alpha, \tau) = \left(\frac{1+2\tau}{3}\right)^{3/2} \left[P_1(\tau) + P_2(\tau)\sin^2\alpha\right], \tag{4.25}$$

$$P_1(\tau) = \frac{5}{6(1-\tau)^2} \left[14\tau + 25 - \frac{3(9\tau+4)}{\sqrt{|1-\tau|}} p_0(\tau)\right], \tag{4.26}$$

$$P_2(\tau) = \frac{5}{12\tau(1-\tau)^2} \left[\frac{3\tau(23\tau+16)}{\sqrt{|1-\tau|}} p_0(\tau) - 28\tau^2 - 91\tau + 2\right]. \tag{4.27}$$

Here, the function $p_0(\tau)$ is given by equation (2.64). In an isotropic plasma with $\tau = 1$ we have $P_1(1) = P_2(1) = 1$.

In Fig. 4.3 the normalized friction coefficient $P(\alpha, \tau)$ for the additional stopping power S_1 is plotted as a function of τ for $\alpha = 0$ (solid line), $\alpha = \pi/6$ (dotted line), $\alpha = \pi/3$ (dashed line), $\alpha = \pi/2$ (dot-dashed line). The general behavior of $P(\alpha, \tau)$ is similar to the friction coefficient of the plasma without magnetic field (see Fig. 4.1). But the correction $P(\alpha, \tau)$ can be also negative at small τ and α, which corresponds to a slight decrease of the stopping power, equation (4.22).

4.2.2 High Projectile Velocities

When the projectile ion moves with large velocity ($v_i \gg \bar{v}_{th}$), equation (4.23), yields

$$S_1 \simeq -\frac{Z^2 e^2 \omega_p^2}{4v_i^2} \left[1 + \cos^2\alpha + \frac{v_{th\|}^2}{v_i^2} C(\alpha, \tau)\right], \tag{4.28}$$

where

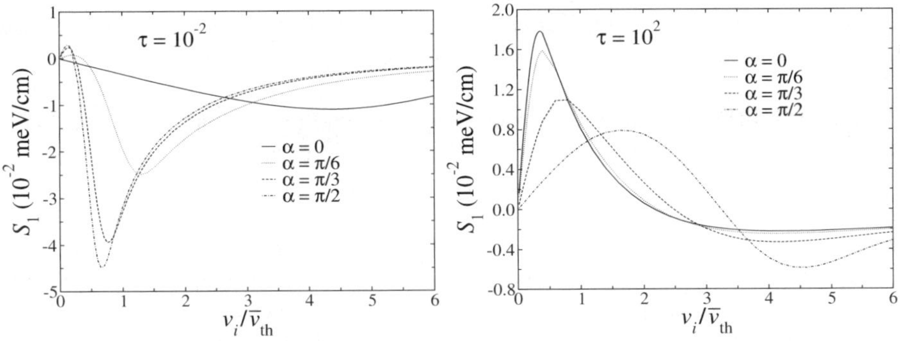

Fig. 4.4. Additional stopping power S_1 (in 10^{-2} meV/cm) in plasma ($n_e = 10^8\,\mathrm{cm}^{-3}$, $\bar{T} = 0.1\,\mathrm{eV}$, $\tau = 10^{-2}$ (left), $\tau = 10^2$ (right)) with weak magnetic field (see equation 4.23) as a function of projectile velocity v_i (in units of $\bar{v}_{\mathrm{th}}$) for $\alpha = 0$ (solid line), $\alpha = \pi/6$ (dotted line), $\alpha = \pi/3$ (dashed line), $\alpha = \pi/2$ (dot-dashed line). After [90].

$$C(\alpha, \tau) = 12\tau\left(1 - 3\cos^2\alpha\right) + \frac{7(1-\tau)}{2}\left(15\cos^4\alpha - 12\cos^2\alpha + 1\right). \qquad (4.29)$$

This result at $\tau = 1$ is in agreement with the results of Honda et al [61] and May and Cramer [83]. Although S_1 in equation (4.28) is proportional to the plasma density, the full correction term $\eta^2 S_1$ does not depend on the plasma density.

In Fig. 4.4 we show the velocity dependence of the function S_1 for $\tau = 10^{-2}$ (left) and $\tau = 10^2$ (right). The different curves are $\alpha = 0$ (solid line), $\alpha = \pi/6$ (dotted line), $\alpha = \pi/3$ (dashed line), $\alpha = \pi/2$ (dot-dashed line). For small and medium projectile velocities the weak magnetic field decreases the total stopping power for small τ and increases it in the high τ limit. For high projectile velocities the magnetic field always reduces the stopping power independent of the temperature anisotropy, see equation (4.28).

4.3 Stopping in Plasmas With Strong Magnetic Field

We now turn to the case when a projectile ion moves in an anisotropic electron plasma with a strong magnetic field, which is on one hand, sufficiently weak to allow a classical description ($\hbar\Omega_e < k_{\mathrm{B}}T_\perp$ or $\hbar/mv_{\mathrm{th}\perp} < a_e$), and, on the other hand, comparatively strong so that the cyclotron frequency of the plasma electrons exceeds the plasma frequency $\Omega_e \gg \omega_p$. This limits the values of the magnetic field, the transverse temperature and the plasma density. From these conditions we can obtain

$$3 \times 10^{-7} n_e^{1/2} < B < 10^4 T_\perp\,, \qquad (4.30)$$

where n_e is measured in cm^{-3}, $T_\perp$ in eV, and B in T. Conditions (4.30) are always true in the range of parameters $n_e < 10^{15}\,\mathrm{cm}^{-3}$, $B < 10\,\mathrm{T}$, $T_\perp > 10^{-3}\,\mathrm{eV}$. Then the transverse motion of the electrons is completely quenched and the stopping power depends only on the longitudinal electron temperature $T_{\|}$. The dependence on the transverse temperature will be only introduced by the cut–off parameter equation (2.59).

In the limit of a sufficiently strong magnetic field the second term in (2.49) is proportional to a_e and can be omitted. Then $X(t) = t^2 \cos^2\beta$ and in equation (2.48) the term proportional to $1/\Omega_e$ can also be omitted. Changing the integration variable in (2.48), $t \to t/|\cos\beta|$, we find that $\mathcal{G} + i\mathcal{F} = g(\zeta_1) + if(\zeta_1)$, where $\zeta_1 = \omega/|k_{\|}|v_{\text{th}\|}$ with g, f from equations (4.3). Thus the dielectric function of the strongly magnetized plasma coincides with the dielectric function of the isotropic field-free plasma, where k, v_{th} and λ_{D} are replaced by $|k_{\|}|$, $v_{\text{th}\|}$ and $\lambda_{\text{D}\|}$, respectively.

Consider equation (2.60) in the limit of strong magnetic field. Assuming that the variable k is scaled in units of $\lambda_{\text{D}\|}^{-1}$ from (2.60) we obtain

$$S_{\text{inf}} = \frac{2Z^2 e^2}{\pi^2 \lambda_{\text{D}\|}^2} \int_0^{\xi} k^3 \text{d}k \int_0^1 \text{d}\mu \int_0^{\pi} \text{d}\phi \frac{\cos\Theta f(\zeta_1)}{\left[k^2 + g(\zeta_1)\right]^2 + f^2(\zeta_1)}, \tag{4.31}$$

where now $\zeta_1 = x\cos\Theta/\mu$ with $x = v_i/v_{\text{th}\|}$. After integration over k equation (4.31) yields

$$S_{\text{inf}} = \frac{Z^2 e^2}{2\pi^2 \lambda_{\text{D}\|}^2} \int_0^1 \text{d}\mu \int_0^{\pi} \text{d}\phi \cos\Theta Q_0\left(x\frac{\cos\Theta}{\mu}, \xi\right). \tag{4.32}$$

Here the function Q_0 is given by equation (4.6). For further simplification of equation (4.32) we introduce the new variable of integration $y = \cos\Theta/\mu$. After integrating with respect to ϕ the stopping power in the presence of a strong magnetic field is finally

$$S_{\text{inf}}(v_i, \alpha) = \frac{Z^2 e^2}{8\pi \lambda_{\text{D}\|}^2} Q(x, \alpha), \tag{4.33}$$

where

$$Q(x, \alpha) = \sin^2\alpha \int_{-\infty}^{+\infty} \frac{Q_0(xy, \xi)\, y\text{d}y}{(y^2 - 2y\cos\alpha + 1)^{3/2}}. \tag{4.34}$$

Previously (see, e.g., [25, 31, 88, 89, 115]) only the case of $\alpha = 0$ (the motion of the projectile ion parallel to the magnetic field) has been investigated. In this case the integral in equation (4.34) diverges, while the prefactor $\sin^2\alpha$ tends to zero, cf. (3.197). Introducing the new variable of integration in equation (4.34), $y' = (y - \cos\alpha)/\sin\alpha$, we obtain for vanishing angle α

$$Q(x, \alpha \to 0) = 2Q_0(x, \xi). \tag{4.35}$$

Thus expression (4.33) with (4.35) reproduces the known results for the stopping power on an ion which moves along the direction of the magnetic field. Below we discuss its low- and high-velocity limits.

4.3.1 Small Projectile Velocities

In the low velocity limit ($v_i \ll v_{\text{th}\|}$) equation (4.34) becomes

$$Q(x,\alpha) \simeq 2x\left\{ \sqrt{2\pi}\psi(\xi)\left[\sin^2\alpha \ln\left(\frac{2}{x\sin\alpha}\right) + 1 - 2\sin^2\alpha\right] + C_1(\xi)\sin^2\alpha \right\}, \tag{4.36}$$

where

$$C_1(\xi) = \int_0^1 \frac{\mathrm{d}y}{y^2}\left[Q_0(y,\xi) - \sqrt{2\pi}\psi(\xi)y\right] + \int_1^\infty \frac{\mathrm{d}y}{y^2} Q_0(y,\xi). \tag{4.37}$$

The function ψ is defined by equation (4.11). Since we deal with small ion beam-plasma coupling $\mathcal{Z} \ll 1$ we have, $\xi \gg 1$ in equations (4.36) and (4.37) and the function $C_1(\xi)$ simplifies to

$$C_1(\xi) \simeq \sqrt{2\pi}\ln(2/\gamma_0)\ln\xi + 2.01, \tag{4.38}$$

where $\gamma_0 = \mathrm{e}^\gamma$ and $\gamma = 0.5772$ is Euler's constant.

We note that the friction coefficient S_{inf}/v_i from equations (4.33) and (4.36) contains a term which depends logarithmically on v_i and which vanishes for $\alpha \to 0$. In Sect. 4.4 we show that this behavior is a characteristic feature of the LR stopping power at low velocities for a finite strength of the magnetic field.

4.3.2 High Projectile Velocities

In the case of high projectile velocities ($v_i \gg v_{\text{th}\|}$) the general expression (4.34) becomes

$$Q(x,\alpha) \simeq \frac{4\pi}{x^2}\left\{ \sin^2\alpha\left[\ln\left(\frac{2x}{\sin\alpha}\right) + C_2(\xi) - 2\right] + 1\right\}, \tag{4.39}$$

where

$$C_2(\xi) = \frac{1}{2\pi}\int_0^1 Q_0(y,\xi)y\mathrm{d}y + \int_1^\infty \frac{\mathrm{d}y}{y}\left[\frac{y^2}{2\pi}Q_0(y,\xi) - 1\right] \tag{4.40}$$

which gives for $\xi \gg 1$ $C_2(\xi) \simeq \ln\xi$. The stopping power for strong magnetic fields shows, in the low- and the high-velocity limits (equations (4.36) and (4.39)) an enhancement for ions moving transverse to the magnetic field compared to the case of the longitudinal motion ($\alpha = 0$). This effect is in agreement with PIC simulation results [127]. In contrast to the field-free case for a strong magnetic field and for ion parallel motion ($\alpha = 0$) with $v_i \gg v_{\text{th}\|}$ (equations (4.33) and (4.39)) $S_{\text{inf}} \simeq Z^2e^2\omega_p^2/2v_i^2$ is independent of k_{max}. The cut–off k_{max} necessary at low ion velocities is, however, less well defined here than for the field-free case, where the cut–off (2.59) was deduced from the binary collision picture. Now, the electrons are forced to move parallel to $\boldsymbol{B}$. Since we assumed the motion of the ion in this direction as well the ion and an electron just pass each other along a straight line. For symmetry reasons the total momentum transfer and the stopping power is zero. Purely binary interactions contribute nothing and the stopping of the ion is only due

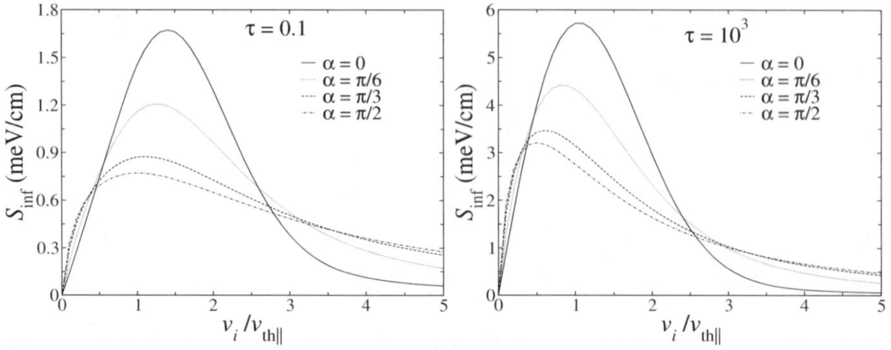

Fig. 4.5. Stopping power S_{inf} (in meV/cm) in plasma ($n_e = 10^6\,\text{cm}^{-3}$, $T_{\|} = 10^{-4}$ eV, $\tau = 0.1$ (left), $\tau = 10^3$ (right)) with strong magnetic field as a function of projectile velocity v_i (in units of $v_{\text{th}\|}$) for $\alpha = 0$ (solid line), $\alpha = \pi/6$ (dotted line), $\alpha = \pi/3$ (dashed line), $\alpha = \pi/2$ (dot-dashed line). After [90].

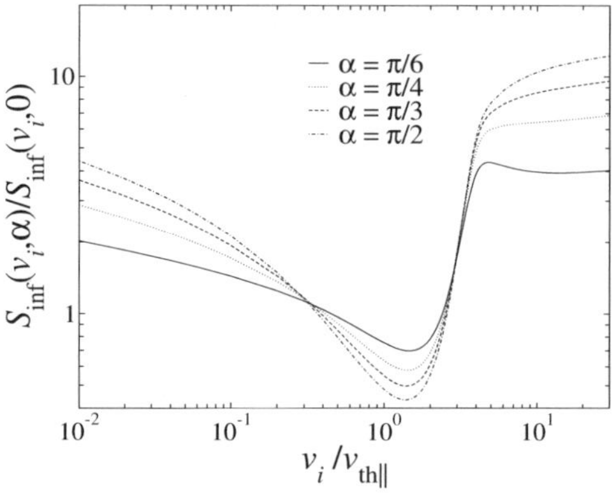

Fig. 4.6. The ratio $S_{\text{inf}}(v_i, \alpha)/S_{\text{inf}}(v_i, 0)$ as a function of projectile velocity v_i (in units of $v_{\text{th}\|}$) for $T_{\|} = 10^{-4}$ eV, $\tau = 10^3$, $\alpha = \pi/6$ (solid line), $\alpha = \pi/4$ (dotted line), $\alpha = \pi/3$ (dashed line), $\alpha = \pi/2$ (dot-dashed line). After [90].

to the collective response of the plasma, that is, due to modes with long wavelengths $k < 1/\lambda_{D\|}$. This suggests taking $k_{\max}$ of the order of $1/\lambda_{D\|}$, but further investigations are clearly needed here for a more precise description in this particular case.

In Fig. 4.5, the stopping power S_{inf} is plotted as a function of projectile velocity (in units of $v_{\text{th}\|}$) for $n_e = 10^6\,\text{cm}^{-3}$, $T_{\|} = 10^{-4}$ eV , $T_{\perp} = 10^{-5}$ eV (left), $T_{\perp} = 0.1$ eV (right), and for four different values of angle α : $\alpha = 0$ (solid line), $\alpha = \pi/6$ (dotted line), $\alpha = \pi/3$ (dashed line) and $\alpha = \pi/2$ (dot-dashed line). The enhancement of $S_{\text{inf}}(v_i, \alpha)$ with respect to $S_{\text{inf}}(v_i, 0)$ in the low and in high velocity limit by increasing of the angle α is documented in Fig. 4.6, for $T_{\|} = 10^{-4}$ eV, $T_{\perp} = 0.1$ eV, $n_e = 10^6\,\text{cm}^{-3}$, $\alpha = \pi/6$ (solid line), $\alpha = \pi/4$ (dotted line), $\alpha = \pi/3$ (dashed line) and $\alpha = \pi/2$ (dot-dashed line). For medium projectile velocities $v_i \simeq v_{\text{th}\|}$ the stopping power is always higher for small α, cf. with Fig. 4.5.

4.4 Stopping in the Low-Velocity Limit at Arbitrary Field Strengths

We now proceed with a projectile ion at low velocities and an arbitrary magnetic field. In the presence of a magnetic field the LR friction coefficient contains a term which diverges like $\ln(v_{\mathrm{th}\|}/v_i)$ in addition to the usual constant one (see e.g. Sect. 4.1). Below we investigate the case of the interaction of a heavy (no cyclotron motion) projectile ion with a magnetized anisotropic electron–ion plasma. For this consideration it is convenient to use the Bessel function representation of the dielectric function which has been given, e.g., by Ichimaru [67] (see Appendix A, equation (A.7)) and to write the real and imaginary parts of equation (A.7) (see also (2.48)) separately,

$$\mathcal{G}(\omega) = 1 - \sum_\nu \varrho_\nu \frac{\sqrt{2}\omega}{|k_\||v_{\mathrm{th}\nu\|}} \Lambda_0(z_\nu)\,\mathrm{Di}\left(\frac{\omega}{\sqrt{2}|k_\||v_{\mathrm{th}\nu\|}}\right) \tag{4.41}$$
$$- \sum_\nu \varrho_\nu \frac{\sqrt{2}}{|k_\||v_{\mathrm{th}\nu\|}} \sum_{n=1}^{\infty} \Lambda_n(z_\nu)\left\{\omega\left[\mathrm{Di}\left(\frac{\omega+n\Omega_\nu}{\sqrt{2}|k_\||v_{\mathrm{th}\nu\|}}\right) + \mathrm{Di}\left(\frac{\omega-n\Omega_\nu}{\sqrt{2}|k_\||v_{\mathrm{th}\nu\|}}\right)\right]\right.$$
$$\left. + n\Omega_\nu\left(\frac{1}{\tau_\nu}-1\right)\left[\mathrm{Di}\left(\frac{\omega-n\Omega_\nu}{\sqrt{2}|k_\||v_{\mathrm{th}\nu\|}}\right) - \mathrm{Di}\left(\frac{\omega+n\Omega_\nu}{\sqrt{2}|k_\||v_{\mathrm{th}\nu\|}}\right)\right]\right\},$$

$$\mathcal{F}(\omega) = \sqrt{\frac{\pi}{2}} \sum_\nu \varrho_\nu \left\{\frac{\omega}{|k_\||v_{\mathrm{th}\nu\|}} \Lambda_0(z_\nu)\exp\left(-\frac{\omega^2}{2k_\|^2 v_{\mathrm{th}\nu\|}^2}\right)\right. \tag{4.42}$$
$$+ \frac{2}{|k_\||v_{\mathrm{th}\nu\|}} \sum_{n=1}^{\infty} \Lambda_n(z_\nu)\exp\left(-\frac{\omega^2+n^2\Omega_\nu^2}{2k_\|^2 v_{\mathrm{th}\nu\|}^2}\right)$$
$$\left.\times\left[\omega\cosh\left(\frac{n\Omega_\nu\omega}{k_\|^2 v_{\mathrm{th}\nu\|}^2}\right) + n\Omega_\nu\left(\frac{1}{\tau_\nu}-1\right)\sinh\left(\frac{n\Omega_\nu\omega}{k_\|^2 v_{\mathrm{th}\nu\|}^2}\right)\right]\right\}.$$

Here $\varrho_\nu = \lambda_{\mathrm{D}\|}^2/\lambda_{\mathrm{D}\nu\|}^2$. The other notations in equations (4.41) and (4.42) are explained in Appendix A. Note that in equations (4.41) and (4.42) we explicitly separate the terms with $n = 0$ and $n \neq 0$.

For the friction coefficient we have to consider the SP S, given by equation (2.60) in the low-velocity limit and thus the functions $\mathcal{G}(\omega)$ and $\mathcal{F}(\omega)$ given by equations (4.41) and (4.42), when $\omega = \boldsymbol{k}\cdot\boldsymbol{v}_i$. Now we have to write the Taylor expansion of equations (4.41) and (4.42) for small $\omega = \boldsymbol{k}\cdot\boldsymbol{v}_i$. However, the first term of equation (4.42) exhibits a singular behavior in the limit of $\omega = \boldsymbol{k}\cdot\boldsymbol{v}_i \to 0$ where the $k_\|$ integration diverges logarithmically for small $k_\|$. We must therefore keep $\omega = \boldsymbol{k}\cdot\boldsymbol{v}_i$ finite in that integration to avoid such a divergence. The anomalous contribution which arises from the first term of equation (4.42) is $S_{\mathrm{I}} = S_{\mathrm{an}} + S_{\mathrm{I}}'$ (see Appendix B for details), where

$$S_{\mathrm{an}} = \left(\frac{2}{\pi}\right)^{1/2} \frac{Z^2 e^2}{8\lambda_{\mathrm{D}\|}^2} \sin^2\alpha \sum_\nu \varrho_\nu \frac{v_i}{v_{\mathrm{th}\nu\|}}\left[\ln\left(\frac{8v_{\mathrm{th}\nu\|}^2}{v_i^2\sin^2\alpha}\right) - \gamma - 1\right]\mathcal{K}(\varsigma_\nu) \tag{4.43}$$

with

$$\mathcal{K}(\varsigma_\nu) = 2\int_0^{\xi} \frac{\Lambda_0(k^2\varsigma_\nu)k^3\mathrm{d}k}{\left[k^2 + \mathcal{H}_0(k)\right]^2} \tag{4.44}$$

is the anomalous term while the other one,

$$S_{\mathrm{I}}' = \left(\frac{2}{\pi}\right)^{1/2} \frac{Z^2\not{k}^2}{\lambda_{\mathrm{D}\|}^2} \sum_\nu \varrho_\nu \frac{v_i}{v_{\mathrm{th}\nu\|}} \int_0^1 \frac{\mathrm{d}\mu}{\mu} \tag{4.45}$$

$$\times \left[\left(\mu^2\cos^2\alpha + \frac{1-\mu^2}{2}\sin^2\alpha\right)\int_0^{\xi} \frac{\Lambda_0(z_\nu)k^3\mathrm{d}k}{\left[k^2 + \mathcal{H}(k,\mu)\right]^2} - \frac{1}{4}\sin^2\alpha\,\mathcal{K}(\varsigma_\nu)\right]$$

is linear with respect v_i and contributes to the usual SP. In (4.43)–(4.45) we have introduced the following notations $\eta_\nu = \Omega_\nu/\omega_{p\nu}$, $\varsigma_\nu = \tau_\nu/\eta_\nu^2\varrho_\nu$, $\Lambda_0(z) = \mathrm{e}^{-z}I_0(z)$ and γ is the Euler's constant. The other quantities are explained in Appendix B.

Consider now the limit of a vanishing magnetic field. In general the stopping power (2.53) in the limit of vanishing magnetic field or ion velocity strongly depends on the order of these limits. In a weak magnetic field the additional stopping power is even, i.e. quadratic with respect to B, due to the symmetry of the system (see (4.22)). As discussed in Sect. 4.2 in the case when $B \to 0$ there is a restriction on the ion velocity and v_i cannot be arbitrary small. In the present case of small ion velocities this is a restriction on the magnetic field and the obtained stopping powers (4.43) and (4.45) are valid for sufficiently large parameters η_ν (i.e., magnetic field strengths), so that $v_i/v_{\mathrm{th}\nu\|} \lesssim \eta_\nu$. In the limit of vanishing magnetic field $\eta_\nu \lesssim v_i/v_{\mathrm{th}\nu\|}$ the S_{an} vanishes as shown in Sect. 4.2.1. Hence the anomalous term equation (4.43) with (4.44) arises from the presence of the magnetic field and is not restricted to anisotropic plasmas.

For an isotropic plasma ($\tau_\nu = 1$) and for a sufficiently weak magnetic field $\eta_\nu\sqrt{\varrho_\nu} < \xi$ (or $\Omega_\nu < k_{\max}v_{\mathrm{th}\nu\|}$), equation (4.44) takes the form [66]

$$\mathcal{K}(\varsigma_\nu) \simeq \int_0^{\infty} \frac{\Lambda_0(x)x\mathrm{d}x}{(x+\varsigma_\nu)^2} = \mathrm{e}^{\varsigma_\nu}\left[(1+\varsigma_\nu)K_0(\varsigma_\nu) - \varsigma_\nu K_1(\varsigma_\nu)\right], \tag{4.46}$$

where K_0 and K_1 are the modified Bessel functions of the second kind. In the case of very strong magnetic field $\eta_\nu > \xi\sqrt{\tau_\nu/\varrho_\nu}$ (or $\Omega_\nu > k_{\max}v_{\mathrm{th}\nu\perp}$) the function $\mathcal{K}(\varsigma_\nu)$ reads $\mathcal{K}(\varsigma_\nu) \simeq \psi(\xi)$, where $\psi(\xi)$ is defined in (4.11).

The physical origin of such an anomalous friction coefficient may be traced to the spiral motion of the electrons and ions along the magnetic field lines. An analysis shows that the anomalous term does not vanish switching off the collective excitations in the dielectric function (see, e.g., Sect. 4.6 for details). This term being proportional to $\propto \mathrm{Im}\,\varepsilon(\boldsymbol{k},\omega)$ arises at small $k_\|$ in the $k_\|$–integration. Recalling that the spectral density of the charge fluctuations in plasma is $\propto \mathrm{Im}\,\varepsilon(\boldsymbol{k},\omega)$ [67, 72] the anomalous friction arises as a result of the long–wavelength fluctuations (i.e., small $k_\|$). The plasma particles naturally tend to couple strongly with these fluctuations along the magnetic field. In addition, when such fluctuations are characterized by slow variation in time (i.e., small $\omega = \boldsymbol{k}\cdot\boldsymbol{v}_i$), the contact time or the rate of energy exchange between the particles and the fluctuations will be further enhanced. In a plasma, such low–frequency fluctuations are provided by the slow projectile ion. The above coupling can therefore be an efficient mechanism of energy exchange between the plasma particles and the projectile ion. In the limit of $v_i \to 0$, the frequency $\omega = \boldsymbol{k} \cdot \boldsymbol{v}_i \to 0$ tends to zero as well. The contact time thus becomes infinite and the friction coefficient diverges. On the other hand one must query how far such a strong coupling is described adequately in a linearized treatment, this will be further discussed in Sect. 4.6.

The anomalous friction coefficient (see equation (4.43)) vanishes, however, when the ion moves along the magnetic field ($\alpha = 0$). Then the friction coefficient is solely given by the second term of equation (4.42) as well as by equation (4.45). The contributions of these terms to the stopping power lead to the usual friction law in a plasma and reads for arbitrary angles α

$$S'_{\mathrm{I}} + S_{\mathrm{II}} \simeq \left(\frac{2}{\pi}\right)^{1/2} \frac{Z^2 e^2}{\lambda^2_{\mathrm{D}\|}} \sum_{\nu} \varrho_{\nu} \frac{v_i}{v_{\mathrm{th}\nu\|}} \int_0^1 \frac{\mathrm{d}\mu}{\mu} \tag{4.47}$$

$$\times \left\{ \left(\mu^2 \cos^2\alpha + \frac{1-\mu^2}{2}\sin^2\alpha\right) \int_0^{\xi} \frac{\mathcal{P}_{\nu}(k,\mu)\, k^3 \mathrm{d}k}{[k^2 + \mathcal{H}(k,\mu)]^2} - \frac{1}{4}\sin^2\alpha \mathcal{K}(\varsigma_{\nu}) \right\}$$

with

$$\mathcal{P}_{\nu}(k,\mu) = \Lambda_0(z_{\nu}) + 2\sum_{n=1}^{\infty} \Lambda_n(z_{\nu}) \exp\left(-\frac{n^2\eta_{\nu}^2\varrho_{\nu}}{2k^2\mu^2}\right)\left[1 + \left(\frac{1}{\tau_{\nu}} - 1\right)\frac{n^2\eta_{\nu}^2\varrho_{\nu}}{k^2\mu^2}\right] \tag{4.48}$$

and $\mathcal{H}(k,\mu)$ as defined by equation (B.5). Note that the anomalous term (4.43) is not divergent at large k and thus does not require an upper cut–off $k_{\max}$ (2.59).

4.5 High-Velocity SP in a Magnetized Plasma

In this section we consider the SP of an ion moving in a magnetized plasma in the high–velocity regime. We shall assume that due to the high–frequencies involved the electrons give the main contribution to the stopping power and the influence of the ionic component is neglected. We study the influences (i) of the magnetic field and the Coulomb coupling on the SP of heavy ions with rectilinear trajectories (heavy ions), (ii) the effect of the finite curvature of the trajectory of the ion beam due to the cyclotron motion (light ions). As has been stressed in Sect. 2.4 the dielectric formalism is valid for the case of weak coupling between the ion and the plasma, in high–velocity limit given by the parameter $\mathcal{Z} = \sqrt{3}\,|Z|\,\Gamma^{3/2}\,(v_{\mathrm{th}}/v_i)^3$ [134]. Here v_{th} is the electron thermal velocity, $\Gamma = e^2/(k_{\mathrm{B}} T a_{\mathrm{WS}})$ is the plasma parameter, where $a_{\mathrm{WS}} = (4\pi n_e/3)^{-1/3}$ is the Wigner–Seitz radius, and T is the temperature of the electron plasma. Since we assume the high–velocity ion beam the dielectric formalism in linear response becomes accurate. Since the early 1960's several theoretical calculations of the SP in a high–velocity regime with the ion velocity v_i much higher than the electron thermal velocity v_{th} have been presented in [3,61,88,90,97,122]. These papers are generally focused on the case of weakly coupled plasmas, $\Gamma \ll 1$. Here we consider the general situation with $\Gamma \sim 1$ [97].

Within the linear response formalism the SP and the longitudinal dielectric function of the plasma, $\varepsilon(\boldsymbol{k}, \omega) = k_i k_j \varepsilon_{ij}(\boldsymbol{k}, \omega)/k^2$, are related by the equations (2.53) and (2.57), where $\varepsilon_{ij}(\boldsymbol{k}, \omega)$ denotes the dielectric tensor of the magnetized plasma [67,72,77]. To account for the strong Coulomb coupling between plasma electrons we use the cold plasma approximation with the dielectric tensor $\varepsilon_{ij} = \varepsilon_{ij}(\omega)$ constructed in [101] by means of the classical theory of moments. If the magnetic field is chosen to be parallel to the z–direction, the components of this dielectric tensor read $\varepsilon_{xz} = \varepsilon_{zx} = \varepsilon_{yz} = \varepsilon_{zy} = 0$, $\varepsilon_{xy} = -\varepsilon_{yx}$, $\varepsilon_{xx} = \varepsilon_{yy} = \varepsilon_{\perp}(\omega)$, $\varepsilon_{zz} = \varepsilon_{\|}(\omega)$. Thus

$$\varepsilon(\boldsymbol{k}, \omega) \equiv \varepsilon(\omega, \beta) = \varepsilon_{\|}(\omega) \cos^2 \beta + \varepsilon_{\perp}(\omega) \sin^2 \beta \,, \tag{4.49}$$

where β denotes the angle between the wave vector $\boldsymbol{k}$ and the magnetic field,

$$\varepsilon_{\|}(\omega) = 1 + \frac{\omega_p^2}{\Omega_{\|}^2 - \omega(\omega + i\nu_e)}\,, \qquad \varepsilon_{\perp}(\omega) = 1 + \frac{\omega_p^2\left[\omega(\omega + i\nu_e) - \Omega_{\perp}^2\right]}{\Omega_e^2 \omega^2 - \left[\omega(\omega + i\nu_e) - \Omega_{\perp}^2\right]^2}\,. \tag{4.50}$$

Here ν_e is the effective collision frequency accounting the damping of plasma waves. In the high–velocity limit the damping effects can be neglected and consequently the imaginary part of the inverse dielectric function can be described by a sharp energy loss at the plasma excitation frequencies[1]. The positive parameters $\Omega_{\perp}$ and $\Omega_{\|}$ take into account the Coulomb correlations between the particles, and are expressed via the second frequency moment of the magnetized plasma conductivity tensor Hermitian part [101], so that

$$\Omega_{\perp}^2 = \frac{\omega_p^2}{2} \sum_{\boldsymbol{q} \neq 0} S_{ei}(\boldsymbol{q}) \frac{q_{\perp}^2}{q^2}\,, \qquad \Omega_{\|}^2 = \omega_p^2 \sum_{\boldsymbol{q} \neq 0} S_{ei}(\boldsymbol{q}) \frac{q_{\|}^2}{q^2}\,, \tag{4.51}$$

$S_{ei}(\boldsymbol{q})$ being the partial electron–ion static structure factor, and $q_{\perp}$ ($q_{\|}$) the component of the vector $\boldsymbol{q}$ on the direction perpendicular (parallel) to the external magnetic field. We shall not specify here the frequencies $\Omega_{\perp}$ and $\Omega_{\|}$. We mention only that in the ideal plasma limit both $\Omega_{\perp}$ and $\Omega_{\|} \to 0$. For our purpose, it is sufficient to make an assumption that the frequencies $\Omega_{\perp}$ and $\Omega_{\|}$ are not affected by the magnetic field. Moreover, to simplify the situation we consider a hydrogen plasma. In this approximation the frequencies $\Omega_{\perp}$ and $\Omega_{\|}$ coincide,

$$\Omega_{\|}^2 = \Omega_{\perp}^2 = h_{ei}(0) \frac{\omega_p^2}{3} = \frac{\omega_p^2}{3} \sum_{q \neq 0} S_{ei}(q)\,, \tag{4.52}$$

and are directly related to the zero–separation value of the electron–ion correlation function $h_{ei}(0)$ [101].

[1] The effect of damping, however, is considered in Sect. 4.5.3 to avoid the divergence of the SP due to resonance excitations.

In the limit of small damping, $\nu_e \to 0$, the imaginary part of the inverse longitudinal dielectric function is then given by the expression

$$\mathrm{Im}\frac{-1}{\varepsilon(\boldsymbol{k},\omega)} \to \pi\frac{|\omega|}{\omega}\delta\left[\varepsilon(\omega,\beta)\right] . \tag{4.53}$$

The dispersion relation for the longitudinal magnetized plasma modes (traditionally called as the plasma resonances) is given by the equation $P(\omega) = \cos^2\beta$, where

$$P(\omega) = \frac{\varepsilon_\perp(\omega)}{\varepsilon_\perp(\omega) - \varepsilon_\parallel(\omega)} = \frac{(\omega^2 - \Omega_-^2)(\omega^2 - \Omega_+^2)(\Omega_\perp^2 - \omega^2)}{\omega^2\omega_p^2\Omega_e^2} . \tag{4.54}$$

Here Ω_- and Ω_+ are the solutions of the equation $\varepsilon_\perp(\omega) = 0$ and play the role of the plasma upper hybrid frequencies in the case of strong coupling,

$$\Omega_\pm^2 = \Omega_\perp^2 + \frac{1}{2}\left(\omega_H^2 \pm \sqrt{\omega_H^4 + 4\Omega_e^2\Omega_\perp^2}\right) , \tag{4.55}$$

and $\omega_H^2 = \omega_p^2 + \Omega_e^2$ is the upper hybrid frequency for ideal magnetized plasma. From equation (4.54) it is seen a cubic equation with respect to ω^2 can be obtained. It determines three resonance frequencies. This is in contrast to ideal magnetized plasmas, where only two resonances exist if the ion motion is neglected (see, e.g., [77]). For the transverse propagation (at $\beta = \pi/2$) the frequencies of the plasma resonances, $\Omega_1(\beta)$, $\Omega_2(\beta)$ and $\Omega_3(\beta)$, are equal to Ω_-, $\Omega_\perp$, and Ω_+, respectively. For the case of longitudinal propagation (at $\beta = 0$), the frequencies $\Omega_1(\beta)$, $\Omega_2(\beta)$ and $\Omega_3(\beta)$ of these resonances are ω_1, ω_2 and ω_3, respectively. They are given explicitly by $\omega_1 = \widetilde{\omega}_1$, $\omega_2 = \min(\widetilde{\omega}_{2,3})$, $\omega_3 = \max(\widetilde{\omega}_{2,3})$, where

$$\widetilde{\omega}_2 = \sqrt{\omega_p^2 + \Omega_\perp^2} , \qquad \widetilde{\omega}_{\substack{3\\1}} = \frac{1}{2}\left(\sqrt{\Omega_e^2 + 4\Omega_\perp^2} \pm \Omega_e\right) . \tag{4.56}$$

The frequencies (so called cut–off frequencies) $\omega_{1,2,3}$ and $\Omega_\pm$, $\Omega_\perp$ determine the boundaries between the domains of propagation for different longitudinal plasma modes. In the limit of weakly coupled plasma ω_1 vanishes and $\omega_2 = \omega_-$, $\omega_3 = \omega_+$, where $\omega_- = \min[\omega_p; \Omega_e]$ and $\omega_+ = \max[\omega_p; \Omega_e]$ are the usual cut–off frequencies. Schematically the frequencies of the plasma resonances are shown in Fig. 4.7 (left). One can check that the inverse dielectric function defined in (4.53) with expression (4.49) satisfies the frequency sum rule.

4.5.1 Heavy Ions With Rectilinear Trajectories

Having introduced and specified the dielectric function and the behavior of the plasma longitudinal modes we may now proceed to analyze the stopping power of a projectile heavy ion (no cyclotron orbit curvature) passing a plasma for different incident angles α and magnetic field strengths. Our starting point is the equation (2.53) which with the dielectric function (4.49) with (4.53) may be represented in the form

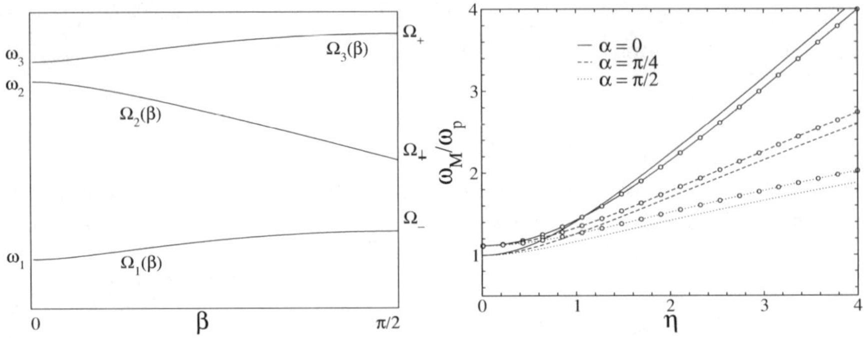

Fig. 4.7. Left panel. The resonance frequencies $\Omega_1(\beta)$, $\Omega_2(\beta)$ and $\Omega_3(\beta)$, as a function of β. Right panel. The mean excitation frequency, ω_M (in units of ω_p) for vanishing coupling (the lines without symbols) and for $w = 0.5$ (the lines with circles). $\alpha = 0$ (solid lines), $\alpha = \pi/4$ (dashed lines), $\alpha = \pi/2$ (dotted lines). After [97].

$$S = \frac{Z^2 e^2}{\pi v_i^2 \sin\alpha} \int_0^{\pi} \mathrm{d}\beta \int_0^{\infty} \delta\left[\varepsilon(\omega,\beta)\right] \omega \mathrm{d}\omega \int_0^{k_{\max}} \frac{\mathrm{d}k}{k} \int_0^{2\pi} \mathrm{d}\phi \delta\left[\cos\phi - \Phi(k,\omega,\beta)\right], \tag{4.57}$$

where

$$\Phi(k,\omega,\beta) = \frac{\omega/kv_i - \cos\beta\cos\alpha}{\sin\beta\sin\alpha}. \tag{4.58}$$

The ϕ–integral requires that $|\Phi(k,\omega,\beta)| < 1$. This condition restricts the integration domain for the k–integral. We perform in equation (4.57) the integrations over ϕ, k and β. The straightforward calculation will lead to

$$S = \frac{Z^2 e^2}{\pi v_i^2} \sum_{j=1,2} \int_{C_j} \frac{X_j(\omega)\,\omega \mathrm{d}\omega}{\sqrt{\left|\varepsilon_{\perp}(\omega)\left\{\left[\varepsilon_{\perp}(\omega) - \varepsilon_{\parallel}(\omega)\right]\sin^2\alpha - \varepsilon_{\perp}(\omega)\right\}\right|}}, \tag{4.59}$$

where the ω–integrations are performed in the domains $\omega \geq 0$ and $0 \leq P(\omega) \leq \sin^2\alpha$ and $\sin^2\alpha \leq P(\omega) \leq 1$ in C_1 and C_2, respectively. Here

$$X_1(\omega) = \sum_{\sigma=\pm} \ln\left[S_{\sigma}(\omega) + \sqrt{S_{\sigma}^2(\omega) - 1}\right], \tag{4.60}$$

$$X_2(\omega) = \begin{cases} 0, & S_+(\omega) > 1 \\ \frac{\pi}{2} - \arcsin\left(S_+(\omega)\right), & -1 \leq S_+(\omega) \leq 1 \\ \pi, & S_+(\omega) < -1 \end{cases} \tag{4.61}$$

$$S_{\pm}(\omega) = \frac{1}{\sin\alpha\sqrt{1 - P(\omega)}}\left[\frac{k_{\max} v_i}{\omega}\left(\sin^2\alpha - P(\omega)\right) \pm \sqrt{P(\omega)}\,|\cos\alpha|\right]. \tag{4.62}$$

In the first term of equation (4.59) it is required that $S_{\pm}(\omega) > 1$ otherwise the corresponding logarithmic function in equation (4.60) vanishes. In the second term of equation (4.59) we require the inequality $S_+(\omega) \leq 1$. At $S_+(\omega) > 1$ the second term of equation (4.59) vanishes. All these restrictions on the values of $S_{\pm}(\omega)$ additionally reduce the ω–integration domains in (4.59). Equation (4.59) is the main result for the SP of the heavy ion in high–velocity limit. It is valid in the limit when $k_{\max} v_i > \omega_p$. For two particular values of the incident angle, $\alpha = 0$ and $\alpha = \pi/2$, with ion motion along and perpendicular to the magnetic field, respectively, equation (4.59) is simplified,

$$S_{\parallel} = \frac{Z^2 e^2}{v_i^2} \int_{C_{\parallel}} \frac{\omega d\omega}{|\varepsilon_{\perp}(\omega)|} \tag{4.63}$$

with $C_{\parallel} \in \left[\omega \geq 0, \omega^2/k_{\max}^2 v_i^2 \leq P(\omega) \leq 1\right]$ and

$$S_{\perp} = \frac{2Z^2 e^2}{\pi v_i^2} \int_{C_{\perp}} \frac{\omega d\omega}{\sqrt{|\varepsilon_{\perp}(\omega)\, \varepsilon_{\parallel}(\omega)|}} \ln\left[S(\omega) + \sqrt{S^2(\omega) - 1}\right] \tag{4.64}$$

with $S(\omega) = (k_{\max} v_i/\omega)\sqrt{1 - P(\omega)}$ and $C_{\perp} \in \left[\omega \geq 0,\ 0 \leq P(\omega) \leq 1 - \omega^2/k_{\max}^2 v_i^2\right]$.

For arbitrary incident angles α some general results can be extracted out of expression (4.59). Consider the limit of extremely large velocities. Neglecting all terms in (4.59) that vanish as $O(v_i^{-2})$ we obtain the following leading term for the stopping power at high velocities

$$S = \frac{Z^2 e^2 \omega_p^2}{v_i^2} \left[\ln\left(\frac{k_{\max} v_i}{\omega_p}\right) - F\right], \tag{4.65}$$

where the constant F is given by

$$F(\eta, \alpha) = \frac{2}{\pi \omega_p^2} \int_{C_1} \frac{\ln\left(\omega/\omega_p\right) \omega d\omega}{\sqrt{\varepsilon_{\perp}(\omega)\left\{\left[\varepsilon_{\perp}(\omega) - \varepsilon_{\parallel}(\omega)\right] \sin^2\alpha - \varepsilon_{\perp}(\omega)\right\}}}. \tag{4.66}$$

Here $\eta = \Omega_e/\omega_p$ is the measure of the electron cyclotron frequency and the integration domain C_1 is defined above. For the dielectric function given by expressions (4.49) and (4.50) the integration domain C_1 includes three subdomains: $\Omega_1(\alpha) \leq \omega \leq \Omega_-$, $\Omega_{\perp} \leq \omega \leq \Omega_2(\alpha)$, and $\Omega_3(\alpha) \leq \omega \leq \Omega_+$, where $\Omega_{1,2,3}(\alpha)$ are the solutions of the equation $P(\omega) = \sin^2\alpha$. Equation (4.65) is the generalization of the Bohr's and Akhiezer's expressions for the SP in the high–velocity limit obtained for ideal plasmas without and with magnetic field [3, 4], respectively.

For deriving equation (4.65) we have used the identities $f_1 = 1$ and $f_2 = 0$, where

$$f_1 = \frac{2}{\pi \omega_p^2} \int_{C_1} \frac{\omega d\omega}{\sqrt{D_1(\omega)}}, \tag{4.67}$$

$$f_2 = \frac{1}{\omega_p^2} \int_{C_2} \frac{\omega d\omega}{\sqrt{D_2(\omega)}} - \frac{2}{\pi \omega_p^2} \int_{C_1} \frac{\omega d\omega}{\sqrt{D_1(\omega)}} \ln\left[\frac{\sin\alpha \sqrt{1 - P(\omega)}}{2\left(\sin^2\alpha - P(\omega)\right)}\right]. \tag{4.68}$$

Here $D_1(\omega)$ is the function under the square root in (4.66), $D_2(\omega) = -D_1(\omega)$. Note that $D_1(\omega) \geq 0$ and $D_2(\omega) \geq 0$ in C_1 and C_2, respectively.

For proving these identities we use the following obvious relations

$$\left.\frac{\wp\left(\sqrt{P(\omega)}\right)}{\sqrt{D_1(\omega)}}\right|_{C_1} = 2\int_0^{\sin\alpha} \frac{d\mu}{\sqrt{\sin^2\alpha - \mu^2}} \wp(\mu)\,\delta\left[\varepsilon(\omega,\mu)\right], \tag{4.69}$$

$$\left.\frac{\wp\left(\sqrt{P(\omega)}\right)}{\sqrt{D_2(\omega)}}\right|_{C_2} = 2\int_{\sin\alpha}^{1} \frac{d\mu}{\sqrt{\mu^2 - \sin^2\alpha}} \wp(\mu)\,\delta\left[\varepsilon(\omega,\mu)\right], \tag{4.70}$$

where $\wp(\mu)$ is an arbitrary function of μ. The dielectric function $\varepsilon(\omega,\mu)$ is given by (4.49) with $\mu = \cos\beta$. The left hand side of (4.69) and (4.70) are non-zero only in C_1 and C_2, respectively. Substituting expression (4.69) into (4.67) with $\wp(\mu) = 1$, changing the integration order and using the frequency sum rule we obtain $f_1 = 1$. Similarly expression (4.68) becomes

$$f_2 = \int_{\sin\alpha}^{1} \frac{d\mu}{\sqrt{\mu^2 - \sin^2\alpha}} - \frac{2}{\pi}\int_0^{\sin\alpha} \frac{d\mu}{\sqrt{\sin^2\alpha - \mu^2}} \ln\left[\frac{\sin\alpha\sqrt{1-\mu^2}}{2\left(\sin^2\alpha - \mu^2\right)}\right]. \tag{4.71}$$

Since both terms in the right hand side of (4.71) are equal to $\ln\left[(1 + |\cos\alpha|)/\sin\alpha\right]$, we finally arrive at $f_2 = 0$.

For the ideal plasma the constant $F(\eta,\alpha)$ vanishes in the limit $\eta \to 0$, from which one retrieves the field–free result. For a strongly coupled plasma and a vanishing magnetic field equation (4.66) yields $F(\eta,\alpha) = (1/2)\ln(1 + w^2)$, where $w = \Omega_\perp/\omega_p$. At arbitrary incident angle α and at strong ($\eta \gg 1$) magnetic fields equation (4.66) reads

$$F(\eta,\alpha) = \ln\eta + \frac{\sin^2\alpha}{4}\left[2\ln\frac{w + \sqrt{w^2 + \sin^2\alpha}}{2\eta} + \left(\frac{\sin\alpha}{w + \sqrt{w^2 + \sin^2\alpha}}\right)^2\right]. \tag{4.72}$$

When the ion moves along the magnetic field ($\alpha = 0$) the integrals in (4.66) can be evaluated exactly. The result is

$$F(\eta,0) = \frac{1}{q_+^2 - q_-^2}\left[\left(w^2 - q_-^2\right)\ln q_- + \left(q_+^2 - w^2\right)\ln q_+\right], \tag{4.73}$$

where $q_\pm = \Omega_\pm/\omega_p$.

Some general results can be extracted out of the expressions (4.65) and (4.66) concerning the influence of the Coulomb coupling and the magnetic field on the high–velocity SP. First, the constant $F(\eta,\alpha)$ is positive at arbitrary η and α and grows with the magnetic field strength and correlation intensity characterized by w. An exception is the ion parallel motion ($\alpha = 0$). In this case the SP increases with w for a strong magnetic field. The SP of a plasma is then reduced with increasing magnetic field and the coupling parameter. However we note that for a strong magnetic field the leading logarithmic term in expression (4.65) with (4.72) is $\ln[(k_{\max}v_i/\omega_p)(\omega_p/\Omega_e)^y]$, where $y = (1 + \cos^2\alpha)/2$. The argument of this function should be a large quantity which restricts the strength of the magnetic field,

$k_{\max}v_i > \omega_p(\Omega_e/\omega_p)^y$. Moreover, the leading term in equation (4.65) may be represented as $\ln(k_{\max}v_i/\omega_M)$, where $\omega_M = \omega_p\, e^{F(\eta,\alpha)}$ is the mean excitation energy. It is seen that the dynamic screening length becomes $\lambda_M = v_i/\omega_M$ and decreases both with increasing magnetic field and the Coulomb coupling. This suggests that in high–velocity limit the dynamic cut–off parameter $\langle s_{\max}\rangle$ introduced in Sect. 2.4 (see the second relation in (2.59)) must be replaced by λ_M. The third point is that the high–velocity SP monotonically increases with α, having its minimum at parallel motion, $\alpha = 0$, and its maximum at perpendicular motion $\alpha = \pi/2$. All these features are shown in Fig. 4.7 (right panel) where the frequency ω_M is plotted as a function of the magnetic field for different incident angle α and coupling parameter. The lines with and without circles correspond to $w = 0.5$ and $w = 0$ respectively.

4.5.2 Weakly Coupled Plasma with Strong Magnetic Fields

As have been discussed above equations (4.65) and (4.66) are not valid for strong magnetic fields when $\Omega_e \gtrsim k_{\max}v_i \gg \omega_p$. To analyze the stopping power in this regime we should start from more general expression (4.59). Here we consider two particular examples, ion motion along ($\alpha = 0$) and transverse ($\alpha = \pi/2$) to the magnetic field in a weakly coupled ($w = 0$) plasma when the stopping power can be evaluated from (4.63) and (4.64), respectively.

In a weakly coupled plasma there are only two plasma resonances. In the limit $w \to 0$ the first plasma mode with frequency $\Omega_1(\beta)$ vanishes and the frequencies of the other two modes are given by

$$\widetilde{\Omega}_{\pm}^2(\beta) = \frac{1}{2}\left(\omega_H^2 \pm \sqrt{\omega_H^4 - 4\Omega_e^2\omega_p^2\cos^2\beta}\right). \tag{4.74}$$

Here $\widetilde{\Omega}_-(\beta)$ and $\widetilde{\Omega}_+(\beta)$ correspond to the frequencies $\Omega_2(\beta)$ and $\Omega_3(\beta)$ introduced in the previous section. We note that the expressions (4.63) and (4.64) in the limit of ideal plasma is more general than the Akhiezer's result [3] since they involve not only the main logarithmic contribution but also the terms which are proportional to $O(v_i^{-2})$. Some particular regimes for the magnetic field strength are considered below.

(i) *Longitudinal motion of the ion.* If the incident ion moves parallel to the magnetic field expression (4.63) at $k_{\max}v_i > \Omega_e\omega_p/\omega_H$ yields

$$S_{\parallel} = \frac{Z^2 e^2\omega_p^2}{v_i^2}\left[\ln\left(\frac{\omega_-\Lambda_+}{\omega_H\Lambda_-}\right) + \frac{\Omega_e^2}{2}\left(\frac{1}{\omega_-^2} + \frac{1}{\Lambda_+^2} - \frac{1}{\Lambda_-^2}\right)\right], \tag{4.75}$$

where $\Lambda_+ = \max[k_{\max}v_i; \omega_+]$, $\Lambda_- = \min[k_{\max}v_i; \omega_-]$. The cut–off frequencies ω_- and ω_+ have been introduced in Sect. 4.5. Consider two particular cases: For strong magnetic fields with $\omega_p < k_{\max}v_i < \Omega_e$ equation (4.75) reads

$$S_{\parallel} = \frac{Z^2 e^2\omega_p^2}{v_i^2}\left[\frac{1}{2} - \ln\left(\frac{\omega_H}{\Omega_e}\right)\right]. \tag{4.76}$$

This expression is in agreement with our previous result, see Sect. 4.3.2 where we have used the full dielectric function with thermal motion of the electrons. In contrast to the field–free case, for $\alpha = 0$ and strong magnetic fields, $S_{\parallel} \simeq Z^2 e^2 \omega_p^2 / 2v_i^2$ is independent of $k_{\max}$.

In the high–velocity limit $k_{\max} v_i > \omega_+$, the stopping power is

$$S_{\parallel} = \frac{Z^2 e^2 \omega_p^2}{v_i^2} \left[\ln \left(\frac{k_{\max} v_i}{\omega_H} \right) + \frac{\Omega_e^2}{2 (k_{\max} v_i)^2} \right] . \tag{4.77}$$

It is easy to see that the leading logarithmic term in (4.77) agrees with expressions (4.65) and (4.66). However the last expression involves an additional term which decreases as v_i^{-2}.

(ii) *Transverse motion of the ion*. For the ion transverse motion we consider the case of strong magnetic field, $\omega_p < k_{\max} v_i < \omega_H$. The other limiting case, $k_{\max} v_i > \omega_H$, can be extracted from (4.65) and (4.66).

In the high–velocity limit with $\omega_p \ll k_{\max} v_i < \omega_H$ equation (4.64) becomes

$$S_{\perp} = \frac{Z^2 e^2 \omega_p^2}{v_i^2} \left[F_1(\eta) \ln \left(\frac{k_{\max} v_i}{\omega_p} \right) - F_2(\eta) \right] . \tag{4.78}$$

Here $p_{\pm} = \widetilde{\Omega}_{\pm}(\mu)/\omega_p$, $\widetilde{\Omega}_{\pm}(\mu)$ are given by (4.74) where $\cos\beta$ is replaced by μ,

$$F_1(\eta) = \frac{2}{\pi} \int_0^1 d\mu \frac{\eta^2 \mu^2 - p_-^2(\mu)}{\sqrt{(1-\mu^2)\left[(1+\eta^2)^2 - 4\eta^2\mu^2\right]}} , \tag{4.79}$$

$$F_2(\eta) = \frac{2}{\pi} \int_0^1 d\mu \frac{\eta^2 \mu^2 - p_-^2(\mu)}{\sqrt{(1-\mu^2)\left[(1+\eta^2)^2 - 4\eta^2\mu^2\right]}} \ln \left(\frac{p_-(\mu)}{2\sqrt{1-\mu^2}} \right) . \tag{4.80}$$

In a strong magnetic field limit ($\eta \gg 1$) $F_1(\eta) = 1/2$ and $F_2(\eta) = (1/2)\ln(1/C)$, where $C = 2/\mathrm{e} \simeq 0.735$. In this case the stopping power is

$$S_{\perp} = \frac{Z^2 e^2 \omega_p^2}{v_i^2} \ln \left(C \frac{k_{\max} v_i}{\omega_p} \right)^{1/2} . \tag{4.81}$$

In contrast to the case $\alpha = 0$ the leading logarithmic term for the ion transverse motion depends here on $k_{\max}$. However, this term is modified as a consequence of a strong magnetic field. This term also differs from the high–velocity expression (4.65) with (4.72) derived for arbitrary α. The physical reason for the modification of this term is the following. In high–velocity limit both low– and high–frequency magnetized plasma modes (see expression (4.74)) are excited and both of them contribute to the SP. The contribution of each mode is equal to the half of the main logarithmic term. However in the intermediate velocity regime considered here only the low–frequency mode is effectively excited. The contribution of this mode leads to the stopping power (4.81).

4.5.3 Light Ions, The Effect of the Cyclotron Rotation

We now focus our attention on the high–velocity SP for light projectile ions and study the influence of the cyclotron rotation on the SP [88, 122, 126]. We consider the case when the ion moves perpendicular to the magnetic field with $\alpha = \pi/2$. Since we assume a non–relativistic ion the synchrotron radiation is neglected. The velocity, coordinate and the time–averaged SP of the ion are given by expressions (2.54), (2.55) and (2.57) respectively, where $v_{i\parallel} = 0$ and $v_{i\perp} = v_i$. In the high–velocity limit we use the cold plasma approximation, expression (4.49). In this case equation (2.57), written in the cylindrical geometry with the axis along the magnetic field, becomes

$$\langle S \rangle = \frac{4Z^2 e^2}{\pi a_c} \sum_{n=1}^{\infty} n \int_0^{k_{\perp\max}} J_n^2(k_\perp a_c) k_\perp \mathrm{d}k_\perp \int_0^{\infty} \mathrm{d}k_\parallel \mathrm{Im}\left[\frac{-1}{k_\parallel^2 \varepsilon_\parallel(n\Omega_c) + k_\perp^2 \varepsilon_\perp(n\Omega_c)}\right]. \tag{4.82}$$

Here $k_{\perp\max}$ is a maximal cut–off for $k_\perp$–integration and a_c is the cyclotron radius of the projectile ion. Performing the $k_\parallel$–integration we finally obtain

$$\langle S \rangle = \frac{2Z^2 e^2}{\pi a_c^2} \sum_{n=1}^{\infty} n Q_n(\zeta)\, \mathrm{Im}\left[\frac{-1}{\varepsilon_\parallel(n\Omega_c) T(n\Omega_c)}\right], \tag{4.83}$$

where $\zeta = k_{\perp\max} a_c$ and

$$Q_\nu(\zeta) = \pi \int_0^{\zeta} J_\nu^2(y)\, \mathrm{d}y, \tag{4.84}$$

$$T(\omega) = \sqrt{\frac{|\mathcal{A}(\omega)| + \mathrm{Re}\mathcal{A}(\omega)}{2}} + \mathrm{i}\frac{|\mathcal{A}''(\omega)|}{\mathcal{A}''(\omega)} \sqrt{\frac{|\mathcal{A}(\omega)| - \mathrm{Re}\mathcal{A}(\omega)}{2}} \tag{4.85}$$

with $\mathcal{A}(\omega) = \varepsilon_\perp(\omega)/\varepsilon_\parallel(\omega)$, $\mathcal{A}''(\omega) = \mathrm{Im}[\mathcal{A}(\omega)]$.

The expression (4.83) is general. It involves all cyclotron harmonics and is valid for arbitrary complex dielectric functions $\varepsilon_\perp(\omega)$ and $\varepsilon_\parallel(\omega)$ which may involve the damping effect. Below we will see that the SP diverges near characteristic resonances. Therefore we include the damping effect in order to remove such singular behavior of the SP. When the damping vanishes we find from (4.83)

$$\langle S \rangle = \frac{2Z^2 e^2}{\pi a_c^2} \sum_n \frac{n Q_n(\zeta)}{\sqrt{|\varepsilon_\parallel(n\Omega_c) \varepsilon_\perp(n\Omega_c)|}}. \tag{4.86}$$

Here the summation over harmonic numbers n is performed in the domain where $\mathcal{A}(n\Omega_c) < 0$ ($\varepsilon_\perp(n\Omega_c)$ and $\varepsilon_\parallel(n\Omega_c)$ must have different signs). The stopping power given by equation (4.86) shows a divergence at magnetic field strengths for which $\varepsilon_\parallel(n\Omega_c) = 0$ or $\varepsilon_\perp(n\Omega_c) = 0$. These conditions correspond to the excitation of the plasma and upper hybrid oscillations in a plasma by rotating ion. Physically, this divergence is due to a resonant coupling of two oscillators with the frequencies Ω_c, i.e., the incident projectile ion, and ω_p (or ω_H), i.e., the plasma (upper hybrid) waves, respectively. When we neglect the damping effects, these resonances have an infinite amplitude. For the dielectric functions given by equation (4.50) with $\nu_e = 0$ and $\Omega_\perp = 0$ (neglecting the damping and Coulomb coupling) the summation in (4.86) is performed in the domains $1 \leq n \leq n_1$ and $n_2 \leq n \leq n_3$, where

$$n_1 = \begin{cases} [c], & \gamma \geq c \\ [\gamma], & \gamma < c \end{cases}, \quad n_2 = \begin{cases} [\gamma] + 1, & \gamma \geq c \\ [c] + 1, & \gamma < c \end{cases}, \quad n_3 = \left[\sqrt{\gamma^2 + c^2} \right] \tag{4.87}$$

with $\gamma = \omega_p/\Omega_c$ and $c = \Omega_e/\Omega_c = M/|Z|m$. Here $[x]$ denotes the rounding of the number x downwards to the nearest integer. When γ or $\sqrt{\gamma^2 + c^2}$ are integers the resonances occur at the harmonic numbers $n = \gamma$ or $n = \sqrt{\gamma^2 + c^2}$. This will show the oscillatory behavior of the SP with respect to the parameter γ (with infinitely large amplitudes), where the SP has maxima at $\gamma_N = N = 1, 2, ...$ (plasma oscillations) and $\gamma_N = \sqrt{N^2 - c^2}$ (upper hybrid oscillations) with an integer N such that $N \geq [c] + 1$. For an ion with $c \gg 1$ the periodicity of the upper hybrid resonances is much larger than those for the plasma resonances and they can be easily resolved in γ. For an electron (or positron) projectile with $c = 1$ both resonances may coalesce into a single entity and they cannot be resolved in γ anymore. Representing the parameter γ in the form $\gamma = 2\pi a_c/\lambda_{p,H}$, where $\lambda_{p,H} = 2\pi v_i/\omega_{p,H}$ is the wavelength of the plasma (upper hybrid) waves, the resonance condition can be alternatively formulated in another form: The resonances occur when the cyclotron circle with the length $2\pi a_c$ contains integer number of plasma (upper hybrid) waves. In addition, the *antiresonance* conditions may also be formulated (see, e.g., [126]) for intermediate values of $2\pi a_c/\lambda_{p,H} \simeq 1.5, 2.5, ...$, where the SP drops down. These near-cancellation effects are due to destructive interferences in the self–interaction of the projectile ion with its own induced charge density.

In contrast to the previous sections we have used the cylindrical geometry and introduced the cut–off parameter $k_{\perp\max}$ for $k_\perp$–integration for deriving equations (4.82) and (4.83). This allows to perform $k_\parallel$–integration and yields more convenient expressions for the numerical evaluation of the SP. A similar expression for the SP can be found in a spherical geometry which again leads to a resonance behavior of $\langle S \rangle$. For instance, for vanishing damping and within spherical geometry the SP is again given by expression (4.86) with the summation domain $\mathcal{A}(n\Omega_c) < 0$ but the argument of the function $Q_n(\zeta)$ contains an extra factor $\sqrt{1 - P(n\Omega_c)}$ [122], where $P(\omega)$ has been introduced by (4.54). In addition, now $\zeta = k_{\max} a_c$ with the cut–off parameter $k_{\max}$ introduced in a spherical geometry (see Sect. 2.4). Note that in this case due to the appearance of the factor $\sqrt{1 - P(n\Omega_c)}$ in the argument of $Q_n(\zeta)$ the summand in (4.86) behaves as $|\varepsilon_\parallel(n\Omega_c)|^n$ and vanishes near plasma resonances $\varepsilon_\parallel(n\Omega_c) = 0$. Therefore the SP receives contribution only from upper hybrid excitations $\varepsilon_\perp(n\Omega_c) = 0$.

For a numerical evaluation of the SP (4.83) as well as for the study a number of particular cases the behavior of the function $Q_\nu(\zeta)$ determined by (4.84) is important. Consider briefly the properties of function $Q_\nu(\nu\zeta)$. To find the asymptotic value of that function at $\nu \gg 1$ and $\zeta > 1$, we divide the domain of integration in (4.84) into two parts with $y < \nu$ and $\nu < y < \nu\zeta$ and use the uniform asymptotic expansion of the Bessel functions of the first kind $J_\nu(\nu y)$ for $y > 1$ and $\nu \gg 1$ [52]. Thus, we find

$$Q_\nu(\nu\zeta) \simeq q_\nu + \ln\left(\zeta + \sqrt{\zeta^2 - 1}\right) + \frac{1}{6}\Gamma\left(\frac{1}{3}\right)\left(\frac{3}{2\nu}\right)^{1/3} + \mathrm{O}\left(\frac{1}{\nu}\right), \tag{4.88}$$

where $\Gamma(z)$ is the Euler's function and

$$q_\nu = \pi \int_0^\nu J_\nu^2(y)\mathrm{d}y\,. \tag{4.89}$$

Numbers q_ν are less than 1, and slowly fall off as $q_\nu \sim \nu^{-1/3}$ with increasing ν. Here we point out some values of q_ν: $q_1 \simeq 0.225$, $q_{20} \simeq 0.096$, $q_{100} \simeq 0.057$.

At $\zeta < \nu$, the argument of the Bessel function is smaller than the index. In this case, the Bessel function is exponentially small, and at a fixed value of ζ, $Q_\nu(\zeta)$ vanishes exponentially as ν increases. Therefore, the series entering equation (4.83) is cut at $n_{\max} \simeq \zeta$ and taking into account the damping effect the SP is determined by harmonics having $n < n_{\max}$.

Consider equation (4.83) in the case of strong magnetic fields ($\gamma \ll 1$) with $\nu_e \neq 0$ and $\Omega_\perp = 0$. Note that this is a domain far from the resonances in the SP. Two cases must be mentioned here.

(i) $c = \Omega_e/\Omega_c$ is no integer. In this case we find from (4.83)

$$\langle S \rangle \simeq \frac{Z^2 e^2 \omega_p^2}{\pi v_i^2} \frac{\nu_e}{\Omega_c} \sum_{n=1}^{\infty} \frac{1}{n^2} Q_n(\zeta) \left[1 + \frac{2n^4}{(n^2 - c^2)^2}\right]. \tag{4.90}$$

From (4.90) it is seen that the SP decreases with increasing magnetic field.

(ii) $c = 1$ (electron or positron test particle). From equation (4.83) we obtain the leading term for the SP

$$\langle S \rangle \simeq \frac{Z^2 e^2 \omega_p^2}{\pi v_i^2} \frac{\Omega_e}{2\nu_e} Q_1(\zeta) \left[\frac{2}{(1+\sigma^2)\left(\sqrt{1+\sigma^2}+1\right)}\right]^{1/2} \tag{4.91}$$

with $\sigma = \omega_p^2/2\nu_e\Omega_e$. For a further increase of the magnetic field with $\sigma \ll 1$ the numerical factor within the square root in (4.91) tends to unity. In this case the SP may increase with increasing the magnetic field. These two asymptotic examples show the strong dependence of the SP on the mass of a test particle in the case when the magnetic field is sufficiently strong.

Consider briefly the zero curvature result, equations (4.65) and (4.66) with $\alpha = \pi/2$, which can be recovered from (4.86) in the limit of a heavy ion, $M \to \infty$. Since $\Omega_c \to 0$ in this limit we can transform the summation in equation (4.86) into an integration with respect to $\omega = n\Omega_c$. In the next step we use the asymptotic relation (4.88) at $\nu \to \infty$ and $\zeta \gg 1$, $Q_\nu(\nu\zeta) \to \ln(2\zeta)$. Using these expressions one arrives at equation (4.65) with (4.66) where $\alpha = \pi/2$ and $k_{\max} = 2k_{\perp\max}$.

Figure 4.8 shows the SP given by equation (4.83) as a function of the dimensionless parameter γ for a proton test particle. Note that for a proton the parameter $c \simeq 1836.1$ and, for instance, the first three upper hybrid resonances occur at

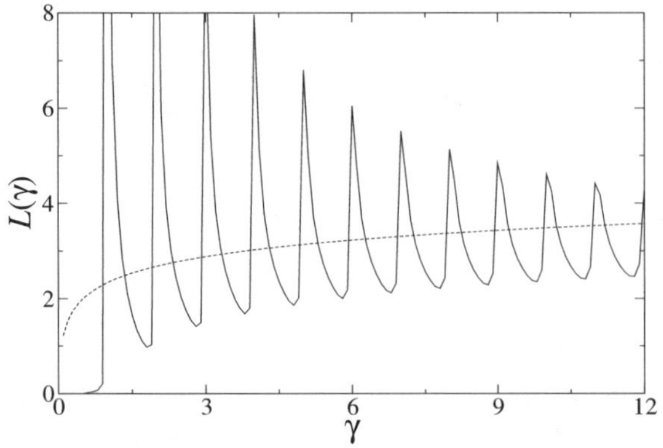

Fig. 4.8. Dependence of the stopping power normalized as $S = (Z^2e^2\omega_p^2/v_i^2)L(\gamma)$ on the dimensionless parameter $\gamma = \omega_p/\Omega_c$ for proton projectile in high-velocity limit, assuming a finite curvature of the test particle and introducing some arbitrary cut–off, $k_{\perp\max}v_i/\omega_p = 200$. The damping frequency is $\nu_e/\omega_p = 0.01$. The dashed line shows the asymptotic result for zero curvature, i.e., $M \to \infty$ and $\Omega_c \to 0$ obtained from equations (4.65) and (4.66) with $\alpha = \pi/2$.

$\gamma_1 \simeq 57.5$, $\gamma_2 \simeq 83.6$ and $\gamma_3 \simeq 103.2$, respectively. Therefore these resonances are outside of the domain $0 < \gamma \leq 12$ shown in Fig. 4.8. It is seen that the SP oscillates as a function of the magnetic field around the zero curvature result. One expects that the resonance peaks are damped out with γ, i.e., at small magnetic fields approaching asymptotically the zero curvature result. In addition both curves approach asymptotically to the field–free value $\ln(k_{\max}v_i/\omega_p)$ at the vanishing magnetic field, i.e. $\gamma \to \infty$.

4.6 Reduced LR (RLR) Treatment

For a theoretical description of the energy loss of ions in a plasma two standard approaches, namely BC, Chap. 3 and LR, Chap. 4 have been discussed in the previous sections. Both treatments, LR and BC, can be regarded as complementary to each other and both of them are of physical interest. Within the LR treatment (dielectric theory) the stopping power can receive a dynamic contribution from collective plasma excitations. It requires a cut–off at small distances where hard collisions between ion and electrons cannot be treated any more as a weak perturbation. Within the BC picture the interaction between the plasma electrons is only treated approximatively by an effective interaction or an upper cut–off for the impact parameters, to account for screening. In this case the stopping power of an ion is the result of the energy transfer in successive binary collisions. In the limit of a non–interacting electron gas ($U_{ee} \to 0$, where U_{ee} is the interaction potential between the electrons), the LR and BC treatments should therefore provide the same result for the stopping power. But even in the absence of a magnetic field both approaches give slightly different results. Exactly the same results can be achieved if physically reasonable cut–offs are used in the Coulomb logarithms [107, 120]. In the presence of a magnetic field the situation is dramatically changed. Here the agreement between the LR and BC treatments breaks down for intermediate and low ion velocities as discussed in [93, 94, 124]. The disagreement is larger for an ion motion along the magnetic field than transverse to it. In addition, the BC treatment also predicts an energy gain (negative stopping power) for very low ion velocities [124] (see, e.g. (3.143)).

In this section we consider BC between ion and electrons in the presence of an arbitrary magnetic field and demonstrate that full agreement between BC and LR treatments in the limit of a non–interacting electron plasma is guaranteed for a regularized interaction which is both of finite range and less singular than the Coulomb interaction at the origin. The regularized and screened potential (3.3) considered in Chap. 3 is an example of such interaction.

As basis for further considerations, we recall briefly the main aspects of the linear response theory for the ion–plasma interaction in the presence of an external magnetic field discussed in more details in Sect. 2.4. Here, for simplicity, we consider the electron plasma and all indices referring to the electrons are omitted. In addition, we introduce two kind of SPs, S_{BC} and S_{LR}, calculated on the basis of BC and LR, respectively. The stopping power S_{LR} of an ion is defined in Sect. 2.4, equation (2.53). The collective excitations (i.e. magnetized plasma modes) contributing to the stopping power are contained in $\varepsilon(\boldsymbol{k}, \boldsymbol{k}\cdot\boldsymbol{v}_i)$. In general, equation (2.53) cannot be evaluated in closed form except for the limiting cases $B \to 0$ and $B \to \infty$. For connecting now the dielectric picture discussed so far in the previous Sections of this Chapter to the BC approach we assume weakly interacting electrons, $e^2 n_e^{1/3} \to 0$ (or $\omega_p \to 0$), with

$$\mathrm{Im}\frac{-1}{\varepsilon(\boldsymbol{k},\omega)} = \frac{\mathrm{Im}\,\varepsilon(\boldsymbol{k},\omega)}{|\varepsilon(\boldsymbol{k},\omega)|^2} \simeq u_C(\boldsymbol{k})\,\mathrm{Im}\,\chi^{(0)}(\boldsymbol{k},\omega)\,. \tag{4.92}$$

In this approximation which we call Reduced Linear Response (RLR) approach for the SP, the stopping power thus reads

$$S_{\mathrm{RLR}} = \frac{Z^2 e^2}{v_i}\int \mathrm{d}\boldsymbol{k}\,|u(\boldsymbol{k})|^2\,(\boldsymbol{k}\cdot\boldsymbol{v}_i)\,\mathrm{Im}\,\chi^{(0)}(\boldsymbol{k},\boldsymbol{k}\cdot\boldsymbol{v}_i)\,, \tag{4.93}$$

where $-Ze^2 u(\boldsymbol{k})$ is the Fourier transformed ion–electron interaction potential introduced in Sect. 3.2. Now the stopping power does not receive any contribution from the dynamic collective plasma modes, but some collective contributions (i.e. screening) can be easily reintroduced by replacing $u(\boldsymbol{k})$ with an screened ion–electron interaction potential. Then equation (4.93) amounts to neglecting the electron–electron interaction in the target except for the screening of the ion. Note that the approximation (4.93) is equivalent replacing the denominator in (4.92) $|\varepsilon(\boldsymbol{k},\omega)|^2$ by its static value at $\omega = 0$ and replacing the bare Coulomb interaction $u_C(\boldsymbol{k})$ by the screened one according to

$$u(\boldsymbol{k}) \to \frac{u_C(\boldsymbol{k})}{\varepsilon(\boldsymbol{k},0)}\,. \tag{4.94}$$

On the other hand to mimic collectivity in (4.93) the dynamic screening length (the second relation in (2.59)) can be introduced in the interaction potential as discussed in Sect. 2.4 and [134].

To illustrate the problem and the interrelation of the LR, RLR and BC treatments, we consider the cases without and with an infinitely strong magnetic field. We show that even in the absence of a magnetic field these approaches yield slightly different results when using the standard averaging procedure. The discrepancy grows with the strength of the magnetic field.

4.6.1 RLR, LR and BC Treatments Without Magnetic Field

Without magnetic field the imaginary part of the susceptibility of an isotropic Maxwellian plasma can be found from (2.47), (4.2) and (4.3), where for the isotropic plasma we take $Y = 1$,

$$\mathrm{Im}\,\chi^{(0)}(\boldsymbol{k},\omega) = \frac{(2\pi)^{5/2}}{4\lambda_D^2}\frac{\omega}{k v_{th}}\exp\left(-\frac{\omega^2}{2k^2v_{th}^2}\right). \tag{4.95}$$

The expression (4.93) for the RLR stopping power in the case of the Coulomb interaction, $u(\boldsymbol{k}) = u_C(\boldsymbol{k})$, and taking into account the dynamic screening (i.e. k_{min} and k_{max} from (2.59)) yields

$$S_{RLR} = \mathcal{S}_0\frac{g^2}{x^2}L_{th}(x)\left[\mathrm{erf}\left(\frac{x}{\sqrt{2}}\right) - \sqrt{\frac{2}{\pi}}\,x\,\mathrm{e}^{-x^2/2}\right], \tag{4.96}$$

where $\mathrm{erf}(x)$ is the error function, $x = v_i/v_{th}$, $\mathcal{S}_0 = (k_B T/\ell)^2$ and

$$L_{th}(x) = \ln\frac{k_{max}}{k_{min}} = \ln\frac{\left(1+x^2\right)^{3/2}}{g} = \ln\frac{1}{\mathcal{Z}(x)} \tag{4.97}$$

is a Coulomb logarithm. Here, the coupling parameter $\mathcal{Z}$ is defined by expression (2.42), and the parameter g is expressed through the Coulomb coupling parameter Γ, $g = \sqrt{3}\,|Z|\,\Gamma^{3/2}$. In the Coulomb logarithm L_{th}, the averaged lower and upper cut–offs, equation (2.59), have been used. Similarly, for the screened Coulomb potential $u(\boldsymbol{k}) = u_D(\boldsymbol{k})$ with the static screening length λ_D expression (4.93) yields again the RLR result (4.96), where, however, the Coulomb logarithm $L_{th}(x)$ is replaced by $(1/2)\psi(\xi)$ (see (4.11)) with $\xi = k_{max}\lambda_D = (1+x^2)/g$. Since $g \ll 1$ for an ideal plasma the Coulomb logarithm L_{th} can be considerably larger than $(1/2)\psi(\xi)$, i.e. the Coulomb logarithm with static screening. That is, the reduction of the SP (4.93) by ignoring the collective effects in the dielectric function can be compensated in part by introducing a dynamic, velocity dependent screening in the interaction potential.

Within the perturbative BC treatment, Chap. 3, we need to consider only the second order energy transfer during an electron–ion collision, as the first order energy transfer is proportional to the impact parameter $\boldsymbol{s}$ and vanishes after averaging over $\boldsymbol{s}$. The angular averaged second order energy transfer is given by equation (3.59)

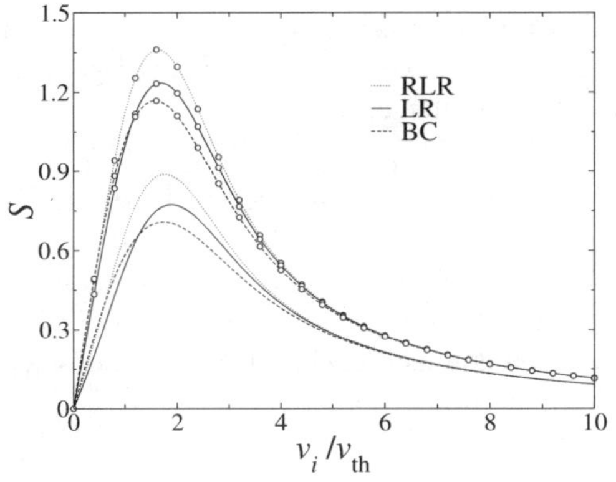

Fig. 4.9. Stopping powers (in units of $g^2 \mathcal{S}_0$) within the RLR with dynamic screening (4.96) (dotted lines), LR (2.53) (solid lines) and BC (4.99) (dashed lines) treatments as a function of the ion velocity v_i (in units of v_{th}) in a plasma without magnetic field for $g = 0.1$ (lines without circles) and $g = 0.01$ (lines with circles). After [93].

(with $\rho \to 0$ for Coulomb interaction). The energy loss of the ion due to the successive collisions in a homogeneous electron plasma is obtained by integrating equation (3.59) over an area element $\mathrm{d}^2\boldsymbol{s}$ perpendicular to the relative current density $n_e \boldsymbol{v}_r$ with $\boldsymbol{v}_r = \boldsymbol{v}_e - \boldsymbol{v}_i$ and averaging over the unperturbed electron distribution function f_0

$$S_{\text{BC}} = -\left(\frac{\mathrm{d}\mathcal{E}_i}{\mathrm{d}l}\right)_{\text{BC}} = -\frac{2\pi n_e}{v_i} \int \mathrm{d}\boldsymbol{v}_e f_0(\boldsymbol{v}_e) v_r \int_{s_{\min}}^{s_{\max}} \left\langle \Delta E_i^{(2)} \right\rangle_{\hat{s}} s\mathrm{d}s \,. \tag{4.98}$$

Here the upper cut–off $s_{\max}$ accounts for screening, while $s_{\min}$ is the cut–off below which the perturbative treatment of the Coulomb interaction fails (see equation (2.58)). For Rutherford scattering hard collisions are taken into account by regularizing the s–integral in (4.98) according to $L(s_{\max}, s_{\min}) \to \Lambda(s_{\max}, s_{\min})$, which yields the exact result [68]. Here the conventional L and modified Λ Coulomb logarithms have been introduced by (2.23) and (2.11), respectively.

For an isotropic Maxwellian distribution (2.26) we obtain from (3.59) with $\rho \to 0$ and (4.98) after the replacement $L \to \Lambda$

$$S_{\text{BC}} = \frac{\mathcal{S}_0 g^2}{2x^2\sqrt{2\pi}} \int_0^{\infty} \frac{\mathrm{d}y}{y^2} \ln\left(1 + \frac{y^6}{g^2}\right) \left[(xy-1)\mathrm{e}^{-(y-x)^2/2} + (xy+1)\mathrm{e}^{-(y+x)^2/2}\right] , \tag{4.99}$$

where $y = v_r/v_{\text{th}}$.

In Fig. 4.9, the normalized stopping powers within RLR (dotted lines) and BC (dashed lines) treatments, equations (4.96) and (4.99), are plotted versus the ion velocity. The full LR results, S_{LR}, including the electron–electron interaction are also plotted for comparison (solid lines). All these approaches yield close results except for some deviations in the intermediate velocity range.

To make a contact between LR and BC results, (4.96) and (4.99), we note that the integral in (4.99) divided by the factor $\sqrt{2\pi}$ is identical with the expression in square brackets in (4.96) if the logarithmic factor in (4.99) is taken out of the y–integral at some average value $\langle y\rangle = \langle v_r\rangle/v_{\rm th}$. The averaged stopping power in the BC treatment can thus be rewritten as

$$\bar{S}_{\rm BC} = \frac{1}{2L_{\rm th}(x)} \ln\left(1 + \frac{\langle y^6\rangle}{g^2}\right) S_{\rm RLR}\,. \tag{4.100}$$

This demonstrates that both approaches are equivalent if comparable cut–off procedures, resulting in equal Coulomb logarithms, are used. Equation (4.100) shows that for the conformity between both approaches one must choose $\langle y^6\rangle = (1+x^2)^3 - g^2$. In the limit $g \ll 1$, this condition becomes $\langle y^2\rangle \simeq 1 + x^2$.

4.6.2 RLR, LR and BC Treatments With Strong Magnetic Fields

In the presence of a strong magnetic field, the imaginary part of $\chi^{(0)}(\boldsymbol{k},\omega)$ can be obtained from the general expressions (2.47) and (2.48) (neglecting the contribution of the plasma ions) in the limit of $\Omega_e \to \infty$. As discussed in Sect. 4.3 (see discussion before (4.31)) the final result is given by (4.95) where the wave vector k should be replaced by $|k_{\|}|$, $k \to |k_{\|}|$. Here $k_{\|}$ is the component of $\boldsymbol{k}$ along the magnetic field. Substituting this equation into (4.93) and taking into account the dynamic screening for the Coulomb potential we obtain

$$S_{\rm RLR} = \mathcal{S}_0 \frac{g^2}{2\sqrt{2\pi}} x L_{\rm th}(x) \sin^2\alpha \int_{-\infty}^{\infty} \frac{y^2 e^{-y^2/2} dy}{q^{3/2}(x,y)}\,. \tag{4.101}$$

Here α is the angle between the magnetic field and the ion velocity $\boldsymbol{v}_i$, the Coulomb logarithm $L_{\rm th}$ is the same as in (4.97) and $q(x,y) = y^2 - 2xy\cos\alpha + x^2$, where $y = v_{e\|}/v_{\rm th}$. We recall that for the screened Coulomb potential with static screening the Coulomb logarithm in (4.101) should be replaced by $(1/2)\psi(\xi)$.

The second order energy transfer for an electron–ion Coulomb collision in the presence of a strong magnetic field is obtained from (3.143), where $\bar{\boldsymbol{v}}_r = v_{e\|}\boldsymbol{b} - \boldsymbol{v}_i$, and $v_{e\|}$ is the electron velocity along the magnetic field. The last result has been given in [120] for the case $v_{e\|} = 0$. In the general case, this term also leads to an energy gain for $v_i^2 < v_{e\|}^2$.

The integration of (3.143) over the impact parameter $\bar{s}$ is similar to that used in the previous subsection, with $s_{\rm min}$ and $s_{\rm max}$ from (2.58). However, now v_r is replaced by the relative velocity of the guiding center $\bar{v}_r$. Averaging expression (3.143) over an isotropic Maxwell distribution function f_0, we arrive at

$$S_{\rm BC} = \mathcal{S}_0 \frac{g^2}{4\sqrt{2\pi}} x \sin^2\alpha \int_{-\infty}^{\infty} \frac{dy(x^2 - y^2)e^{-y^2/2}}{q^{5/2}(x,y)} \ln\left[1 + \frac{q^3(x,y)}{g^2}\right]. \tag{4.102}$$

Note that the RLR (4.101) and BC (4.102) stopping powers are related to (3.197) and (3.196), respectively, which, however, have been derived for the regularized potential given by (3.3). The Coulomb logarithm in (4.101) and the logarithmic factor in (4.102) are replaced there by the function $\mathcal{U}_0(\varkappa)$.

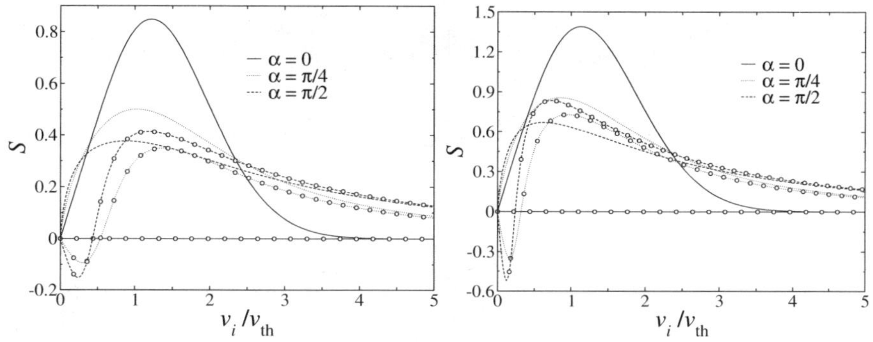

Fig. 4.10. Stopping powers (in units of $g^2\mathcal{S}_0$) within the RLR with dynamic screening (lines without circles) and the BC (lines with circles) treatment as a function of the ion velocity v_i (in units of v_{th}) in a plasma with a strong magnetic field for $g = 0.1$ (left), $g = 0.01$ (right), $\alpha = 0$ (solid lines), $\alpha = \pi/4$ (dotted lines) and $\alpha = \pi/2$ (dashed lines). After [93].

When the logarithmic factor in (4.102) is now taken out with some average value $\langle q^3(x, y)\rangle$ we obtain the stopping power $\bar{S}_{\text{BC}}$ in which the y–integral does not coincide with the y–integral in (4.101) as it has been the case in the absence of a magnetic field, cf. (4.100). Moreover, at low ion velocity $x \to 0$ the stopping power $\bar{S}_{\text{BC}}$ behaves as $1/x$ and tends to infinity. The RLR stopping power (4.101), on the other hand, leads at low ion velocities to a term which behaves $\propto x\ln(1/x)$, for both the full and the reduced LR treatments [90]. This anomalous SP has been investigated in details in Sect. 4.4. This is a quite unexpected behavior compared to the well–known linear velocity dependence without magnetic field [38, 105, 134].

The stopping powers (4.101) and (4.102) depend on the angle α. For small $\alpha \to 0$, expression (4.101) yields

$$S_{\text{RLR}}(\alpha \to 0) = \mathcal{S}_0 \frac{g^2}{\sqrt{2\pi}} L_{\text{th}}(x) x\, \mathrm{e}^{-x^2/2}\,, \tag{4.103}$$

whereas for BC, the stopping power vanishes as

$$S_{\text{BC}}(\alpha \to 0) = \mathcal{S}_0 \frac{g^2}{4\sqrt{2\pi}} x \sin^2\alpha \int_{-\infty}^{\infty} \frac{\mathrm{d}y (x^2 - y^2)\mathrm{e}^{-y^2/2}}{|y - x|^5} \ln\left[1 + \frac{(y - x)^6}{g^2}\right]. \tag{4.104}$$

This result coincides with the exact behavior (no perturbation treatment) of the BC stopping power for vanishing α. In the presence of a strong magnetic field, the electrons move parallel to the magnetic field. For reasons of symmetry, no velocity can be transferred to positively charged ions which also move parallel to the field, see, e.g., equations of motion (3.202) with $\tilde{v}_{i\perp} = 0$.[2] The energy transfer and hence the stopping power within BC treatment must therefore vanish.

In Fig. 4.10 the stopping powers within RLR (4.101) (the lines without circles) and BC (4.102) (the lines with circles) treatments are plotted for plasmas in a strong

[2] It should be noted that this may also be true for negatively charged ions ($Z < 0$). For instance, equations of motion (3.202) for the Coulomb potential yield vanishing velocity

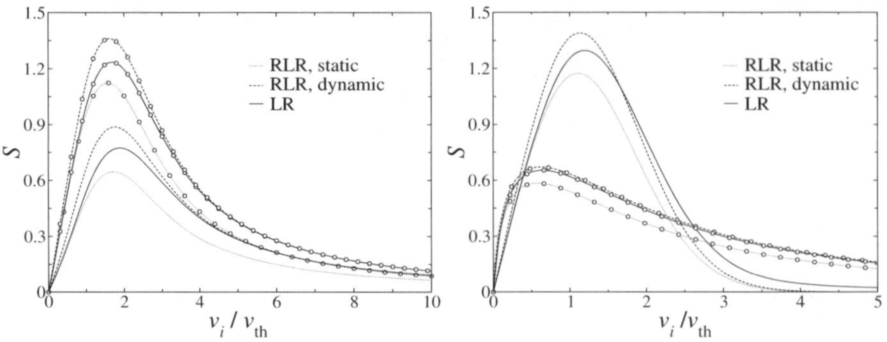

Fig. 4.11. Stopping powers (in units of $g^2 \mathcal{S}_0$) within the RLR with the Coulomb potential and dynamic screening (dashed lines), within the RLR with Debye potential and static screening (dotted lines) and the LR (solid lines) treatments as a function of the ion velocity v_i (in units of v_{th}) in a plasma without magnetic field (left) and with strong magnetic field (right). Left panel, the lines with and without circles correspond to $g = 0.01$ and $g = 0.1$, respectively. Right panel, for $g = 0.01$, $\alpha = 0$ (lines without circles) and $\alpha = \pi/2$ (lines with circles).

magnetic field for three values of α: $\alpha = 0$ (solid lines), $\alpha = \pi/4$ (dotted lines), $\alpha = \pi/2$ (dashed lines). The difference between the two treatments is noteworthy especially in the low and intermediate ion velocity limits. It is related to the different cut–off procedures, i.e. $k_{\min}$, $k_{\max}$ in RLR and $s_{\min}$, $s_{\max}$ in BC. In particular, the large stopping power predicted by the RLR for $\alpha = 0$ is unrealistic, since it vanishes within an exact (nonlinear) BC treatment, as discussed above. This is not healed by including collective effects, see Fig. 4.11, where we compare the reduced and full LR stopping powers without magnetic field ((4.96) and (4.7), respectively) and for an infinitely strong magnetic field ((4.101) and (4.33), respectively). The RLR results for the screened Coulomb interaction and with static screening are also plotted for comparison (dotted lines). (Note that in the left panel of Fig. 4.11 the curves corresponding to the LR and RLR with dynamic screening results are the same as in Fig. 4.9). Figure 4.11 shows by introducing the dynamic screening in the RLR that the role of collective excitations as predicted in this linear treatment is not as important here as in the limiting case considered in [88, 120]. We come back to this point at the end of this Section. As all dynamic collective excitations are absent in a RLR treatment with static screening, the resulting stopping power is always smaller than

---

and energy transfers if $\bar{s} > s_0 = 2|Z|e^2/m\bar{v}_{r\|}^2$, where $\bar{v}_{r\|}$ is the initial relative velocity. In opposite case with $\bar{s} < s_0$ the velocity transfer is $\Delta v_{\|} = -2\bar{v}_{r\|}$ which corresponds to a reversion of the initial motion, i.e. to a backscattering event. Then the energy transfer is $\Delta E_i = 2mv_{i\|}\bar{v}_{r\|}$.

those for the LR.[3] This difference vanishes in the low–velocity limit where only the single-particle excitations contribute to the SP, see Fig. 4.11.

In the low–velocity limit when $x \ll 1$, the BC stopping power is linear in x and negative,

$$S_{\mathrm{BC}}(x \to 0) = -\mathcal{S}_0 \frac{g}{8\sqrt{2\pi}} x \sin^2\alpha \int_0^\infty \frac{\mathrm{d}y}{y^2}\, \mathrm{e}^{-gy} \ln\left(1 + 8gy^3\right) . \tag{4.105}$$

This corresponds to an energy gain of an ion at slow transverse motion. For a finite magnetic field strength and anisotropic (non–equilibrium) velocity distributions typical for electron cooling, such energy gains have indeed been observed in non–perturbative numerical simulations of binary collisions [138] as well as in a numerical solution of the non–linearized Vlasov–Poisson equation [127]. The perturbing ion drives the equilibration between the longitudinal and transverse electron temperatures. There results, in the average, a shrinking of the cyclotron radii of the electrons and an energy gain of the ion caused by the released transverse electron energy. For an infinitely strong magnetic field and an effective one–dimensional electron motion, however, this mechanism does not work. Both the unexpected $\propto x\ln(1/x)$ behavior in the LR and the energy gain in (second order) BC therefore indicate a breakdown of a perturbation treatment for $B \to \infty$ and small ion velocities. This is also supported by considering again the case where the ion moves along the magnetic field. As discussed above, the energy transfer in binary collisions is zero for positively charged ions due to symmetry. But for negatively charged ions, the electron can either pass over the potential well, which gives again no energy transfer, or it is reflected with a momentum transfer of two times its initial momentum, see also the example and discussion given in Sect. 6.2.3. Thus all scattering events contributing to the stopping power are non–perturbative in this case. It is evident, that this situation cannot be treated by either the LR or the second order BC which both depend on the square of the ion–charge. However, the cut–off procedures employed in the (perturbative) BC (see Sects. 2.2 and 3.10) lead to stopping powers which are much closer to numerical simulations [127, 138] than the LR predictions [124] for positively charged ions and finite magnetic fields.

---

[3] It should be emphasized that these results do not contradict to the "equipartition rule". Originally this rule first formulated by Lindhard and Winther in [80] states the equality of the k–integrals in the SP (i.e., only the one–dimensional integral $k^2\mathrm{d}k$ in the volume element $\mathrm{d}\boldsymbol{k}$ in (2.53)) for collective and single–particle excitations. Although these excitations contribute equally in this integral, nevertheless the SPs for both excitations may considerably differ from each other, see, e.g., Fig. 4.11 and [92,98] for a review. Moreover, the "equipartition rule" has been formulated for the field–free case and for fully degenerate plasmas with a sharp boundary between the spectra of collective and single–particle excitations. In a classical plasma with a Maxwellian distribution of the particles this boundary is not as well defined as for a degenerate plasma. In addition the presence of a magnetic field introduces in general an infinite number of plasma collective modes (see, e.g., [72]) and the equipartition rule may not necessarily hold in the present context. Some examples of the violation of this rule even for $\boldsymbol{B} = 0$ are considered in [92, 98].

To shed some more light on the differences between BC and LR and the role of the dynamic excitations discussed so far in this Section we now turn to some comparison with numerical results from the PIC method (Sect. 2.5) and the CTMC treatment (Sect. 2.3). The most pronounced discrepancies have been found for a strong magnetic field in the case of small angles α and low ion velocities (see Fig. 4.10), i.e. for strong ion–electron coupling $\mathcal{Z} \gtrsim 1$ (2.42) or $g = \sqrt{3}\,|Z|\,\Gamma^{3/2} \gtrsim 1$. We thus focus in Fig. 4.12 on some examples in this regime and consider the typical parameters ($B = 6$ T, $T_e = 4.2$ K, $n_e = 7 \times 10^6$ cm^{-3}) for the cooling of highly charged ions with strongly magnetized electrons in a trap. More details on electron cooling in traps will be discussed in Sect. 6.2.

Compared are in Fig. 4.12 the numerical solution of the nonlinear Vlasov-Poisson equations (2.65), (2.66) obtained by PIC simulations as outlined in Sect. 2.5 (curves with filled circles) with the results of a non-perturbative BC treatment (curves with open circles) given by CTMC simulations (see Sect. 2.3) and the perturbative LR (dashed curves) and RLR (dotted curves) approaches. The numerical treatments have been performed with the strong, but finite magnetic field $B = 6$ T, while for LR and RLR an infinitely strong magnetic field with a one-dimensional electron motion along $\boldsymbol{b}$ was assumed. It was checked by CTMC simulations that a strong finite field of $B = 6$ T yields, within numerical fluctuations, in fact the same results as obtained by restricting the electron motion to be directed along the magnetic field. In the PIC simulations a pure Coulomb interaction is used, as it is in the employed LR expression for infinitely strong magnetic field (expressions (4.33), (4.34) with the required cut–off to exclude the violent close scattering events $\xi = k_{\max}\lambda_D = (1 + x^2)/g$, where $x = v_i/v_{th}$). The BC (CTMC) and the RLR have been calculated on basis of a dynamic screened potential given by u_D (3.2), (3.6) with a velocity dependent screening length $\lambda(v_i) = \lambda_D\,(1 + x^2)^{1/2}$. The RLR stopping power is then given by (4.101) after replacing the L_{th} (appropriate for the Coulomb interaction) by $\psi(\xi)/2$ (appropriate for the screened potential) and $\xi = k_{\max}\lambda(v_i) = (1 + x^2)^{3/2}/g$, i.e.

$$S_{\mathrm{RLR}} = \mathcal{S}_0 \frac{g^2}{4\sqrt{2\pi}}\, x\,\psi(\xi(x))\, \sin^2\alpha \int_{-\infty}^{\infty} \frac{y^2 e^{-y^2/2} dy}{q^{3/2}(x, y)}\,, \qquad (4.106)$$

where $\psi(\xi)$ is defined through (4.11). We first focus on the non-perturbative treatments. The difference between PIC (full circles) and BC (CTMC) (open circles) reveals the collective dynamic polarization contribution to the stopping power which cannot be approximated by an effective velocity dependent spherical interaction, $u_D(\lambda(v_i))$, as assumed for the BC. This contribution turns out to be particular large for large coupling ($g \gtrsim 1$) and small angles α and diminishes with decreasing coupling and increasing angle (from top left to bottom right). For $\alpha \to 0$, there remain, however, such dynamic contributions also in the limit $g \ll 1$. As already argued previously in several places the energy transfer in a BC treatment with a spherical symmetric interaction must vanish for symmetry reasons for the quasi one-dimensional electron motion in a strong magnetic field. But this argument does not apply in the general case where the electron-electron interaction is included as the polarization

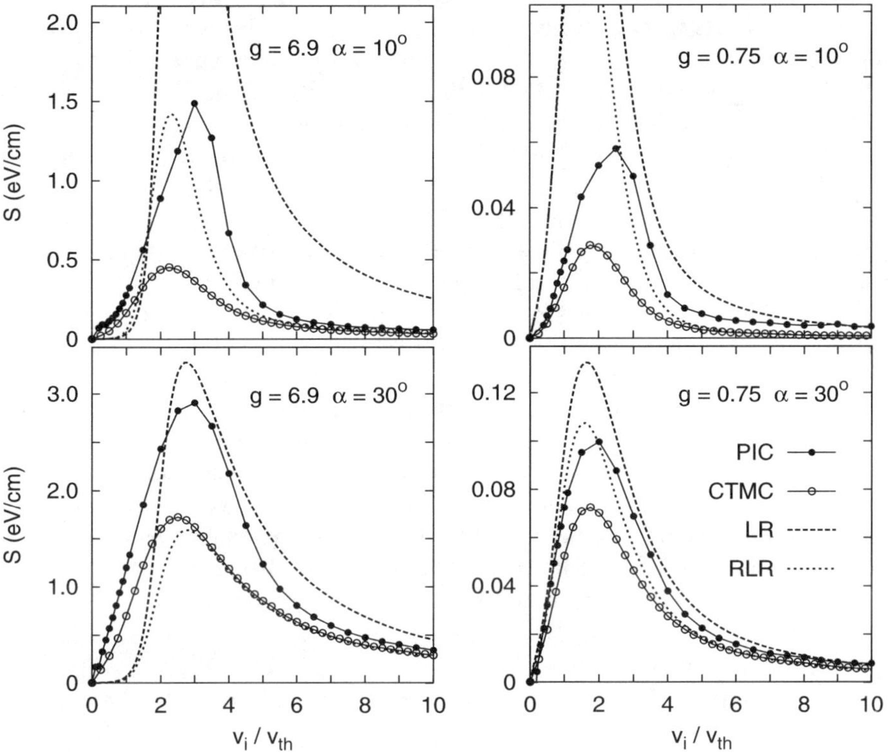

Fig. 4.12. Stopping power S as function of the ion velocity v_i in units of $v_{\rm th}$ for the two angles $\alpha = 10^{\rm o}$ and $\alpha = 30^{\rm o}$ and for $B = 6$ T, $T_e = 4.2$ K, $n_e = 7 \times 10^6$ cm^{-3} which are typical conditions for electron cooling in a trap. Compared are results from the PIC method (Sect. 2.5) (filled circles), from CTMC simulations (Sect. 2.3) (open circles) with the LR (4.33), (4.34) (dashed curves) and RLR (4.106) (dotted curves) treatments. For the CTMC and RLR a screened interaction u_D (3.2) with a velocity dependent screening length $\lambda(v_i)$ (see text) was used. The coupling parameters $g = 6.9$ (left panel) and $g = 0.75$ (right) correspond to ion charge states $Z = 92$ (i.e bare Uranium) and $Z = 10$, respectively, or for $g = 0.75$ to $Z = 92$ and a higher temperature $T_e = 18.4$ K instead of $T_e = 4.2$ K.

cloud generated here by a moving ion is usually non-spherical. And this results in a finite stopping power also for parallel ion motion in a strong magnetic field. The effects of this dynamic polarization contribution in the perturbative treatment is quite similar and can be identified by comparing the LR (dashed curves) with the RLR (dotted curves) results. Here the influence of the dynamic collectivity is also reduced with decreasing coupling and increasing angle, except for $\alpha \to 0$ and large velocities $v_i \gg v_{\rm th}$ where the LR goes like $1/v_i^2$, see (4.39), while the RLR vanishes exponentially with $\exp(-v_i^2/2v_{\rm th}^2)$, cf. (4.103).

But very interesting is here of course the comparison of PIC with LR and BC (CTMC) with RLR, respectively, i.e. of the non-perturbative with the corresponding perturbative approaches. As expected the perturbative treatments fail at strong coupling ($g \gtrsim 1$ and $v_i/v_{\text{th}} \lesssim 1$), see the low velocity regime in the left panel of Fig. 4.12, but tend towards the non-perturbative numerical results for decreasing coupling ($g < 1$ and/or $v_i/v_{\text{th}} \gg 1$). For ions moving along the magnetic field lines or with small angles, the LR significantly overestimates the stopping power when compared to the PIC. At low ion velocities this is not healed by a decreasing coupling strength. Unfortunately there are no PIC results at very small coupling ($g \to 0$) available as this method suffers increasingly from numerical noise in this limit.

Altogether the general trends and observations, which are supported by PIC calculations in other parameter regimes [127, 128], are as follows: Dynamic collective contributions to the stopping power, which cannot be simply incorporated into a BC or RLR treatment by employing an effective interaction with velocity dependent screening, get unimportant for weak coupling, i.e. in the domain where perturbative approaches are valid, except in cases of a strong magnetic field and ions moving with small angles to $\boldsymbol{b}$. Under the latter conditions these dynamic contribution dominate the energy loss and cannot be described within a BC treatment. The perturbative dielectric approach, i.e. the LR, on the other hand here also fails and overestimates the stopping power in particular in the low velocity regime $v_i \approx v_{\text{th}}$.

4.7 Conformity Between Reduced LR and BC approaches

The results obtained so far strongly suggest that the large discrepancies between the LR and the BC seen at strong magnetic fields are peculiar to the Coulomb interaction which requires cut–offs. It is the main concern of this Section to show that these discrepancies are in fact a consequence of the different cut–off procedures in the LR and BC and that the standard cut–off recipes for the non–magnetized case are not guaranteed to work in the presence of an external magnetic field. To this end we replace the Coulomb interaction by an effective, regularized interaction potential, which decays faster than r^{-1} at large distances and increases slower than r^{-1} at small ones. The introduction of such a regularized potential can be viewed as an alternative implementation of cut–offs. It is justified by the same line of arguments. At large distances, the bare Coulomb interaction is screened by the polarization of the electrons. At small distances a perturbative treatment of the Coulomb interaction leads to divergencies. One thus attempts to approximate the finite cross section of a non–perturbative treatment by either a cut–off or a regularization of the interaction.

Our starting point is the second order φ–averaged energy transfer (3.95). Next, the φ–averaged ion energy change, $\langle \Delta E_i^{(2)} \rangle_\varphi$, is integrated over the impact parameters $\bar{\boldsymbol{s}}$ in the full 2D space. For calculation of this $\bar{\boldsymbol{s}}$–integral we split all vector variables $\boldsymbol{C}$ in (3.95) into components parallel $C_\parallel^{(r)}$ and perpendicular $\boldsymbol{C}_\perp^{(r)}$ to the relative velocity $\bar{\boldsymbol{v}}_r$ as in (3.131), where $\bar{\boldsymbol{n}}_r = \bar{\boldsymbol{v}}_r/\bar{v}_r$ is the unit vector along $\bar{\boldsymbol{v}}_r$. Note that the component of $\boldsymbol{k}$ perpendicular to the magnetic field, $\boldsymbol{k}_\perp$, is now a function of $k_\parallel^{(r)}$

and $\boldsymbol{k}_\perp^{(r)}$. Introducing an effective transport cross section, $\sigma(\boldsymbol{v}_r, \boldsymbol{v}_i)$, and performing the $\bar{s}$ and τ-integrations in (3.95) we obtain

$$\sigma(\bar{\boldsymbol{v}}_r, \boldsymbol{v}_i) \equiv \int \mathrm{d}^2\bar{\boldsymbol{s}} \left\langle \Delta E_i^{(2)} \right\rangle_\varphi \tag{4.107}$$
$$= -\frac{(2\pi)^4 Z^2 \not{e}^4}{2m\bar{v}_r} \int \mathrm{d}\boldsymbol{k}\, |u(\boldsymbol{k})|^2 (\boldsymbol{k}\cdot\boldsymbol{v}_i) \sum_{n=-\infty}^{+\infty} J_n^2(k_\perp a)\, g_n(\boldsymbol{k}) ,$$

where

$$g_n(\boldsymbol{k}) = \frac{1}{\pi \mathrm{i}} \left\{ \frac{k_\parallel^2}{(\zeta_n(\boldsymbol{k}) + \mathrm{i}0)^2} + \frac{k_\perp^2}{2\Omega_e} \left[\frac{1}{\zeta_{n-1}(\boldsymbol{k}) + \mathrm{i}0} - \frac{1}{\zeta_{n+1}(\boldsymbol{k}) + \mathrm{i}0} \right] \right\} . \tag{4.108}$$

Here $\zeta_n(\boldsymbol{k}) = \boldsymbol{k}\cdot\bar{\boldsymbol{v}}_r + n\Omega_e$. The second order singularity in (4.108) (the first term) must be understood as

$$\frac{1}{(\zeta_n + \mathrm{i}0)^2} \rightarrow \frac{1}{\zeta_n(\zeta_n + \mathrm{i}0)} = \frac{1}{\zeta_n^2} - \frac{\pi \mathrm{i}}{\zeta_n}\delta(\zeta_n) . \tag{4.109}$$

It is easy to see that the contribution of the imaginary part of $g_n(\boldsymbol{k})$ to the effective cross section, (4.107), vanishes. The contribution of the real part yields

$$\sigma(\bar{\boldsymbol{v}}_r, \boldsymbol{v}_i) = -\frac{(2\pi)^4 Z^2 \not{e}^4}{2m\bar{v}_r} \int \mathrm{d}\boldsymbol{k}\, |u(\boldsymbol{k})|^2 (\boldsymbol{k}\cdot\boldsymbol{v}_i) \sum_{n=-\infty}^{+\infty} J_n^2(k_\perp a) \tag{4.110}$$
$$\times \left\{ k_\parallel^2 \delta'(\zeta_n(\boldsymbol{k})) + \frac{k_\perp^2}{2\Omega_e} \left[\delta(\zeta_{n+1}(\boldsymbol{k})) - \delta(\zeta_{n-1}(\boldsymbol{k})) \right] \right\} ,$$

where $\delta'(x)$ defines the derivative of the δ–function with respect to the argument.

For the Coulomb interaction $u(k) = u_C(k)$, the full 2D integration over the $\bar{s}$-space results in a logarithmic divergence of the $\boldsymbol{k}$-integration in (4.110). To cure this, we introduce cut–off parameters $k_{\min}$ and $k_{\max}$ as it was done in the linear response formulation (see (2.59)).

For applications to the energy loss of ions moving in a magnetized homogeneous plasma we average the ion energy change during successive binary collision over the distribution function of the electrons f_0. The standard procedure for averaging over distribution function is given by (4.98), where the s–integral and v_r must be replaced by $(1/2\pi)\sigma(\bar{\boldsymbol{v}}_r, \boldsymbol{v}_i)$ and $\bar{v}_r$, respectively. Substituting (4.110) for the effective transport cross section σ into (4.98), we obtain

$$S_{\mathrm{BC}} = \frac{Z^2 \not{e}^2}{v_i} \int \mathrm{d}\boldsymbol{k}\, |u(\boldsymbol{k})|^2 (\boldsymbol{k}\cdot\boldsymbol{v}_i)\, \Xi(\boldsymbol{k}, \boldsymbol{k}\cdot\boldsymbol{v}_i) , \tag{4.111}$$

where

$$\Xi(\boldsymbol{k}, \omega) = 2\pi^3 \omega_p^2 \sum_{n=-\infty}^{+\infty} \int \mathrm{d}\boldsymbol{v}_e f_0(\boldsymbol{v}_e) J_n^2(k_\perp a) \tag{4.112}$$
$$\times \left\{ k_\parallel^2 \delta'(\xi_n) + \frac{k_\perp^2}{2\Omega_e} \left[\delta(\xi_{n+1}) - \delta(\xi_{n-1}) \right] \right\} ,$$

and $\xi_n = k_{\|} v_{e\|} + n\Omega_e - \omega$.

With the relations $J^2_{n-1}(z) - J^2_{n+1}(z) = (4n/z)\, J_n(z) J'_n(z)$, for the Bessel functions (see, e.g., [52]), the function $\Xi(\boldsymbol{k}, \omega)$ becomes

$$\Xi(\boldsymbol{k}, \omega) = 2\pi^3 \omega_p^2 \sum_{n=-\infty}^{+\infty} \int \mathrm{d}\boldsymbol{v}_e f_0(\boldsymbol{v}_e) \times \left\{ k_{\|} J_n^2(k_\perp a) \frac{\partial}{\partial v_{e\|}} \delta(\xi_n) + \delta(\xi_n) \frac{n\Omega_e}{v_{e\perp}} \frac{\partial}{\partial v_{e\perp}} J_n^2(k_\perp a) \right\}. \tag{4.113}$$

After a partial integration of (4.113) we arrive at $\Xi(\boldsymbol{k}, \omega) = \mathrm{Im}\chi^{(0)}(\boldsymbol{k}, \omega)$, where $\chi^{(0)}(\boldsymbol{k}, \omega)$ is the susceptibility of magnetized electrons given by (2.47) and (A.6) with the electronic component only. The comparison of (4.111) and (4.93) then yields $S_{\mathrm{BC}} = S_{\mathrm{RLR}}$. Thus, starting from BC treatment we obtain the result derived in a reduced LR treatment. This shows the complete conformity between both approaches.

In the presence of a strong magnetic field the RLR expression for the stopping power (4.93) with an arbitrary spherically symmetric interaction potential $u(r)$ yields

$$S_{\mathrm{RLR}}(x) = \mathrm{S}_0 \frac{g^2 \mathcal{U}_0}{2\sqrt{2\pi}} x \sin^2 \alpha \int_{-\infty}^{\infty} \frac{y^2 \mathrm{e}^{-y^2/2} \mathrm{d}y}{q^{3/2}(x, y)} = S_{\mathrm{BC}}(x). \tag{4.114}$$

Here

$$\mathcal{U}_0 = \frac{(2\pi)^4}{4} \int_0^{\infty} k^3 u^2(k) \mathrm{d}k \tag{4.115}$$

and $u(k)$ is the Fourier transformed interaction potential. Now there appears in (4.114) the numerical factor $\mathcal{U}_0$ instead of the Coulomb logarithm (4.97) in (4.101). Equation (4.115) also gives a criterion in Fourier space for the regularized potential which is equivalent to the conditions considered above. Indeed, it must behave like $u(k) \sim k^{-2-p}$ at $k \to 0$ and $k \to \infty$ with negative or positive values, $p < 0$ and $p > 0$, respectively. Note that for the Coulomb potential $p = 0$ in both limits and the lower and upper cut–offs must be introduced in (4.115).

5 Quantum Theory of SP in Magnetized Plasmas

5.1 Dielectric Theory

In Chap. 4 we have considered the energy loss of a test ion in a magnetized plasma where the motion of the projectile ion as well as the plasma have been treated classically. Here we use a quantum mechanical description of the beam–plasma interaction rather than a classical dielectric function. Again we assume a weak coupling , $\mathcal{Z} \ll 1$, between projectile ion and plasma, where the coupling parameter $\mathcal{Z}$, within quantum description is given by $\mathcal{Z} = |Z|e^2/\hbar\bar{v}_r$.(see, e.g., [134]). Hence the dielectric formalism in linear response becomes accurate in the limit of high test particle velocities. We describe the plasma in the random-phase approximation (RPA) and are therefore restricted to the weak–coupling limit of the interparticle interactions. Unlike its classical counterpart the RPA dielectric function due to the wave nature of particles guarantees the convergence of the $\boldsymbol{k}$–integral for short–range interactions and avoids the cut–off procedure. Furthermore, we assume that the electrons give the main contribution to the stopping power.

The general theory for the SP has been developed in [3, 8] which, in principle, allows to account for the quantum nature of the incident test particle as well. In particular, in [8] the SP is expressed by the dynamical charge–charge structure factor. The latter is obtained from the longitudinal dielectric function $\varepsilon(\boldsymbol{k}, \omega)$ of the plasma, by employing the fluctuation dissipation theorem [76]. However this approach does not allow to calculate the dielectric function of the plasma explicitly. Instead we use the self–consistent field method developed originally for unmagnetized plasmas by Ehrenreich and Cohen [39] as well as by Goldstone and Gottfried [49]. This was generalized to the case of magnetized plasmas by Zyryanov [142] (see also [143]), Mermin and Canel [84] as well as Celli and Mermin [30]. The derivations in [30, 84] are very close to those given by Zyryanov and Kalashnikov [142, 143] except that the authors of [30, 84] include the interaction of the electron spin with the magnetic field explicitly.

We consider a quantum electron gas in a uniform magnetic field $\boldsymbol{B}$, initially in thermal equilibrium, which is subsequently perturbed by a weak self–consistent potential $\phi(\boldsymbol{r}, t)$ of the projectile ion. We assume a heavy ion such that a classical description of its motion with a rectilinear trajectory is applicable. The classical description of an individual electron is valid as long as the Landau energy satisfies $\hbar\Omega_e \ll E_0$, where E_0 is the mean energy of the random motion of electrons. In the case $\hbar\Omega_e \gtrsim E_0$ the quantum effects play an important role, but one can still treat the self–consistent electric field $\phi(\boldsymbol{r}, t)$ created by a moving ion by classical theory.

Consider the single–particle equation of motion for the density operator

$$\mathrm{i}\hbar\frac{\partial\hat{\rho}}{\partial t} = [\hat{H}, \hat{\rho}] \tag{5.1}$$

with $\hat{H} = \hat{H}_0 + \hat{H}_1$, where $\hat{H}_0$ is the unperturbed Hamiltonian for a single electron in a magnetic field and $\hat{H}_1 = -e\phi(\boldsymbol{r}, t)$ is the time–dependent perturbation. We seek an approximate solution of the equation of motion in which $\phi(\boldsymbol{r}, t)$ is considered as a perturbation. Within the perturbation theory $\hat{\rho}$ can be represented as $\hat{\rho} = \hat{\rho}_0 + \hat{\rho}_1$, where $\hat{\rho}_1$ is the first–order perturbation to the unperturbed density operator $\hat{\rho}_0$. The operator $\hat{\rho}_1$ can be treated as a linear functional of the self–consistent potential, $\hat{\rho}_1 \sim \phi(\boldsymbol{r}, t)$. We start with the zero–order unperturbed equation of motion, $\mathrm{i}\hbar(\partial\hat{\rho}_0/\partial t) = [\hat{H}_0, \hat{\rho}_0]$. Since the density of an electron gas is constant in equilibrium before the perturbation, the equilibrium density operator commutes with $\hat{H}_0$, $[\hat{H}_0, \hat{\rho}_0] = 0$. We take its diagonal matrix elements in a representation of eigenstates $|\alpha\rangle = \psi_\alpha(\boldsymbol{r})$ of $\hat{H}_0$ to be given by the Fermi–Dirac distribution function $f(E_\alpha)$ with $\hat{\rho}_0|\alpha\rangle = f(E_\alpha)|\alpha\rangle$. Here $\hat{H}_0|\alpha\rangle = E_\alpha|\alpha\rangle$, E_α are the eigenvalues of the Hamiltonian $\hat{H}_0$, i.e. for free particles. The corresponding eigenstates are labeled by α. In the case of magnetized electrons in the homogenous magnetic field along the z-axis, $\alpha = \{n, \sigma, q_y, q_z\}$ with the Landau level $n = 0, 1, 2...$, spin variable $\sigma = \pm 1/2$ and the components of the momentum, q_y and q_z.

Next, we consider the first–order linearized equation of motion for the density operator $\hat{\rho}_1$,

$$\mathrm{i}\hbar\frac{\partial\hat{\rho}_1}{\partial t} = [\hat{H}_1, \hat{\rho}_0] + [\hat{H}_0, \hat{\rho}_1]\,. \tag{5.2}$$

We assume that all corrections vanish at $t \to -\infty$, i.e. before the perturbation. Following Ehrenreich and Cohen [39], we use (5.2) to find the equation of motion of the matrix elements, $\mathcal{N}_{\alpha\beta}(t) = \langle\alpha|\hat{\rho}_1(\boldsymbol{r}, t)|\beta\rangle$, of the operator $\hat{\rho}_1$, which describes the transitions between the stationary states, α and β, of the system,

$$\left(\mathrm{i}\hbar\frac{\partial}{\partial t} + E_\beta - E_\alpha\right)\mathcal{N}_{\alpha\beta}(t) = -e\left[f(E_\beta) - f(E_\alpha)\right]\int \mathrm{d}\boldsymbol{k}\phi(\boldsymbol{k}, t)\,S_{\alpha\beta}(\boldsymbol{k}) \tag{5.3}$$

with

$$S_{\alpha\beta}(\boldsymbol{k}) = \langle\alpha|\mathrm{e}^{\mathrm{i}\boldsymbol{k}\cdot\boldsymbol{r}}|\beta\rangle\,. \tag{5.4}$$

Here $\phi(\boldsymbol{k}, t)$ is the Fourier transformed self–consistent potential. For deriving (5.3) we have used the relations

$$\langle\alpha|\hat{H}_0\hat{\rho}_1|\beta\rangle = E_\alpha\mathcal{N}_{\alpha\beta}(t)\,, \quad \langle\alpha|\hat{\rho}_0\hat{H}_1|\beta\rangle = -ef(E_\alpha)\langle\alpha|\phi(\boldsymbol{r}, t)|\beta\rangle\,. \tag{5.5}$$

Making the Fourier transformation with respect to t and using the initial condition $\mathcal{N}_{\alpha\beta}(t) \to 0$ at $t \to -\infty$ from (5.3) we find the solution for the matrix $\mathcal{N}_{\alpha\beta}(\omega)$,

$$\mathcal{N}_{\alpha\beta}(\omega) = e\frac{f(E_\beta) - f(E_\alpha)}{E_\alpha - E_\beta - \hbar\omega - \mathrm{i}0}\int \mathrm{d}\boldsymbol{k}\phi(\boldsymbol{k}, \omega)\,S_{\alpha\beta}(\boldsymbol{k})\,, \tag{5.6}$$

where the infinitesimal i0 guarantees the vanishing of $\mathcal{N}_{\alpha\beta}(t)$ at $t \to -\infty$.

Next, we express $\phi(\boldsymbol{k}, \omega)$ in terms of the matrix element of the operator $\hat{\rho}_1$ using the Poisson equation

$$\varepsilon_0 k^2 \phi(\boldsymbol{k}, \omega) = \rho_i(\boldsymbol{k}, \omega) - e n_1(\boldsymbol{k}, \omega) . \tag{5.7}$$

Here $\rho_i(\boldsymbol{k}, \omega)$ and $-e n_1(\boldsymbol{k}, \omega)$ are the Fourier transformed charge densities of the projectile ion and plasma electrons, respectively. The first–order perturbation $n_1(\boldsymbol{r}, t)$ to the equilibrium density of the plasma can be evaluated from

$$n_1(\boldsymbol{r}, t) = \mathrm{Tr}\left[\delta(\boldsymbol{r} - \boldsymbol{r}')\hat{\rho}_1(\boldsymbol{r}', t)\right] = \sum_{\alpha,\beta} \psi_\beta^*(\boldsymbol{r})\psi_\alpha(\boldsymbol{r})\mathcal{N}_{\alpha\beta}(t) , \tag{5.8}$$

where the trace is evaluated by means of the eigenfunctions $|\alpha\rangle$. Thus, using (5.8) we arrive at

$$n_1(\boldsymbol{k}, \omega) = \frac{1}{(2\pi)^3}\sum_{\alpha,\beta} S_{\alpha\beta}^*(\boldsymbol{k})\mathcal{N}_{\alpha\beta}(\omega) . \tag{5.9}$$

Finally, substituting (5.9) with (5.6) into (5.7) we obtain an integral equation for the Fourier transformed self–consistent potential

$$\phi(\boldsymbol{k}, \omega) + V(k)\int \mathrm{d}\boldsymbol{k}'\phi(\boldsymbol{k}', \omega)\sum_{\alpha,\beta} S_{\alpha\beta}^*(\boldsymbol{k})S_{\alpha\beta}(\boldsymbol{k}') \tag{5.10}$$

$$\times \frac{f(E_\beta) - f(E_\alpha)}{E_\alpha - E_\beta - \hbar\omega - \mathrm{i}0} = \frac{\rho_i(\boldsymbol{k}, \omega)}{\varepsilon_0 k^2} .$$

Here $u_C(k) = 1/2\pi^2 k^2$ and $V(k) = e^2 u_C(k)$ is the Fourier transformed electron–electron Coulomb interaction potential. The result given by (5.10) is general because up to now we have not specified the initial state of the plasma. For a general unperturbed Hamiltonian $\hat{H}_0$ one would have to resort to additional approximations to solve (5.10). However, when $\hat{H}_0$ represents a translationally invariant system, the kernel of (5.10) will contain a delta function of $\boldsymbol{k} - \boldsymbol{k}'$ which enables the equation to be solved algebraically[1].

To carry this through in the present case we must make use of the explicit unperturbed eigenfunctions and eigenvalues of $\hat{H}_0$, i.e., for an electron gas in a uniform magnetic field. If we take $\boldsymbol{B}$ to be in the z direction and use for the vector potential $\boldsymbol{A}$ the gauge $A_x = A_z = 0$ and $A_y = Bx$, then

$$\hat{H}_0 = \frac{1}{2m}(\hat{\boldsymbol{p}} + e\boldsymbol{A})^2 + \hbar\Omega_e\hat{\sigma}_z , \tag{5.11}$$

where $\hat{\sigma}_z$ is the spin operator. The unperturbed electron wave function in the Landau state $\alpha = \{n, \sigma, q_y, q_z\}$ is [73]

[1] To solve the integral equation (5.10) Zyryanov [142], for instance, makes an approximation which is equivalent to the RPA. We note that this is unnecessary approximation because for the translational invariant system (5.10) is indeed an algebraic equation. We show this explicitly by calculating the matrix elements of $S_{\beta\alpha}(\boldsymbol{k})$, see expressions (5.16) and (5.127).

$$\psi_\alpha(\boldsymbol{r}) = \frac{1}{L\lambda_B^{1/2}\left(2^n n!\sqrt{\pi}\right)^{1/2}} \mathrm{e}^{\mathrm{i}(q_y y + q_z z)} \mathrm{e}^{-\xi^2/2} H_n(\xi) u_\sigma \tag{5.12}$$

with $\xi = x/\lambda_B + q_y\lambda_B$ and the magnetic length $\lambda_B = (\hbar/m\Omega_e)^{1/2}$. A factor L^{-1} was introduced for normalization reasons (L is the normalization length). In (5.12) u_σ is the spin wave function, H_n is the Hermite polynomial, and the eigenvalues of the free particles E_α are given by

$$E_{n\sigma}(q_z) = \frac{\hbar^2 q_z^2}{2m} + \hbar\Omega_e\left(n + \sigma + \frac{1}{2}\right), \tag{5.13}$$

where $\sigma = \pm 1/2$ is the spin quantum number. The wave functions (5.12) are normalized to $\delta_{\sigma\sigma'}\delta_{nn'}\delta_{q_y;q_y'}\delta_{q_z;q_z'}$, where the Kronecker symbol $\delta_{q;q'}$, in the continuum limit is related to the Dirac delta function as $\delta_{q;q'} = (2\pi/L)\,\delta(q' - q)$. With the wave functions (5.12) we calculate the matrix elements for $S_{\beta\alpha}(\boldsymbol{k})$ [84]

$$S_{\beta\alpha}(\boldsymbol{k}) = \mathrm{i}^{n'-n}\delta_{\sigma\sigma'}\delta_{k_y+q_y;q_y'}\delta_{k_z+q_z;q_z'}\mathrm{e}^{-\mathrm{i}k_x\lambda_B^2(q_y+k_y/2)}\mathrm{e}^{\mathrm{i}(n-n')\theta}F_{nn'}(\zeta) \tag{5.14}$$

with $\beta = \{n', \sigma', q_y', q_z'\}$ as another Landau state, $\zeta = k_\perp^2\lambda_B^2/2$, $\tan\theta = k_y/k_x$ and $k_\perp$ is the component of $\boldsymbol{k}$ perpendicular to $\boldsymbol{b} = \boldsymbol{B}/B$. The function $F_{nn'}(\zeta)$ $(n, n' \geq 0)$ is given by [84]

$$F_{nn'}(\zeta) = \left(\frac{n!}{n'!}\right)^{1/2} \zeta^{(n'-n)/2}\mathrm{e}^{-\zeta/2}L_n^{n'-n}(\zeta) \tag{5.15}$$

with the property of symmetry $F_{nn'}(\zeta) = (-1)^{n-n'}F_{n'n}(\zeta)$, and $L_n^{n'}(\zeta)$ are the generalized Laguerre polynomials.

Now using (5.14) we explicitly calculate the integral involved in (5.10)

$$\int_{-\infty}^{\infty} \mathrm{d}q_y \mathrm{d}q_y' S_{\beta\alpha}(\boldsymbol{k}')S_{\beta\alpha}^*(\boldsymbol{k}) = \frac{2\pi}{\lambda_B^2}\delta_{\sigma\sigma'}\delta_{q_z';q_z+k_z}\delta_{\boldsymbol{k}';\boldsymbol{k}}F_{nn'}^2(\zeta). \tag{5.16}$$

In (5.16) a delta function $\delta(\boldsymbol{k} - \boldsymbol{k}')$ allows to reduce the integral equation (5.10) to an algebraic one. Solving this equation with respect to $\phi(\boldsymbol{k}, \omega)$ we arrive at

$$\phi(\boldsymbol{k}, \omega) = \frac{\rho_i(\boldsymbol{k}, \omega)}{\varepsilon_0 k^2 \varepsilon(\boldsymbol{k}, \omega)}, \tag{5.17}$$

where

$$\varepsilon(\boldsymbol{k}, \omega) = 1 + \frac{4\pi\not{e}^2}{k^2}\sum_{n;n'=0}^{\infty} F_{nn'}^2(\zeta) \times \sum_{\sigma=\pm 1/2}\int_{-\infty}^{\infty}\frac{\mathrm{d}q_z}{(2\pi\lambda_B)^2}\frac{f(E_{n'\sigma}(q_z+k_z)) - f(E_{n\sigma}(q_z))}{E_{n\sigma}(q_z) - E_{n'\sigma}(q_z+k_z) - \hbar\omega - \mathrm{i}0} \tag{5.18}$$

is the longitudinal dielectric function of the magnetized quantum plasma. (For the calculation of the full dielectric tensor we refer to [30, 111, 143]). Note that for a point–like projectile ion with $\rho_i(\boldsymbol{r}, t) = Ze\delta(\boldsymbol{r} - \boldsymbol{r}_i(t))$ the expression (5.17) after inverse Fourier transformation is the same as (2.46) except that the dielectric function in (5.17) is given by (5.18), i.e., for quantum magnetized plasma.

The dielectric function (5.18) is not appropriate to calculate the stopping power in the limiting case of weak and strong magnetic fields. An alternative integral representation for the dielectric function (5.18) considered e.g. by Horing [62] is given in Appendix C.

We make some remarks about the dielectric function (5.18). In obtaining this equation we have not imposed restrictions on the distribution function $f(E)$. Furthermore, equation (5.18) gives two branches of the spectrum of excitations as for the field–free case (see, e.g., [79, 80]). The poles from the denominator in (5.18) give the branch of single–particle excitations, and the root of the integral equation $\varepsilon(\boldsymbol{k}, \omega) = 0$ gives the branch of collective oscillations.

Next we consider the imaginary part of the dielectric function. The δ–function (see expression (C.1)) allows us to perform the q_z–integration in (5.18), and we readily obtain

$$\mathrm{Im}\,\varepsilon(\boldsymbol{k}, \omega) = \frac{1}{k^2|k_z|\lambda_B^2 a_B} \sum_{\sigma=\pm 1/2} \sum_{n;n'=0}^{\infty} F_{nn'}^2(\zeta) \left[f\left(E_{n\sigma}\left(Q_-\right)\right) - f\left(E_{n\sigma}\left(Q_+\right)\right) \right] \quad (5.19)$$

with

$$Q_\pm = \frac{k_z}{2} + \frac{m}{\hbar k_z} \left[\Omega_e(n' - n) \pm \omega \right] . \quad (5.20)$$

Here $a_B = \hbar^2/me^2$ is the Bohr radius. The real part of $\varepsilon(\boldsymbol{k}, \omega)$ can be evaluated from the Kramers–Kronig relation

$$\mathrm{Re}\,\varepsilon(\boldsymbol{k}, \omega) = 1 + \frac{1}{\pi} P \int_{-\infty}^{+\infty} \mathrm{d}\omega' \frac{\mathrm{Im}\,\varepsilon(\boldsymbol{k}, \omega')}{\omega' - \omega} , \quad (5.21)$$

where the integral is to be understood in the sense of a Cauchy principal value. The result reads as

$$\mathrm{Re}\,\varepsilon(\boldsymbol{k}, \omega) = 1 + \frac{1}{\pi k^2 k_z \lambda_B^2 a_B} \sum_{\sigma=\pm 1/2} \sum_{n;n'=0}^{\infty} F_{nn'}^2(\zeta) \left[G_{n\sigma}\left(Q_-\right) + G_{n\sigma}\left(Q_+\right) \right] , \quad (5.22)$$

where the functions $G_{n\sigma}(Q)$ is defined by

$$G_{n\sigma}(Q) = P \int_{-\infty}^{+\infty} \mathrm{d}q \frac{f\left(E_{n\sigma}(q)\right)}{Q - q} . \quad (5.23)$$

Equations (5.19) and (5.22) are valid at arbitrary degeneracy of the plasma and may serve as a starting point for the analysis of $\varepsilon(\boldsymbol{k}, \omega)$ in different parameter regimes. In the next Sections we find some explicit expressions for the dielectric function in these regimes.

In the low–frequency limit the real part of the dielectric function (5.22) involves the statical screening properties of the plasma. In this limit we have for the real part $\mathrm{Re}\,\varepsilon(\boldsymbol{k},0) = 1 + k_s^2/k^2$, where $k_s = 1/\lambda_s$ is the inverse screening length. In general the quantity k_s may be found from the first derivative of the electron density with respect to the chemical potential, i.e., $k_s^2 = 4\pi e^2(\partial n_e/\partial\mu)$. For a nondegenerate plasma one finds that the screening length shows an isotropic behavior and is given by the Debye radius, $\lambda_s = \lambda_{\mathrm{D}}$. For a degenerate plasma and vanishing magnetic field λ_s is approximated by the static Thomas–Fermi length. In a degenerate plasma with a magnetic field the modified Thomas–Fermi length can be computed using the explicit equation of states (see Sect. 5.2.2, equation (5.33)). At finite temperatures analytical expressions for the screening length can be found in [63].

Consider briefly the high–frequency and hydrodynamic (or alternatively the long–wavelength limit $k \to 0$) limits of the dielectric function. In both cases the imaginary part (5.19) of the dielectric function for the Fermi–Dirac distribution function is exponentially small. For the real part we obtain in the high–frequency limit from (5.22) and (5.23)

$$\varepsilon(\boldsymbol{k},\omega) \simeq 1 - \frac{2m}{n_e\hbar k^2}\frac{\omega_p^2}{\omega^2}\frac{1}{(2\pi\lambda_B)^2}\sum_{\sigma=\pm 1/2}\sum_{n=0}^{\infty}\int_{-\infty}^{\infty} \mathrm{d}q f(E_{n\sigma}(q)) \times \sum_{n'=0}^{\infty} F_{nn'}^2(\zeta)\left[\frac{\hbar k_z^2}{2m} + \Omega_e(n'-n)\right], \tag{5.24}$$

where ω_p and n_e are the plasma frequency and density, respectively. The n, n'–sums in (5.24) are evaluated using expressions (D.4), (D.9) and (D.14). This yields $\hbar k^2/2m$. Substituting this value into (5.24) and taking into account the normalization condition for the distribution function (see Sect. 5.2, expression (5.26)) we finally arrive at $\varepsilon(\boldsymbol{k},\omega) \simeq 1 - \omega_p^2/\omega^2$. This is a well known high–frequency approximation of the dielectric function (see, e.g., [74]).

Next consider the hydrodynamic limit of the dielectric function (5.22). (For more detailed calculations and a number of another limiting cases we refer a reader to [13, 30, 62, 84, 142, 143]). To do this we separate out from the double sum over n and n' the terms with $n' = n$ and $n' \neq n$. Then we obtain in the long–wavelength limit $k \to 0$ from (5.22)

$$\varepsilon(\boldsymbol{k},\omega) \simeq 1 + \frac{\omega_p^2}{n_e(2\pi\lambda_B)^2}\sum_{\sigma=\pm 1/2}\sum_{n=0}^{\infty}\int_{-\infty}^{+\infty} \mathrm{d}q f\,(E_{n\sigma}(q)) \times \left[-\frac{\cos^2\beta}{\omega^2}F_{nn}^2(\zeta) + \frac{\sin^2\beta}{\zeta}\sum_{n'\neq n}^{\infty} F_{nn'}^2(\zeta)\frac{n'-n}{\Omega_e^2(n'-n)^2-\omega^2}\right]. \tag{5.25}$$

Here β is the angle between $\boldsymbol{k}$ and $\boldsymbol{B}$. In the second term of (5.25) in the limit $\zeta \to 0$ only the terms with $n' = n \pm 1$ contribute to the dielectric function (see equation (5.15)). Thus keeping only these terms and using the behavior of the functions $F_{nn'}$ at $\zeta \to 0$, $F_{nn}^2(\zeta) \to 1$ and $F_{n;n+1}^2(\zeta) - F_{n;n-1}^2(\zeta) \to \zeta$ we get the dielectric function (4.49) with (4.50) (ignoring the damping and Coulomb correlations) obtained within the hydrodynamical approximation.

5.2 Equation of State for Quantum Magnetized Plasmas

The calculation of the SP in a quantum magnetized plasma in different parameter regimes requires some details about the dielectric function as well as the equation of state, $\mu = \mu(n_e, T)$, where μ, n_e and T are the chemical potential, density and temperature of electrons, respectively. The relation between the chemical potential $\mu\,(n_e, T)$, the electron density n_e, the temperature T and the magnetic field B is established by the normalization condition for the Fermi–Dirac distribution function. Recalling that the density of states in the interval $\mathrm{d}q_z$ for the given Landau level $\{n, \sigma\}$ is $(2\pi\lambda_B)^{-2}\mathrm{d}q_z$ [75] this condition reads

$$\frac{1}{(2\pi\lambda_B)^2}\sum_{n=0}^{\infty}\sum_{\sigma=\pm1/2}\int_{-\infty}^{\infty}\mathrm{d}q_z f\left(E_{n\sigma}\left(q_z\right)\right) = \frac{y\sqrt{2}}{\pi^2 \lambda\!\!\!^{-3}}\sum_{n=0}^{\infty} g_n F_{-1/2}(\mu_n) = n_e\,. \tag{5.26}$$

The arguments of the Fermi–Dirac function $f(E)$ are given by the eigenvalues of the free particles (5.13). Here $g_n = 1 - \frac{1}{2}\delta_{n0}$, $\mu_n = \mu/k_{\mathrm{B}}T - 2ny$, $y = \hbar\Omega_e/2k_{\mathrm{B}}T$, $\lambda\!\!\!^{-} = \hbar/mv_{\mathrm{th}}$ and $v_{\mathrm{th}} = (k_{\mathrm{B}}T/m)^{1/2}$ are the thermal wavelength and velocity of the electron, respectively. In (5.26) we have introduced the Fermi functions $F_\nu(z)$ with $\nu > -1$ which are defined by

$$F_\nu(z) = \int_0^{\infty}\frac{t^\nu \mathrm{d}t}{1+\mathrm{e}^{t-z}}\,. \tag{5.27}$$

For applications the asymptotic behavior of the Fermi function is useful. At $z \to \infty$ and $z \to -\infty$ this function behaves as $F_\nu(z) \simeq z^{\nu+1}/(\nu+1)$ and $F_\nu(z) \simeq \mathrm{e}^z\Gamma(\nu+1)$, respectively.

Note that the limit of vanishing magnetic field of the normalization condition (5.26) can be easily recovered. Since $\Omega_e \to 0$ in this limit we can transform the n-summation in (5.26) into an integration with respect to n. Next we introduce the new variable of the integration according to $n\hbar\Omega_e \to \hbar^2 q_\perp^2/2m$. Using these transformations one arrives at the ordinary normalization condition for the distribution function (see, e.g., [75]).

5.2.1 Critical Temperature

Let us define the critical temperature T_c when the chemical potential vanishes, $\mu\,(n_e, T_c) = 0$. We will show that at $T < T_c$ and $T > T_c$ equation (5.26) allows positive ($\mu > 0$) and negative ($\mu < 0$) solutions for the chemical potential, respectively. Although this temperature behavior of μ is similar to the case with vanishing magnetic field (see, e.g., [75]) the asymptotic values of the chemical potential at $T = 0$ (fully degenerate limit) and $T \to \infty$ (classical limit) as well as the temperature T_c will depend on the magnetic field. As for the unmagnetized case the Fermi energy is defined as $\mathcal{E}_F = \mu(n_e, T \to 0)$. Setting $\mu = 0$ we obtain from (5.26)

$$\sum_{n=0}^{\infty} g_n F_{-1/2}(-2ny_c) = n_e\frac{\hbar\,(2\pi\lambda_B)^2}{2\,(2mk_{\mathrm{B}}T_c)^{1/2}}\,, \tag{5.28}$$

where $y_c = \hbar\Omega_e/2k_{\mathrm{B}}T_c$. Alternatively the left hand side of (5.28) can be expressed through the elementary functions. The straightforward calculations yield

$$\sum_{n=1}^{\infty} \frac{(-1)^{n+1}}{\sqrt{n}} \frac{y_c}{\tanh(ny_c)} = \frac{4}{3\sqrt{\pi}} \left(\frac{E_F}{k_{\mathrm{B}}T_c}\right)^{3/2} . \tag{5.29}$$

Here E_F is the Fermi energy in the absence of magnetic field.

Consider the limit of a vanishing magnetic field, $y_c \to 0$. In this case we obtain from (5.29) (see, e.g., [75])

$$\frac{k_{\mathrm{B}}T_c}{E_F} = \left[\frac{3\sqrt{\pi}}{4}\left(1 - \frac{\sqrt{2}}{2}\right)\zeta(3/2)\right]^{-2/3} , \tag{5.30}$$

where $\zeta(z)$ is the Riemann function. So, in the limit of a vanishing magnetic field $k_{\mathrm{B}}T_c \sim E_F$. In the opposite case of a strong magnetic field ($y_c \gg 1$)

$$\frac{k_{\mathrm{B}}T_c}{E_F} \simeq \left(\frac{8}{3C\sqrt{\pi}} \frac{E_F}{\hbar\Omega_e}\right)^2 , \tag{5.31}$$

where

$$C = \sum_{n=1}^{\infty} \frac{(-1)^{n+1}}{\sqrt{n}} \simeq 0.6049 . \tag{5.32}$$

Equation (5.31) shows that in the limit of a strong magnetic field the temperature T_c decreases with B and $k_{\mathrm{B}}T_c \ll E_F$.

5.2.2 Fully Degenerate Electron Plasma

After these preliminary results we now turn to the analysis of the chemical potential and the Fermi energy in the presence of a magnetic field. For a fully degenerate electron plasma ($T = 0$) we have $f(E) = \Theta(\mathcal{E}_F - E)$. Here $\Theta(z)$ is the Heaviside function. Therefore we obtain from (5.26)

$$\sum_{n=0}^{n_F} g_n \sqrt{\frac{\mathcal{E}_F}{\hbar\Omega_e} - n} = \frac{\pi^2}{\sqrt{2}} n_e \lambda_B^3 , \tag{5.33}$$

where $n_F = [\mathcal{E}_F/\hbar\Omega_e]$ is the allowed number of Landau levels and $[z]$ denotes the integer part of z. Hence expression (5.33) expresses implicitly the Fermi energy through the electron density and the magnetic field. In the limit of a vanishing magnetic field the number of allowed Landau levels increases and the summation in (5.33) can be replaced by an integration. This yields the Fermi energy E_F in the field–free case. However in the opposite case of increasing magnetic field the number n_F is strongly reduced. For instance when $\mathcal{E}_F < \hbar\Omega_e$ only one Landau level exists. Consider two examples.

(i) *One Landau level*, $\mathcal{E}_F < \hbar\Omega_e$. In this case $n_F = 0$ and all terms in (5.33) with $n \geq 1$ vanish and only the term with $n = 0$ contributes to the Fermi energy which reads

$$\mathcal{E}_F = \alpha^3 E_F \left(\frac{E_F}{\hbar\Omega_e}\right)^2, \tag{5.34}$$

where $\alpha = (16/9)^{1/3}$. Note that the condition $\hbar\Omega_e > \mathcal{E}_F$ is equivalent to $\hbar\Omega_e > \alpha E_F$. We also note that in this case $\mathcal{E}_F < \alpha E_F$. From (5.34) one finds the simple relation for the Fermi momentum of the electrons $k_F = 2\pi^2 n_e \lambda_B^2$.

Equation (5.34) shows that in the limit of a strong magnetic field the Fermi energy of the electrons is strongly reduced. Since in this limit the critical temperature T_c is reduced (see (5.31)) as well one can conclude that a strong magnetic field decreases the domain of the positive values of μ.

(ii) *Two Landau levels*, $\hbar\Omega_e \leq \mathcal{E}_F < 2\hbar\Omega_e$. In this case $n_F = 1$ and there are only two contributions in (5.33). Introducing the parameter $\xi = \alpha E_F/\hbar\Omega_e$ the Fermi energy is

$$\mathcal{E}_F = \alpha E_F \left(\frac{\xi^3 + 4}{\xi^2 + 2\sqrt{\xi(\xi^3 + 3)}}\right)^2, \tag{5.35}$$

where there must be $1 \leqslant \xi < 1/C_1$ with $C_1 = (2 + \sqrt{2})^{-2/3}$. This condition is equivalent to $C_1 \alpha E_F < \hbar\Omega_e \leq \alpha E_F$. The Fermi momentum for the two level system can be easily found from (5.35).

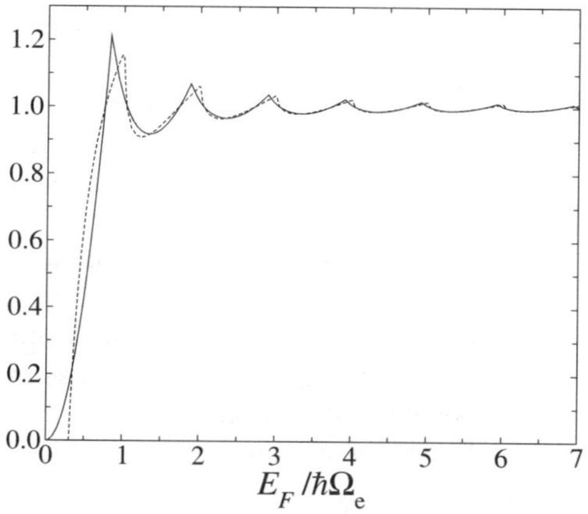

Fig. 5.1. Fermi energy $\mathcal{E}_F$ of the quantum magnetized plasma (in units of E_F, i.e. Fermi energy in field–free case) as a function of $E_F/\hbar\Omega_e$. The solid and dashed lines are found from (5.33) and (5.36), respectively.

For an arbitrary number n_F of Landau levels equation (5.33) can be evaluated using Poisson summation formula (see, e.g., [75]). This way of summing accurately involves the de Haas–van Alphen oscillatory terms in the Fermi energy [75]. Applying this summation formula for the Fermi energy we obtain the following approximate relation

$$\frac{\mathcal{E}_F}{E_F} \simeq 1 - \frac{\sqrt{2}}{4\pi\nu^{3/2}}\left[\sum_{l=1}^{\infty}\frac{1}{l^{3/2}}\sin\left(2\pi l\nu - \frac{\pi}{4}\right) + \frac{\pi\sqrt{2}}{12\nu^{1/2}}\right], \tag{5.36}$$

where $\nu = E_F/\hbar\Omega_e > 1$. The results of the numerical evaluation of (5.33) and (5.36) are shown in Fig. 5.1.

5.2.3 Semiclassical and Classical Limits

The semiclassical limit (in the sense of statistics) is established by the relations $k_B T \gg E_F$ or $e^{\mu/k_B T} \ll 1$ but the external magnetic field is still arbitrary. In this case we can take approximately $f(E) \simeq e^{(\mu-E)/k_B T}$. Then we find from equations (5.13) and (5.26) after summation over all Landau levels

$$e^{\mu/k_B T} = \frac{4}{3\sqrt{\pi}}\left(\frac{E_F}{k_B T}\right)^{3/2}\frac{\tanh(y)}{y} \tag{5.37}$$

and

$$\mu = -k_B T \ln\left[\frac{3\sqrt{\pi}}{4}\left(\frac{k_B T}{E_F}\right)^{3/2}\frac{y}{\tanh(y)}\right], \tag{5.38}$$

where $y = \hbar\Omega_e/2k_B T$ is still an arbitrary parameter. More precisely the semiclassical limit can be formulated using the relations (5.37) and (5.38). It reads $k_B T \gg (\tanh(y)/y)^{2/3} E_F$. Introducing the electron thermal wavelength $\lambda\!\!\!^{-} = \hbar/m v_{th}$ and velocity $v_{th} = (k_B T/m)^{1/2}$ this condition can be alternatively represented in the equivalent form, $(2\pi)^{3/2} n_e \lambda\!\!\!^{-3} \tanh(y)/2y \ll 1$ which requires high temperatures and low densities for an electron plasma. The classical limit is recovered in the limit of small Landau energy compared to the thermal one, $y \to 0$. Equation (5.38) shows that the chemical potential may be strongly modified by an external magnetic field. In addition a magnetic field increases the domain of classical behavior towards higher densities or lower temperatures. In this semiclassical regime the single–electron motion is described by quantum mechanics but the electron distribution is given by a classical Maxwell distribution.

5.3 Dielectric Function, Fully Degenerate Plasma

Consider first the dielectric function in the case of fully degenerate plasma ($T = 0$) when the distribution function is expressed by the Heaviside function $f(E) = \Theta(\mathcal{E}_F - E)$. Using this distribution we calculate the functions $G_{n\sigma}(Q)$ in (5.23) and substitute into (5.22). Thus from (5.22) and (5.19) we obtain for the real and imaginary parts of the dielectric function

$$\operatorname{Re}\varepsilon(\boldsymbol{k},\omega) = 1 + \frac{1}{\pi k^2 k_z a_B \lambda_B^2}\sum_{n'=0}^{\infty}$$
$$\times\left\{\sum_{n=0}^{n_F} F_{nn'}^2(\zeta)\ln\left|\frac{\left[k_z\lambda_B\sqrt{2(\tau-n)}+\frac{k_z^2\lambda_B^2}{2}+n'-n\right]^2-\frac{\omega^2}{\Omega_e^2}}{\left[k_z\lambda_B\sqrt{2(\tau-n)}-\frac{k_z^2\lambda_B^2}{2}+n-n'\right]^2-\frac{\omega^2}{\Omega_e^2}}\right|\right. \tag{5.39}$$
$$\left.+\sum_{n=0}^{n_F-1} F_{nn'}^2(\zeta)\ln\left|\frac{\left[k_z\lambda_B\sqrt{2(\tau-1-n)}+\frac{k_z^2\lambda_B^2}{2}+n'-n\right]^2-\frac{\omega^2}{\Omega_e^2}}{\left[k_z\lambda_B\sqrt{2(\tau-1-n)}-\frac{k_z^2\lambda_B^2}{2}+n-n'\right]^2-\frac{\omega^2}{\Omega_e^2}}\right|\right\},$$

$$\operatorname{Im}\varepsilon(\boldsymbol{k},\omega) = \frac{1}{k^2\,|k_z|\,a_B\lambda_B^2}\sum_{n'=0}^{\infty}$$
$$\times\left\{\sum_{n=0}^{n_F} F_{nn'}^2(\zeta)\left[\Theta\left(\frac{2}{\lambda_B^2}(\tau-n)-Q_-^2\right)-\Theta\left(\frac{2}{\lambda_B^2}(\tau-n)-Q_+^2\right)\right]\right. \tag{5.40}$$
$$\left.+\sum_{n=0}^{n_F-1} F_{nn'}^2(\zeta)\left[\Theta\left(\frac{2}{\lambda_B^2}(\tau-1-n)-Q_-^2\right)-\Theta\left(\frac{2}{\lambda_B^2}(\tau-1-n)-Q_+^2\right)\right]\right\},$$

where $\tau = \mathcal{E}_F/\hbar\Omega_e$, $n_F = [\tau]$ ($[\tau]$ denotes the integer part of τ), $\zeta = k_\perp^2\lambda_B^2/2$ and a_B is the Bohr radius. In (5.40) the quantities $Q_\pm$ are defined by (5.20). Note that in this case the Fermi energy $\mathcal{E}_F$ is determined from (5.33). Equations (5.39) and (5.40) show that as the magnetic field strength decreases, i.e., as n_F increases new terms are added to the n–sums, and this leads to the discontinuous changes of the dispersion relations of the plasma. Consider this briefly by the example of Bernstein modes [15] propagating transverse to the magnetic field with $k_z = 0$. In this case the imaginary part of the dielectric function vanishes and we obtain from (5.39)

$$\varepsilon(\boldsymbol{k},\omega) = 1 + \frac{\omega_p^2}{k^2\lambda_B^2}\frac{\sqrt{2}}{\pi^2 n_e\lambda_B^3}\left[\sum_{n=0}^{n_F}\sqrt{\tau-n}\sum_{n'=-n}^{\infty}F_{n;n'+n}^2(\zeta)\frac{n'}{n'^2\Omega_e^2-\omega^2}\right. \tag{5.41}$$
$$\left.+\sum_{n=0}^{n_F-1}\sqrt{\tau-1-n}\sum_{n'=-n}^{\infty}F_{n;n'+n}^2(\zeta)\frac{n'}{n'^2\Omega_e^2-\omega^2}\right].$$

The dispersion equation $\varepsilon(\boldsymbol{k},\omega) = 0$ with the dielectric function (5.41) has an infinite number of solutions (Bernstein modes [15]) near the harmonics of the cyclotron frequency $l\Omega_e$, where $l = 1, 2, ...$ For instance, the frequencies of the lowest Bernstein modes with $n_F = 0$ ($\tau < 1$) and $n_F = 1$ ($1 \le \tau < 2$) are obtained from (5.41) neglecting the terms with $n' \ge 2$. The straightforward calculations for $n_F = 0$ and $n_F = 1$ yield

$$\omega_{n_F=0}^2(\zeta) \simeq \Omega_e^2 + \omega_p^2 e^{-\zeta}, \tag{5.42}$$

$$\omega_{n_F=1}^2(\zeta) \simeq \Omega_e^2 + \omega_p^2 e^{-\zeta}\left[\sqrt{1-\frac{1}{\tau}}+\frac{(2-\zeta)^2}{2}\right], \tag{5.43}$$

respectively. At $\tau \simeq 1$ the discontinuity between the frequencies (5.42) and (5.43) is determined by the term within square brackets in (5.43).

Equations similar to (5.39) and (5.40) have been previously derived by Zyryanov [142] ignoring the spin of electrons. In contrast to [142] the dielectric function in (5.39) and (5.40) involves the electron spin. In particular, the first and second n–sums in (5.39) and (5.40) are the contributions of the electrons with $\sigma = -1/2$ and $\sigma = +1/2$, respectively, and these contributions do not necessarily coincide as in [142].

5.3.1 Fully Degenerate Plasma in a Strong Magnetic Field

Consider the dielectric function of a fully degenerate plasma in a strong magnetic field with $\hbar\Omega_e > \mathcal{E}_F$ (alternatively $\hbar\Omega_e > \alpha E_F$, where $\alpha = (16/9)^{1/3}$), $\tau < 1$. The Fermi energy $\mathcal{E}_F$ is obtained from (5.34) with the Fermi momentum $k_F = 2\pi^2 n_e \lambda_B^2$. In this case all the electrons are in the lowest Landau state with $n_F = 0$ and have their spins aligned antiparallel to the magnetic field. Thus the second terms in (5.39) and (5.40) vanish and the dielectric function receives contributions only from the first terms with $n = 0$. Using the relation $F_{0n}^2(\zeta) = (\zeta^n/n!)\,\mathrm{e}^{-\zeta}$ (see (5.15)) the real and imaginary parts of the dielectric function become [13, 62]

$$\mathrm{Re}\,\varepsilon(\boldsymbol{k},\omega) = 1 + \frac{1}{\pi a_B \lambda_B^2 k^2 |k_z|}\mathrm{e}^{-\zeta}\sum_{n=0}^{\infty}\frac{\zeta^n}{n!}\ln\left|\frac{\left(|k_z|v_F + \frac{\hbar k_z^2}{2m} + n\Omega_e\right)^2 - \omega^2}{\left(|k_z|v_F - \frac{\hbar k_z^2}{2m} - n\Omega_e\right)^2 - \omega^2}\right|, \tag{5.44}$$

$$\begin{aligned}\mathrm{Im}\,\varepsilon(\boldsymbol{k},\omega) = {} & \frac{1}{a_B \lambda_B^2 k^2 |k_z|}\mathrm{e}^{-\zeta}\sum_{n=0}^{\infty}\frac{\zeta^n}{n!}\left[\Theta\left(k_z^2 v_F^2 - \left(\omega - \frac{\hbar k_z^2}{2m} - n\Omega_e\right)^2\right)\right. \\ & \left. - \,\Theta\left(k_z^2 v_F^2 - \left(\omega + \frac{\hbar k_z^2}{2m} + n\Omega_e\right)^2\right)\right],\end{aligned} \tag{5.45}$$

where $v_F = \hbar k_F/m$ is the Fermi velocity.

In the limit of infinitely strong magnetic field ($\lambda_B \to 0$) only the terms with $n = 0$ contribute to the dielectric function (5.44) and (5.45) which in this case becomes

$$\varepsilon(\boldsymbol{k},\omega) = 1 + \frac{k_z^2}{k^2}\frac{\omega_p^2}{\frac{\hbar^2 k_z^4}{4m^2} - \omega^2} + \mathrm{i}\frac{\pi m \omega_p^2}{\hbar k^2}\left[\delta\left(\omega - \frac{\hbar k_z^2}{2m}\right) - \delta\left(\omega + \frac{\hbar k_z^2}{2m}\right)\right]. \tag{5.46}$$

Note that in a classical limit ($\hbar \to 0$) the real part of the dielectric function (5.46) coincides with the high–frequency classical dielectric function in an infinitely strong magnetic field (see Sect. 4.3). In both cases and in high–frequency limit the plasma frequency is replaced by $\omega_p(k_z/k)$ which is the consequence of the infinitely strong magnetic field.

The dielectric function (5.46) can be treated as a quantum analogy of the plasmon–pole approximation employed to calculate the stopping power in an unmagnetized plasma (see, e.g., [11, 91]). However, due to the strong magnetic field the Fermi velocity in (5.46) vanishes and the three–dimensional wavenumber k in plasmon–pole approximation is replaced here by k_z.

5.3.2 Acoustic Plasma Resonance

In a fully degenerate plasma there exists an acoustic mode (see, e.g., [34, 71, 133]). In a long wavelength limit, $k \to 0$, the frequency of this mode behaves as $\omega(k) \simeq kc_s$, where c_s is some constant and can be treated as the velocity of sound. This quantity can be obtained from the real part of the dielectric function (5.39)

$$\sum_{n=0}^{n_F} g_n \frac{\sqrt{\tau - n}}{V_0^2(\tau - n)\cos^2\beta - c_s^2} = 0\,, \tag{5.47}$$

where $V_0^2 = 2\hbar\Omega_e/m$. Note that this acoustic mode does not exist in transverse direction when $\beta = \pi/2$. Also this acoustic mode does not exist in the presence of strong magnetic field, or alternatively, when there exists only one Landau level with $n_F = 0$. For a degenerate electron plasma with two Landau levels, $n_F = 1$ and

$$c_s^2 = V_0^2 \cos^2\beta \frac{\sqrt{\tau(\tau - 1)}\left(\sqrt{\tau - 1} + 2\sqrt{\tau}\right)}{2\sqrt{\tau - 1} + \sqrt{\tau}}\,. \tag{5.48}$$

Note that for a degenerate electron plasma with two Landau levels the Fermi energy $\mathcal{E}_F$ is determined from (5.35).

This difference in the dispersion from the usual quadratic type $\omega \propto k^2$ for zero magnetic field has its origin in the following: the electron gas which is a single component plasma in the absence of the external magnetic field, behaves, under the field, as a multi–component plasma with electrons at different Landau levels. It is known (see, for example, [5, 72]) that a two–component plasma having charged particles of unequal effective masses (e.g. an electron–hole and electron-ion plasmas) has an acoustic type dispersion for the collective mode. In our case the different Landau levels act as different constraints of motion and hence simulate the effect of unequal effective masses.

5.4 Dielectric Function, Semiclassical Limit

Consider the semiclassical limit for the dielectric function (5.18) for the Fermi–Dirac distribution function for a magnetized electron plasma with temperature T [62]. The relation between the chemical potential $\mu(n_e, T)$ and the electron density n_e is established by the expression (5.38) and the external magnetic field is still arbitrary.

The imaginary part of the dielectric function (5.18) with the distribution function $f(E) \simeq \mathrm{e}^{(\mu-E)/k_B T}$ reads

$$\operatorname{Im} \varepsilon(\boldsymbol{k}, \omega) = \frac{m^2 e^2 \Omega_e}{\hbar^3 k^2 |k_z|} \left(1 + \mathrm{e}^{-2y}\right) \mathrm{e}^{\mu/T} \Im(y, \zeta) , \tag{5.49}$$

where $\Omega_\pm = \omega \pm \hbar k_z^2/2m$,

$$\Im(y, \zeta) = \sum_{n=0}^{\infty} \mathrm{e}^{-2yn} \sum_{s=-n}^{\infty} F^2_{n;s+n}(\zeta) (C_s - D_s) , \tag{5.50}$$

$$C_s = \mathrm{e}^{-2ys} \exp\left[-\frac{(s\Omega_e + \Omega_-)^2}{2k_z^2 v_{\mathrm{th}}^2}\right] , \quad D_s = \exp\left[-\frac{(s\Omega_e + \Omega_+)^2}{2k_z^2 v_{\mathrm{th}}^2}\right] . \tag{5.51}$$

For the evaluation of the function $\Im(y, \zeta)$ we rewrite it in the following form

$$\Im(y, \zeta) = \sum_{s=1}^{\infty} \mathfrak{R}_s(y, \zeta) (C_s - D_s) + \sum_{s=0}^{\infty} \mathrm{e}^{-2ys} \mathfrak{R}_s(y, \zeta) (C_{-s} - D_{-s}) \tag{5.52}$$

with

$$\mathfrak{R}_s(y, \zeta) = \sum_{n=0}^{\infty} \mathrm{e}^{-2yn} F^2_{n;s+n}(\zeta) . \tag{5.53}$$

First, let us consider the function $\mathfrak{R}_s(y, \zeta)$. Using (5.15) we obtain

$$\mathfrak{R}_s(y, \zeta) = \zeta^s \mathrm{e}^{-\zeta} \sum_{n=0}^{\infty} \frac{\kappa^n n!}{(n+s)!} \left[L_n^s(\zeta)\right]^2 , \tag{5.54}$$

where $\kappa = \mathrm{e}^{-2y} < 1$. The sum in (5.54) is well known [52]. It represents the expansion of the modified Bessel functions of the first kind by the generalized Laguerre polynomials

$$\begin{aligned} \mathfrak{R}_s(y, \zeta) &= \frac{1}{\kappa^{s/2}(1-\kappa)} \exp\left(-\zeta \frac{1+\kappa}{1-\kappa}\right) I_s\left(\frac{2\zeta\sqrt{\kappa}}{1-\kappa}\right) \\ &= \frac{\exp\left[y(s+1)\right]}{2\sinh(y)} \mathrm{e}^{-\zeta \coth(y)} I_s\left(\frac{\zeta}{\sinh(y)}\right) . \end{aligned} \tag{5.55}$$

Using the symmetry relation $\mathfrak{R}_{-n}(y, \zeta) = \mathrm{e}^{-2yn} \mathfrak{R}_n(y, \zeta)$, (5.52) can be rewritten as

$$\Im(y, \zeta) = \sum_{n=-\infty}^{\infty} \mathfrak{R}_n(y, \zeta) (C_n - D_n) . \tag{5.56}$$

Substituting the last expression into (5.49) we finally arrive at

$$\begin{aligned} \operatorname{Im} \varepsilon(\boldsymbol{k}, \omega) &= \frac{m\omega_p^2 (2\pi)^{1/2}}{\hbar k^2 |k_z| v_{\mathrm{th}}} \exp\left(-\frac{\hbar^2 k_z^2}{8m^2 v_{\mathrm{th}}^2}\right) \mathrm{e}^{-\zeta \coth(y)} \\ &\times \sinh\left(\frac{\hbar\omega}{2k_B T}\right) \sum_{n=-\infty}^{\infty} \exp\left[-\frac{(\omega - n\Omega_e)^2}{2k_z^2 v_{\mathrm{th}}^2}\right] I_n\left(\frac{\zeta}{\sinh(y)}\right) . \end{aligned} \tag{5.57}$$

This is a semiclassical expression for $\mathrm{Im}\varepsilon(\boldsymbol{k},\omega)$. In a classical limit $y \to 0$, $\zeta \to 0$ and $\zeta/y \to k_\perp^2 a_e^2$, where $a_e = v_{\mathrm{th}}/\Omega_e$ is the cyclotron radius of the electrons. Thus, in this limit (5.57) agrees with the imaginary part of the classical dielectric function (A.7) (neglecting the contribution of plasma ions) derived in Appendix A.

The real part of $\varepsilon(\boldsymbol{k},\omega)$ can be evaluated from the Kramers-Kronig relation (5.21). The straightforward calculations yield

$$\mathrm{Re}\,\varepsilon(\boldsymbol{k},\omega) = 1 + \frac{\sqrt{2}m\omega_p^2}{\hbar k^2\,|k_z|\,v_{\mathrm{th}}}\mathrm{e}^{-\zeta\coth(y)}\sum_{n=-\infty}^{\infty} I_n\left(\frac{\zeta}{\sinh(y)}\right) \times\left[\mathrm{e}^{-ny}\,\mathrm{Di}\left(\frac{p_+}{\sqrt{2}}\right) - \mathrm{e}^{ny}\,\mathrm{Di}\left(\frac{p_-}{\sqrt{2}}\right)\right], \tag{5.58}$$

where $p_\pm = (\Omega_\pm - n\Omega_e)\,/\,|k_z|\,v_{\mathrm{th}}$ and the Dawson integral $\mathrm{Di}\,(z)$ is defined by (4.4). Again, in the classical limit, $\hbar \to 0$, (5.58) agrees with the real part of (A.7) (ignoring in (A.7) the contribution of plasma ions).

Equations (5.57) and (5.58) are valid for arbitrary magnetic fields. In the presence of infinitely strong magnetic fields only the terms with $n = 0$ contribute in (5.57) and (5.58). In this limit the straightforward calculations yield

$$\varepsilon(\boldsymbol{k},\omega) = 1 + \frac{m\omega_p^2}{\hbar k^2}\left\{\frac{1}{\Omega_-}\left[g\left(\frac{\Omega_-}{k_z v_{\mathrm{th}}}\right) - 1\right] - \frac{1}{\Omega_+}\left[g\left(\frac{\Omega_+}{k_z v_{\mathrm{th}}}\right) - 1\right] + \mathrm{i}\left[\frac{1}{\Omega_-}f\left(\frac{\Omega_-}{|k_z|\,v_{\mathrm{th}}}\right) - \frac{1}{\Omega_+}f\left(\frac{\Omega_+}{|k_z|\,v_{\mathrm{th}}}\right)\right]\right\}. \tag{5.59}$$

Here $g(z)$ and $f(z)$ are the real and imaginary parts of the dispersion function of the classical plasma, respectively, (4.3) with (4.4). It is easy to check that in a classical limit $\hbar \to 0$ expression (5.59) yields the classical dielectric function in an infinitely strong magnetic field (see Sect. 4.3). Besides, in the limit of the vanishing temperature (fully degenerate limit), $v_{\mathrm{th}} \to 0$, the dielectric function (5.59) yields the expression (5.46) derived previously.

As we discussed in Sect. 5.2 the strong magnetic field increases the domain of classical behavior of the plasma. Therefore, one can expect that the dielectric function (5.59) can be found starting from the quantum expression (5.18) and assuming the infinitely large magnetic fields when only the Landau levels with $n = 0$ and $\sigma = -1/2$ are occupied. Since in this regime the chemical potential is large and negative (see Sect. 5.2.3) the distribution function is $f(E) \simeq \mathrm{e}^{(\mu-E)/k_{\mathrm{B}}T}$. Substituting this function into (5.18) we arrive after straightforward calculations at (5.59).

5.5 Stopping Power in a Magnetized Quantum Plasma

In this section we calculate on the basis of the theoretical background introduced in Sects. 5.2-5.4 the stopping power (SP) within the LR treatment using the dielectric function of the magnetized quantum plasma (5.18). We consider a heavy projectile ion with $\hbar\Omega_c \ll E_\perp$, where Ω_c and $E_\perp$ are the cyclotron frequency and the transverse energy of the ion, respectively. Consequently the classical motion of the ion with a constant velocity $\mathbf{v}_i$ on a rectilinear trajectory is justified. In this regime the SP may be computed from (2.53) with the dielectric function (5.18). We analyze the SP in dependence on the ion velocity at both low, $v_i \to 0$, and high velocities, $v_i \to \infty$, as well as the influence of the quantizing magnetic field. Note that the dielectric formalism employed in this Chapter is valid if the ion–electron relative velocity satisfies the inequality $\bar{v}_r \gg |Z|e^2/\hbar$. In a low–velocity limit the relative velocity $\bar{v}_r$ may be replaced by the velocity of the random motion of electrons v_0. In an unmagnetized plasma the above inequality requires large temperatures or densities in the case of nondegenerate and degenerate plasmas, respectively. With increasing magnetic field the velocity v_0 may be strongly reduced (see Sect. 5.2.2) and practically it is impossible to fullfill the condition above. Therefore in this case the LR treatment becomes questionable and the calculation of the low–velocity SP may have only a qualitative nature.

5.5.1 Low–Velocity Stopping Power in a Semiclassical Regime

First we consider the SP of a slow ion with $v_i \ll v_{\text{th}}$ in a semiclassical regime when the dielectric function is determined from (5.57) and (5.58). In a classical treatment the SP contains an anomalous term (see Sect. 4.4) which in the low–velocity limit behaves as $\propto v_i \ln(1/v_i)$. We show below that the quantum treatment does not remove the anomaly in the SP.

Let us introduce the Coulomb logarithm $\Lambda_Q(x, \alpha, \eta)$ which is determined by

$$S = \frac{Z^2 e^2 \omega_p^2}{v_{\text{th}}^2} \frac{1}{3} \sqrt{\frac{2}{\pi}} x \Lambda_Q(x, \alpha, \eta) . \tag{5.60}$$

Here α is the angle between $\mathbf{v}_i$ and $\mathbf{B}$, $x = v_i/v_{\text{th}}$, $\eta = \lambda_{\text{D}}/a_e = \Omega_e/\omega_p$, $\lambda\!\!\!^{-} = \hbar/mv_{\text{th}}$, $\varsigma = \lambda\!\!\!^{-}/2\lambda_{\text{D}} = \hbar\omega_p/2k_{\text{B}}T$, where a_e is the cyclotron radius of the electrons. In a low–velocity limit the Coulomb logarithm Λ_Q in (5.60) can be treated as a dimensionless friction coefficient. The evaluation of this quantity in a present case is similar to that adopted in Sect. 4.4. In the imaginary part of the dielectric function (5.57) we separate the terms with $n \neq 0$ and $n = 0$. As shown in Sect. 4.4 the latter is responsible for the anomalous SP. Note that in a semiclassical regime the ratio of the Landau energy to the electron thermal energy, $y = \hbar\Omega_e/2k_{\text{B}}T = \lambda\!\!\!^{-}/2a_e$, is an arbitrary parameter. For the evaluation of the low–velocity SP we distinguish two different cases.

(i) *Classically strong magnetic fields*, $\lambda\!\!\!^{-} \ll a_e \ll \lambda_{\text{D}}$. In this case there are basically three small parameters, x, $1/\eta$ and y. (Note that $\varsigma = y/\eta \ll 1$). Substituting expressions (5.57) and (5.58) into (2.53) in a leading–term approximation and in the low–velocity limit we obtain

$$\Lambda_Q(x,\alpha,\eta) = \frac{3}{8}\sin^2\alpha\left\{\left[\ln(2\eta^2)-\gamma-1\right]\left[\ln\left(\frac{8}{x^2\sin^2\alpha}\right)-\gamma-4\right]\right. \tag{5.61}$$
$$\left.+\frac{\pi^2-18}{6}+C_1\right\}+\frac{1}{4}\left[\ln\left(\frac{8}{\varsigma^4}\eta^2\right)-3\gamma-C_2\right],$$

where γ is the Euler's constant and $C_1 \simeq 0.074$, $C_2 \simeq 0.154$. For deriving (5.61) we have used the relation $\mathcal{K}(z) \simeq \ln(2/z) - \gamma - 1$ for small z, where the function $\mathcal{K}(z)$ is determined from (4.46). In (5.61) the terms proportional to the small parameters x, ς and y have been dropped. The obtained expression for the Coulomb logarithm is valid for $(m/m_i)^{1/3} < x \ll 1$, where m_i is the mass of the plasma ions. Under this condition the contribution of the plasma ions in the SP can be neglected [27, 83].

The different terms in (5.61) have the following origin. The anomalous $\propto \ln(1/x)$ and $\propto \ln(1/\eta)$ terms are the contributions from $n = 0$ in the dielectric function. The other terms result from the summation with respect to $n \geq 1$ in (5.57). Equation (5.61) can be compared with the SP derived in a classical case (Sect. 4.4, expressions (4.43) with (4.47)) neglecting the contribution of the plasma ions. An inspection shows that the both results agree with the choice $k_{\max} \simeq 2/\lambda\!\!\!^{-}$ of the cut–off parameter in a classical treatment [8].

(ii) *Quantum strong magnetic fields*, $a_e \ll \lambda\!\!\!^{-} \ll \lambda_{\mathrm{D}}$. This corresponds to the situation when the Landau energy is larger compared to the thermal energy, $y \gg 1$, and the transverse motion of the electrons, i.e. the terms in (5.57) and (5.58) with $n \neq 0$ can be neglected. In a leading–term approximation and in a limit $x \to 0$ we obtain

$$\Lambda_Q(x,\alpha,\eta) = \frac{3}{8}\sin^2\alpha\left\{\left[\ln\left(\frac{8}{x^2\sin^2\alpha}\right)-4-\gamma\right]\left[\ln\left(\frac{\eta}{\varsigma}\right)-\gamma-1\right]\right. \tag{5.62}$$
$$\left.-\frac{18-\pi^2}{6}\right\}+\frac{3}{4}\left[\ln\left(\frac{\eta}{\varsigma}\right)-\gamma\right].$$

In a low–velocity limit the Coulomb logarithm in a field–free case is $\propto \ln(1/\varsigma)$ (see, e.g., [55]). Therefore, at low velocities the Coulomb logarithms, (5.61) and (5.62), of the plasma increase as a result of an increasing magnetic field strength, in contrast to the high–velocity limit, where the SP decreases, see, e.g., Sects. 4.5.1 and 4.5.2. Besides, the SP with the Coulomb logarithms (5.61) and (5.62) increases with the angle α. In Chap. 4 we have alrady discussed a similar behavior of the classical SP.

Let us discuss the limit of the infinitely strong magnetic field in (5.62). In fact, this expression diverges logarithmically at $\eta \to \infty$. Actually there is a reason which limits the strength of the magnetic field which cannot assumed to be arbitrary large. Equations (5.61) and (5.62) are computed in a low–velocity limit with a strong magnetic field. The resulting SP depends on the order of the limits $v_i \to 0$ and $B \to \infty$. When one deals first with the small velocities the SP for a strong magnetic field $\eta \gtrsim 1$ is valid for $x \lesssim 1/\eta$. Note that this condition together with $a_e \lesssim \lambda\!\!\!^{-}$ restricts the ion velocity stronger, $x \lesssim \varsigma$. In a transition domain with $\eta \sim 1/x$ the Coulomb logarithm (5.62) behaves like $\propto \ln^2 x$ and the anomalous term becomes even more singular.

5.5.2 Stopping power in an Infinitely Strong Magnetic Field, Low–Velocity Limit

In an infinitely strong magnetic field the dielectric function is determined from (5.59). First we consider the SP for an ion motion with arbitrary velocity parallel to the magnetic field, $\alpha = 0$. The straightforward calculations yield

$$S_{\parallel} = \frac{4Z^2 e^2}{\pi \lambda^2} \int_0^{\infty} \left[\frac{\pi}{2} - \arctan \frac{G(s,x) + s^3/\kappa}{F(s,x)} \right] s \mathrm{d}s \,, \tag{5.63}$$

where $\kappa = \varsigma^2/2 = \lambda^2/8\lambda_{\mathrm{D}}^2$, $x = v_i/v_{\mathrm{th}}$, and

$$G(s,x) = \frac{1}{s+x}\left[1 - g(s+x)\right] + \frac{1}{s-x}\left[1 - g(s-x)\right] , \tag{5.64}$$

$$F(s,x) = \sqrt{\frac{\pi}{2}} \left[\mathrm{e}^{-(s-x)^2/2} - \mathrm{e}^{-(s+x)^2/2} \right] . \tag{5.65}$$

In equation (5.64) the function $g(x)$ is the real part of plasma dispersion function (see the first expression of (4.3)). In a classical limit with $\kappa \to 0$ the obtained SP can be compared with the results of Sect. 4.3, expressions (4.33) with (4.35), where Q_0 is determined from (4.6). We discuss this point at the end of this Section.

In a limit of large velocities, $x \gg 1$, equation (5.63) becomes

$$S_{\parallel} \simeq \frac{Z^2 e^2 \omega_p^2}{2 v_i^2} \tag{5.66}$$

in complete agreement with our previous results, see Sects. 4.3.2 and 4.5.2.

Consider now the low–velocity SP with arbitrary α. The inverse dielectric function is obtained from (5.59) which in the low–frequency limit reads

$$\mathrm{Im} \frac{-1}{\varepsilon(\boldsymbol{k},\omega)} \simeq \sqrt{\frac{\pi}{2}} \frac{k^2 \lambda_{\mathrm{D}}^2}{\left[k^2 \lambda_{\mathrm{D}}^2 + \mathcal{G}(k_z \lambda/2)\right]^2} \frac{\omega}{|k_z| v_{\mathrm{th}}} \exp\left(-\frac{\omega^2}{2 k_z^2 v_{\mathrm{th}}^2} - \frac{k_z^2 \lambda^2}{8} \right) . \tag{5.67}$$

Here $\mathcal{G}(z)$ is the statical screening function in a semiclassical regime

$$\mathcal{G}(z) = \frac{1 - g(z)}{z^2} \tag{5.68}$$

which determines the screening properties in a plasma $k_s^2 = k_{\mathrm{D}}^2 \mathcal{G}(k_z \lambda/2)$. Note that in the presence of an infinitely strong magnetic field the plasma is effectively one–dimensional and the screening length depends only on k_z [63]. Moreover, in (5.67) we again keep the exponential factor with the ω^2–term to avoid the logarithmical divergence of the Coulomb logarithm at low velocities which in this regime reads

$$\Lambda_Q(x,\alpha) = \frac{3}{2\pi} \int_0^{\pi} \mathrm{d}\phi \int_0^1 \frac{\mathrm{d}\mu}{\mu} \Phi^2(\mu,\phi,\alpha) \exp\left(-x^2 \frac{\Phi^2(\mu,\phi,\alpha)}{2\mu^2} \right) \Psi(\kappa\mu^2) . \tag{5.69}$$

Here $\Phi(\mu, \phi, \alpha) = \cos\Theta$, $\mu = \cos\beta$, Θ is the angle between the vectors $\boldsymbol{k}$ and $\boldsymbol{v}_i$. In a spherical coordinate system Θ is expressed by the angle α between $\boldsymbol{v}_i$ and $\boldsymbol{B}$ and the spherical coordinates β, ϕ of the vector $\boldsymbol{k}$ (see equation (2.61)). In (5.69) the function $\Psi(u)$ is

$$\Psi(u) = 2\int_0^{\infty} \frac{e^{-s^2/2} s^3 ds}{[s^2 + 2u\mathcal{G}(s)]^2} . \tag{5.70}$$

For an evaluation of the Coulomb logarithm (5.69) the behavior of the function $\Psi(u)$ at small u is important. Equation (5.70) shows that $\Psi(u) \simeq \ln(1/u) - 1 - \gamma$ at $u \to 0$, where γ is the Euler's constant. Due to this logarithmic singularity in $\Psi(u)$ the method of integration employed in Sects. 4.4 and 5.5.1 is not applicable here. Thus we have to deal with the integral

$$Q(x) = \int_0^1 \frac{d\mu}{\mu} \ln\mu \exp\left(-x^2 \frac{\wp_1^2(\mu)}{2\mu^2}\right) \wp_2(\mu) , \tag{5.71}$$

where $\wp_1(\mu)$ and $\wp_2(\mu)$ are some arbitrary functions with $\wp_1(0), \wp_2(0) \neq 0$. In a limit of small x ($x \ll 1$) the function $Q(x)$ behaves as

$$Q(x) \simeq \int_0^1 \frac{d\mu}{\mu} \ln\mu \left[\wp_2(\mu) - \wp_2(0)\right] - \frac{1}{8}\wp_2(0) \left\{\left[\ln\left(\frac{2}{x^2 \wp_1^2(0)}\right) - \gamma\right]^2 + \frac{\pi^2}{6}\right\} . \tag{5.72}$$

For a calculation of (5.69) we decompose the function Ψ into two parts with and without logarithmic singularity. The first part can be evaluated using (5.72). The second part is evaluated employing the method of Sects. 4.4 and 5.5.1. The straightforward calculations yield

$$\begin{aligned} \Lambda_Q(x, \alpha) = \frac{3}{4}\Bigg\{\psi_1(\kappa) + \frac{\sin^2\alpha}{4}\Bigg[\left(\ln\left(\frac{8}{x^2\sin^2\alpha}\right) - 1 - \gamma\right)^2 + \frac{\pi^2 - 18}{2} \\ + 2\left(\ln\frac{1}{\kappa} - 1 - \gamma\right)\left(\ln\left(\frac{8}{x^2\sin^2\alpha}\right) - 1 - \gamma\right) + 2\left(\psi_2(\kappa) - 3\psi_1(\kappa) + 3\right)\Bigg]\Bigg\} , \end{aligned} \tag{5.73}$$

where

$$\psi_1(\kappa) = 2\int_0^{\infty} \frac{e^{-s^2/2} s ds}{s^2 + 2\kappa\mathcal{G}(s)} , \tag{5.74}$$

$$\begin{aligned} \psi_2(\kappa) = \ln\kappa\left[\ln\kappa + 1 + \gamma + \Psi(\kappa)\right] + 2\int_0^{\infty} \frac{ds}{s}\Bigg\{\ln\frac{s^2 + 2\kappa}{s^2} \\ - e^{-s^2/2}\ln\frac{s^2 + 2\kappa\mathcal{G}(s)}{s^2} + 2\kappa\left[\frac{1}{s^2 + 2\kappa} - \frac{e^{-s^2/2}\mathcal{G}(s)}{s^2 + 2\kappa\mathcal{G}(s)}\right] \\ + 4\kappa\ln\kappa\left[\frac{e^{-s^2/2}\mathcal{G}(s)(s^2 + \kappa\mathcal{G}(s))}{(s^2 + 2\kappa\mathcal{G}(s))^2} - \frac{s^2 + \kappa}{(s^2 + 2\kappa)^2}\right]\Bigg\} . \end{aligned} \tag{5.75}$$

The Coulomb logarithm (5.73) can be further simplified for small and large κ.

(i) *Low–density, high–temperature plasma*, $\kappa \ll 1$. In this case $\Psi(\kappa) \simeq \ln(1/\kappa) - 1 - \gamma$, $\psi_1(\kappa) \simeq \ln(1/\kappa) - \gamma$ and $\psi_2(\kappa) = O(\kappa \ln^2 \kappa)$. Neglecting the function $\psi_2(\kappa)$ and substituting the values of $\Psi(\kappa)$ and $\psi_1(\kappa)$ into (5.73) we obtain

$$\Lambda_Q(x,\alpha) = \frac{3}{4}\left\{\ln\frac{1}{\kappa} - \gamma + \frac{\sin^2\alpha}{4}\left[\left(\ln\left(\frac{8}{x^2\sin^2\alpha}\right) - 1 - \gamma\right)^2 + \frac{\pi^2 - 18}{2} + 2\left(\ln\frac{1}{\kappa} - \gamma - 1\right)\left(\ln\left(\frac{8}{x^2\sin^2\alpha}\right) - 4 - \gamma\right)\right]\right\}. \tag{5.76}$$

(ii) *High–density, low–temperature plasma*, $\kappa \gg 1$. In this limit $\Psi(\kappa) \simeq C_3/\kappa^2$, $\psi_1(\kappa) \simeq C_2/\kappa - C_3/\kappa^2$ and

$$\psi_2(\kappa) \simeq \frac{1}{2}(\ln\kappa + \gamma + 1)^2 + \frac{\pi^2 - 2 + 2C_1}{4}, \tag{5.77}$$

where C_1, $C_2 = 13\pi/20$ and $C_3 \simeq 17.07$ are numerical constants with

$$C_1 = -4\int_0^\infty \frac{ds}{s} e^{-s^2/2} \ln G(s) \simeq 1.238\,. \tag{5.78}$$

Now we neglect in equation (5.73) the function $\Psi(\kappa)$ and the second term in $\psi_1(\kappa)$ proportional to κ^{-2}, and substitute the first term of $\psi_1(\kappa)$ and (5.77) into (5.73). The result reads

$$\Lambda_Q(x,\alpha) = \frac{3}{4}\left\{\frac{13\pi}{20\kappa} + \sin^2\alpha\left[\left(\frac{1}{2}\ln\left(\frac{8}{\kappa x^2\sin^2\alpha}\right) - 1 - \gamma\right)^2 + \frac{\pi^2 - 4 + C_1}{4}\right]\right\}. \tag{5.79}$$

Note that in (5.79) we keep the small term proportional to κ^{-1} because for ion parallel motion with $\alpha = 0$ only this term contributes to the Coulomb logarithm.

The regime (i) with $\hbar\omega_p \ll k_B T$ does not necessarily coincide with the classical situation considered in Sect. 4.3.1. This is because the Coulomb logarithm (5.76) is computed assuming that the ratio of the Landau energy to the thermal energy is infinitely large. This indicates that the Coulomb logarithm in the regime (i) still has quantum nature even in the high–temperature limit. The quantum origin of (5.76) for nonzero α is expressed by the velocity dependency of the SP which is now proportional to $\propto \ln^2 x$ in contrast to the anomalous term $\propto \ln x$ in a classical case (4.36). Thus the quantum effects intensify the anomaly in a low–velocity SP unless $\alpha = 0$. Moreover, for a parallel motion of the ion ($\alpha = 0$) the term $\ln^2 x$ in (5.76) vanishes and complete agreement with the classical result (4.36) can be achieved for $k_{\max} = (2mv_{\text{th}}/\hbar)C_0$ with $C_0 \simeq 1.75$. It is seen that this cut–off parameter differs from the Bethe value [118] by the numerical factor C_0.

In a second regime (ii) with $\hbar\omega_p \gg k_B T$ it is impossible to recover the fully degenerate situation with $T = 0$ smoothly since the Coulomb logarithm (5.79) for $\alpha \neq 0$ diverges logarithmically at large κ. This is because the SP depends on the order of the limits $x \to 0$ and $\kappa \to \infty$ which restricts the velocity of the ion $x \lesssim \kappa^{-1/2}$ in (5.79). The specific case of a nearly degenerate plasma is considered in the next Section.

5.5.3 Stopping power in a Strong Magnetic Field in the Nearly Degenerate Regime

The Fermi (see equation (5.34) and Fig. 5.1) and the thermal energies of electrons should not both vanish. Therefore, we assume a small but finite temperature (or, alternatively, a large but finite magnetic field) and show explicitly how the LR treatment of the SP breaks down at low velocities.

The most characteristic feature of the dielectric function (5.46) is the strong damping of the single–particle modes and the stopping power receives contribution only from the collective excitations. Since the effective plasma frequency in the approximation (5.46) is $\omega_p(k_z/k)$ the excitation of collective waves in a transverse direction is strongly restricted. For the SP we consider only two specific cases, namely parallel or transverse motion of the ion with respect to the magnetic field.

For parallel motion of the ion ($\alpha = 0$) we obtain from (2.53) and (5.46)

$$S_{\parallel} = \frac{2Z^2 e^2 \omega_p^2}{v_c^2} \begin{cases} x^2 \,, & x < 1 \\ \frac{1}{x^2 + \sqrt{x^4 - 1}} \,, & x \geq 1 \end{cases} \tag{5.80}$$

where $x = v_i/v_c$ with $v_c^2 = \hbar\omega_p/m$. In a low–velocity limit the SP (5.80) falls as x^2 unlike the linear behavior $\propto x$ considered in Chap. 4 and Sect. 5.5. In the opposite case of high velocities expression (5.80) vanishes as x^{-2} but the SP is by a factor of two larger than equation (4.76) as well as (5.66).

Similarly for a transverse motion of the ion ($\alpha = \pi/2$) we find

$$S_{\perp} = \frac{4Z^2 e^2 \omega_p^2}{\pi v_c^2 x} \int_0^{\mu_0(x)} \frac{\mu^2 d\mu}{\mathcal{D}_+(\mu) + \mathcal{D}_-(\mu)} \mathrm{K}\left(\mathcal{D}(\mu)\right) \,. \tag{5.81}$$

Here μ_0 is the solution of the equation $\mu_0^3 = x^2(1 - \mu_0^2)$, $\mathrm{K}(x)$ is the complete elliptic integral, and

$$\mathcal{D}_{\pm}(\mu) = \sqrt{x^2\left(1 - \mu^2\right) \pm \mu^3} \,, \qquad \mathcal{D}(\mu) = \frac{4\mathcal{D}_+(\mu)\mathcal{D}_-(\mu)}{\left[\mathcal{D}_+(\mu) + \mathcal{D}_-(\mu)\right]^2} \,. \tag{5.82}$$

We also provide the asymptotic solutions $\mu_0(x)$ which are important for treating the low– and high–velocity limits of the SP. At $x \ll 1$ and $x \gg 1$ this function behaves as $\mu_0(x) \simeq x^{2/3}(1 - x^{4/3}/3)$ and $\mu_0(x) \simeq 1 - 1/(2x^2)$, respectively. Using these asymptotic values the low– and high–velocity limits of (5.81) read

$$S_{\perp} = \frac{2Z^2 e^2 \omega_p^2}{3v_c^2} \,, \qquad S_{\perp} = \frac{Z^2 e^2 \omega_p^2}{2v_i^2} \ln\left(C_2 \frac{2mv_i^2}{\hbar\omega_p}\right) \,, \tag{5.83}$$

respectively, where $C_2 = 8\mathrm{e}^{-5/2} \simeq 0.66$. Clearly, the behavior of the SP (5.81) is unphysical at low velocities as it does not vanish for $v_i \to 0$. In the high–velocity limit the SP in (5.83) yields the Bethe result [118] apart from the numerical factors C_2 in a logarithmic function and $1/2$ in a front of the SP. As we discussed in Sect. 4.5.2 the factor $1/2$ appears due to the strong magnetic field. Moreover, the result in (5.83) agrees with the classical SP (4.81) if the classical cut–off parameter $k_{\max}$ is replaced there by $\simeq 2mv_i/\hbar$.

5.6 Binary Collision Treatment, Conformity Between BC and RLR

Consider now the electron-ion binary collisions (BC) in the presence of quantizing magnetic field. We consider an electron and a projectile heavy ion moving in a homogeneous magnetic field $\boldsymbol{B}$. We assume a mass of the ion $M \gg m$ such that a classical description of its motion with a rectilinear trajectory is applicable. The particles interact with the potential $-e\Phi_{\mathrm{ei}}(\boldsymbol{r}-\boldsymbol{v}_i t)$, where $\boldsymbol{r}$ and $\boldsymbol{v}_i t$ are the coordinates of the electron and ion, respectively. For charged particles the function $\Phi_{\mathrm{ei}}(\boldsymbol{r})$ can be expressed, for instance, by screened potential, $\Phi_{\mathrm{ei}}(\boldsymbol{r}) = Ze\exp(-r/\lambda)/4\pi\varepsilon_0 r$ (λ is the screening length) for application in plasmas. Note that Φ_{ei} is related to the potential $U(\boldsymbol{r}) = -Ze^2 u(\boldsymbol{r})$ introduced in Sect. 3.2 as $U(\boldsymbol{r}) = -e\Phi_{\mathrm{ei}}(\boldsymbol{r})$.

Our starting point is the Schrödinger equation

$$\left(\hat{H}_0 + \hat{H}_1(t)\right)\psi = \mathrm{i}\hbar\frac{\partial\psi}{\partial t} \tag{5.84}$$

with the time–dependent Hamiltonian $\hat{H}_1(t) = -e\Phi_{\mathrm{ei}}(\boldsymbol{r}-\boldsymbol{v}_i t)$ and the Hamiltonian of free electron $\hat{H}_0$ (5.11). We note that due to the spin operator $\hat{\sigma}_z$ commutes with the full Hamiltonian in (5.84), and the coefficient of $\hat{\sigma}_z$ in (5.11) is a constant, $\hat{\sigma}_z$ is conserved and the spin and coordinate variables in (5.84) are separable.

Alternatively, one can consider the scattering process with the stationary interaction Hamiltonian $\hat{H}_1 \sim \Phi_{\mathrm{ei}}(\boldsymbol{r})$, where $\boldsymbol{r}$ is the relative radius vector of the particles, such a program was carried out in [58, 100, 103]. There the energy of the particles is conserved. In contrast to previous work we use here the more realistic time–dependent Hamiltonian $\hat{H}(t) = \hat{H}_0 + \hat{H}_1(t)$. Thus, after the interaction the electron is in a superposition state of the eigenvalues of $\hat{H}_0$ (see below) and the energy of an electron has not a definite value since its energy is not conserved [95, 96].

We seek an approximate solution of (5.84) in which the interaction potential is considered as a perturbation. We start with the zero–order unperturbed eigenstates which are described by the zero–order Schrödinger equation

$$\hat{H}_0\psi_\alpha^{(0)} = \mathrm{i}\hbar\frac{\partial\psi_\alpha^{(0)}}{\partial t}, \tag{5.85}$$

where the corresponding time–dependent eigenstates contain an exponential factor, $\psi_\alpha^{(0)}(\boldsymbol{r},t) = \psi_\alpha^{(0)}(\boldsymbol{r})\exp(-\mathrm{i}\Omega_\alpha t)$ with $\Omega_\alpha = E_\alpha/\hbar$ and E_α from (5.13). The stationary wave functions $\psi_\alpha^{(0)}(\boldsymbol{r})$ are given by (5.12) and are supplied by an additional upper index to indicate the order.

We now seek the solution of the perturbed equation (5.84) in the form

$$\psi_\alpha(\boldsymbol{r},t) = \sum_\beta a_{\alpha\beta}(t)\psi_\beta^{(0)}(\boldsymbol{r},t), \tag{5.86}$$

where the expansion coefficients are a function of time and $\beta = \{n', \sigma', q'_y, q'_z\}$. Substituting (5.86) into (5.84), and recalling that the function $\psi_\beta^{(0)}(\boldsymbol{r},t)$ satisfies the equation (5.85) we obtain

$$\dot{a}_{\alpha\beta}(t) = -\frac{\mathrm{i}}{\hbar}\sum_{\gamma} h_{\beta\gamma}(t) a_{\alpha\gamma}(t)\,, \tag{5.87}$$

where

$$h_{\beta\gamma}(t) = \mathrm{e}^{\mathrm{i}\Omega_{\beta\gamma}t}\int \mathrm{d}\boldsymbol{r}\psi_{\beta}^{(0)*}(\boldsymbol{r})\,\hat{H}_1(t)\,\psi_{\gamma}^{(0)}(\boldsymbol{r}) \tag{5.88}$$

are the matrix elements of the perturbation, including the time factor. Here $\hbar\Omega_{\beta\gamma} = E_\beta - E_\gamma$.

The expression (5.87) is an exact equation for the expansion coefficients $a_{\alpha\beta}(t)$. Since we take the wave function of the αth stationary state as unperturbed wave function in (5.86), we find in zero-order (free particles) $a_{\alpha\beta}^{(0)}(t) = \delta_{\alpha\beta}$. The equation for the first-order correction is then given by $\dot{a}_{\alpha\beta}^{(1)}(t) = -(\mathrm{i}/\hbar)h_{\beta\alpha}(t)$. Assuming that all corrections vanish at $t \to -\infty$ and Fourier transforming the ion–electron interaction potential we find from (5.86) for the first-order electron wave function

$$\psi_{\alpha}^{(1)}(\boldsymbol{r},t) = \frac{e}{\hbar}\int \mathrm{d}\boldsymbol{k}\Phi_{\mathrm{ei}}(\boldsymbol{k})\sum_{\beta}\psi_{\beta}^{(0)}(\boldsymbol{r},t)\,S_{\beta\alpha}(\boldsymbol{k})\,\frac{\mathrm{e}^{\mathrm{i}(\Omega_{\beta\alpha}-\omega)t}}{\Omega_{\beta\alpha}-\omega-\mathrm{i}0}\,, \tag{5.89}$$

where $\omega = \boldsymbol{k}\cdot\boldsymbol{v}_i$, and the matrix $S_{\beta\alpha}(\boldsymbol{k})$ is given by (5.4) and (5.14). At large $t \to \infty$ the interaction force between the particles vanishes. Thus after the interaction the electron at large t is described by a superposition of stationary zero-order wave functions with the constant expansion coefficients. The first-order correction to the stationary state $\psi_{\alpha}^{(0)}$ can be found from (5.89). At large t from (5.89) we find $\psi_{\alpha}^{(1)}(\boldsymbol{r},t) = \sum_{\beta}\mathcal{T}_{\alpha\beta}\psi_{\beta}^{(0)}(\boldsymbol{r},t)$, where

$$\mathcal{T}_{\alpha\beta} = a_{\alpha\beta}^{(1)}(\infty) = \frac{2\pi\mathrm{i}e}{\hbar}\int \mathrm{d}\boldsymbol{k}\Phi_{\mathrm{ei}}(\boldsymbol{k})\,S_{\beta\alpha}(\boldsymbol{k})\,\delta(\omega-\Omega_{\beta\alpha})\,. \tag{5.90}$$

The quantities $\mathcal{T}_{\alpha\beta}$ give the first–order correction to the eigenstates $\psi_{\alpha}^{(0)}(\boldsymbol{r},t)$ after completion of an electron–ion collision. We note that the diagonal elements of this matrix are $\mathcal{T}_{\alpha\alpha} \sim \Phi_{\mathrm{ei}}(0)$. Therefore for the bare Coulomb interaction potential a lower cut–off k_{min} must be introduced in (5.89) and (5.90) to avoid the divergence at small k. Here k_{min} must account for screening with $k_{\mathrm{min}} \sim \lambda^{-1}$.

Since the energy of an electron during its interaction with the ion is not conserved the Fermi golden rule is not applicable for the calculation of the energy transfer after a collision between an electron and a nonstationary ion. Therefore we introduce the probability current density of the electron in the state $|\alpha\rangle$ [73] for the calculation of the expected energy transfer

$$\boldsymbol{j}_{\alpha}(\boldsymbol{r},t) = \frac{\mathrm{i}\hbar}{2m}\left(\psi_\alpha\nabla\psi_\alpha^* - \psi_\alpha^*\nabla\psi_\alpha\right) + \frac{e\boldsymbol{A}}{m}\,|\psi_\alpha|^2 \tag{5.91}$$

with the corresponding probability density $\rho_\alpha = |\psi_\alpha|^2$. Then we define the energy transfer of an electron which is initially in the αth Lanadau state as

$$\Delta E_\alpha = -e\int_{-\infty}^{\infty}\mathrm{d}t\int \mathrm{d}\boldsymbol{r}\boldsymbol{j}_\alpha(\boldsymbol{r},t)\cdot\boldsymbol{E}_{\mathrm{ext}}(\boldsymbol{r},t)\,, \tag{5.92}$$

where $\boldsymbol{E}_{\text{ext}}(\boldsymbol{r},t) = -\nabla\Phi_{\text{ei}}(\boldsymbol{r}-\boldsymbol{v}_i t)$ is the electrical field created by a moving ion. Here $-e\boldsymbol{j}_\alpha(\boldsymbol{r},t)$ is the electrical current for an electron in αth state.

The energy loss of the ion per unit length, i.e. the stopping power, in a homogeneous electron plasma is obtained by averaging the energy transfer of an electron per unit time over the unperturbed electron distribution function

$$S_{\text{BC}} = \frac{1}{v_i}\int \mathrm{d}\boldsymbol{r}\boldsymbol{J}(\boldsymbol{r},t)\cdot\boldsymbol{E}_{\text{ext}}(\boldsymbol{r},t) . \tag{5.93}$$

Here we have introduced the averaged electrical current and electrical charge density of the electrons

$$\boldsymbol{J}(\boldsymbol{r},t) = -e\sum_\alpha f(E_\alpha)\boldsymbol{j}_\alpha(\boldsymbol{r},t) , \tag{5.94}$$

$$\rho(\boldsymbol{r},t) = -e\sum_\alpha f(E_\alpha)\rho_\alpha(\boldsymbol{r},t) . \tag{5.95}$$

Let us note that (5.93) appears to differ from the general definition of the stopping power (see, e.g., [134, 136]) in terms of the density matrix. To show the identity of both treatments we consider the relation $\boldsymbol{J}\cdot\boldsymbol{E}_{\text{ext}} = \Phi_{\text{ei}}\nabla\cdot\boldsymbol{J} - \nabla\cdot(\Phi_{\text{ei}}\boldsymbol{J})$. The last term of this equation vanishes after transformation of the volume integral into the surface one according to the Gauss theorem. Using the continuity equation for the probability current and density the first term can be rewritten as

$$\rho\frac{\partial\Phi_{\text{ei}}}{\partial t} - \frac{\partial}{\partial t}(\rho\Phi_{\text{ei}}) = \rho\boldsymbol{v}_i\cdot\boldsymbol{E}_{\text{ext}} - \frac{\partial}{\partial t}(\rho\Phi_{\text{ei}}) . \tag{5.96}$$

The last term in (5.96) is the time derivative of the function $\rho\Phi_{\text{ei}}$ which represents the density of the plasma potential energy $U_{\text{p}}(\boldsymbol{r},t)$ in the electrical field of the external ion. However, the total potential energy $\int U_{\text{p}}(\boldsymbol{r},t)d\boldsymbol{r}$ should be a constant for a homogeneous and infinite plasma and can be dropped from further consideration. Substituting the first term of (5.96) into (5.93) we arrive at

$$S_{\text{BC}} = \frac{\boldsymbol{v}_i}{v_i}\cdot\int \mathrm{d}\boldsymbol{r}\rho(\boldsymbol{r},t)\boldsymbol{E}_{\text{ext}}(\boldsymbol{r},t) = \frac{\boldsymbol{v}_i}{v_i}\cdot\boldsymbol{F}_{\text{tot}} , \tag{5.97}$$

where $\boldsymbol{F}_{\text{tot}} = -\boldsymbol{\mathcal{F}}$ is the total averaged electrical force acting on the plasma. In (5.97) the charge density $\rho(\boldsymbol{r},t)$ represents the eigenvalues of the density matrix operator considered in [134, 136] which guarantees the identity of both treatments.

Similarly one can prove that the energy transfer (5.92) is expressed through the charge density of an electron in the αth state

$$\Delta E_\alpha = -e\int_{-\infty}^{\infty} dt\int \mathrm{d}\boldsymbol{r}\rho_\alpha(\boldsymbol{r},t)\left[\boldsymbol{v}_i\cdot\boldsymbol{E}_{\text{ext}}(\boldsymbol{r},t)\right] . \tag{5.98}$$

Physically the identities of (5.92) with (5.98) and (5.93) with (5.97) are the consequence of the conservation of the total energy of the macroscopic system.

Below we evaluate the general expressions (5.97) and (5.98) within first- and second–order perturbation theory. The first–order energy transfer is proportional to the zero–order probability current density or to the zero–order probability density $\rho_\alpha^{(0)} = |\psi_\alpha^{(0)}|^2$. Using the Fourier representation of the interaction potential the first–order energy transfer reads

$$\Delta E_\alpha = 2\pi \mathrm{i}e \int \mathrm{d}\boldsymbol{k}\Phi_{\mathrm{ei}}(\boldsymbol{k})(\boldsymbol{k}\cdot\boldsymbol{v}_i)\delta(\boldsymbol{k}\cdot\boldsymbol{v}_i)S_{\alpha\alpha}(\boldsymbol{k}) = 0\,. \tag{5.99}$$

Similarly, the first–order SP is proportional to $\rho^{(0)}(\boldsymbol{r},t)$ which is related to the probability density $\rho_\alpha^{(0)}(\boldsymbol{r},t)$ according to (5.95). From (5.97) we find $S_{\mathrm{BC}}^{(1)} = 0$. Then as for classical description, see Sect. 3.5, the first–order energy transfer and SP vanishes and the energy transfer and ion energy loss receive contributions only from higher orders.

Consider now the second–order energy transfer and energy loss which are proportional to the first–order charge density, $\rho_\alpha^{(1)}(\boldsymbol{r},t)$ and $\rho^{(1)}(\boldsymbol{r},t)$, respectively. Here $\rho^{(1)}(\boldsymbol{r},t)$ is related to $\rho_\alpha^{(1)}(\boldsymbol{r},t)$ according to (5.95) and the quantity $\rho_\alpha^{(1)}(\boldsymbol{r},t)$ is given by

$$\rho_\alpha^{(1)}(\boldsymbol{r},t) = \psi_\alpha^{(0)}(\boldsymbol{r},t)\psi_\alpha^{(1)*}(\boldsymbol{r},t) + \psi_\alpha^{(0)*}(\boldsymbol{r},t)\psi_\alpha^{(1)}(\boldsymbol{r},t)\,. \tag{5.100}$$

Using (5.89) and Fourier representation of the interaction potential $\rho_\alpha^{(1)}(\boldsymbol{r},t)$ and $\rho^{(1)}(\boldsymbol{r},t)$ can be written as

$$\rho_\alpha^{(1)}(\boldsymbol{r},t) = \frac{e}{\hbar}\int \mathrm{d}\boldsymbol{k}\Phi_{\mathrm{ei}}(\boldsymbol{k})\mathrm{e}^{-\mathrm{i}\omega t}\sum_\beta\left[\frac{P_{\beta\alpha}(\boldsymbol{r})S_{\alpha\beta}(\boldsymbol{k})}{\Omega_{\beta\alpha}+\omega+\mathrm{i}0} + \frac{P_{\alpha\beta}(\boldsymbol{r})S_{\beta\alpha}(\boldsymbol{k})}{\Omega_{\beta\alpha}-\omega-\mathrm{i}0}\right], \tag{5.101}$$

$$\rho^{(1)}(\boldsymbol{r},t) = -\frac{e^2}{\hbar}\int \mathrm{d}\boldsymbol{k}\Phi_{\mathrm{ei}}(\boldsymbol{k})\mathrm{e}^{-\mathrm{i}\omega t}\sum_{\alpha,\beta}P_{\alpha\beta}(\boldsymbol{r})S_{\beta\alpha}(\boldsymbol{k})\frac{f(E_\alpha)-f(E_\beta)}{\Omega_{\beta\alpha}-\omega-\mathrm{i}0}, \tag{5.102}$$

where we have introduced the matrix $P_{\alpha\beta}(\boldsymbol{r}) = \psi_\alpha^{(0)*}(\boldsymbol{r})\psi_\beta^{(0)}(\boldsymbol{r})$. Using the relation $S_{\beta\alpha}(\boldsymbol{k}) = S_{\alpha\beta}^*(-\boldsymbol{k})$ and substituting (5.101) and (5.102) into (5.98) and (5.97), respectively, and after straightforward calculations one arrives at the following expressions for the second–order energy transfer and stopping power

$$\Delta E_\alpha = -\frac{(2\pi)^2e^2}{\hbar}\int \mathrm{d}\boldsymbol{k}\mathrm{d}\boldsymbol{k}'\Phi_{\mathrm{ei}}(\boldsymbol{k})\Phi_{\mathrm{ei}}(\boldsymbol{k}')(\boldsymbol{k}\cdot\boldsymbol{v}_i)\delta(\omega+\omega') \times \sum_\beta S_{\alpha\beta}(\boldsymbol{k})S_{\beta\alpha}(\boldsymbol{k}')\delta(\Omega_{\alpha\beta}-\omega)\,, \tag{5.103}$$

$$S_{\mathrm{BC}}^{(2)} = -\frac{\mathrm{i}e^2}{\hbar v_i}\int \mathrm{d}\boldsymbol{k}\mathrm{d}\boldsymbol{k}'\Phi_{\mathrm{ei}}(\boldsymbol{k})\Phi_{\mathrm{ei}}^*(\boldsymbol{k}')(\boldsymbol{k}'\cdot\boldsymbol{v}_i)\mathrm{e}^{\mathrm{i}(\omega'-\omega)t} \times \sum_{\alpha,\beta}S_{\beta\alpha}(\boldsymbol{k})S_{\beta\alpha}^*(\boldsymbol{k}')\frac{f(E_\alpha)-f(E_\beta)}{\Omega_{\beta\alpha}-\omega-\mathrm{i}0}, \tag{5.104}$$

where $\omega' = \boldsymbol{k}'\cdot\boldsymbol{v}_i$. Finally using the expression (5.16) and the relation between screened potentials $\Phi_{\mathrm{ei}}(\boldsymbol{k}) = (Ze/4\pi\varepsilon_0)u(\boldsymbol{k})$ we obtain $S_{\mathrm{BC}}^{(2)} = S_{\mathrm{RLR}}$, where S_{RLR} is the RLR stopping power, (4.93) with the susceptibility $\chi^{(0)}(\boldsymbol{k},\omega)$ deduced from the dielectric function of the quantum magnetized plasma, (5.18). This shows that the complete conformity is established only between the BC and RLR approaches since there is no collectivity except screening in either the BC or reduced version of LR treatment. In Sect. 4.7 the same result has been obtained for classical plasmas. However for the latter case the conformity is established only for the regularized interaction potential, which decays faster than r^{-1} at large distances and increases slower than r^{-1} at small ones. Such a regularized potential can be viewed as an alternative implementation of cut–offs. Due to the wave nature of the electrons the quantum mechanical description does not require a cut–off procedure at small distances and hence the conformity between BC and reduced LR approaches is valid for any interaction potential which decays faster than r^{-1}.

Equation (5.103) is the quantum mechanical analogy of the electron second–order energy transfer. Because of conservation of the total energy the ion energy transfer is $-\Delta E_\alpha$. Note that (5.103) was obtained for any $\psi_\alpha^{(0)}(\boldsymbol{r})$ and, therefore, can be applied for different bases of the unperturbed wave functions.

5.7 Correspondence Between Quantum and Classical BC Treatments

In this Section we consider the second–order energy transfer of an electron in αth Landau state (5.103) when the unperturbed Landau levels are represented either in a cartesian (CAR) or a cylindrical (CYL) basis. Within the CAR approach the electron state α is given by the parameters $\alpha = \{n, \sigma, q_y, q_z\}$ [73], $n = 0, 1, 2, \ldots$ denotes the Landau levels, q_z and q_y are the electron momentum components parallel and perpendicular to the external magnetic field, respectively. In this geometry the Schrödinger equation is reduced to the 1D equation for a shifted harmonic oscillator and the corresponding wave functions and eigenvalues are given by (5.12) and (5.13), respectively. In CYL the electron state $\alpha = \{\nu, n, \sigma, q_z\}$ ($\nu = 0, \pm 1, \pm 2 \ldots$ is the angular momentum) and corresponding wave functions differ from CAR [70, 73, 123]. The momentum component q_y in CAR is replaced here by angular momentum ν. Because of the different electron states α in both representations the BC treatment will describe different elementary collision processes which finally results in different energy transfers. We investigate the classical limit of these quantum mechanical expressions obtained from CAR and CYL and compare them with classical results. As the main outcome we will show that using CYL allows a connection to the classical energy transfer as a function of the impact parameter. The use of cylindrical coordinates is hence much more appropriate for the present scattering problem, as it makes the transition to the classical description much more transparent.

5.7.1 Cartesian Basis

In a cartesian basis the vector potential of the magnetic field $\boldsymbol{A}$ has been defined in Sect. 5.1. The eigenstates are labeled by $\alpha = \{n, \sigma, q_y, q_z\}$, and the unperturbed electron wave function in the Landau state α is represented by (5.12) (see [73]). Within the CAR treatment the eigenvalues of the free particles E_α and the matrix $S_{\beta\alpha}(\boldsymbol{k})$ are given by (5.13) and (5.14) with (5.15), respectively. Substituting (5.14) into (5.103) we obtain within CAR

$$\Delta E_\alpha^{\mathrm{CAR}} = \frac{(2\pi)^4 e^2}{\hbar |v_{ix}| L^2} \int \mathrm{d}\boldsymbol{k}\, |\Phi_{\mathrm{ei}}(\boldsymbol{k})|^2 (\boldsymbol{k}\cdot \boldsymbol{v}_i) \sum_{l=-n}^{\infty} F_{n;l+n}^2(\zeta)\, \delta\left(\zeta_l(\boldsymbol{k}) + \frac{\hbar k_z^2}{2m}\right). \tag{5.105}$$

Here v_{ix} is the component of ion velocity along the x-axis, $\zeta_l(\boldsymbol{k}) = l\Omega_e + \boldsymbol{k}\cdot\bar{\boldsymbol{v}}_r$, $\bar{\boldsymbol{v}}_r = v_{e\|}\boldsymbol{b} - \boldsymbol{v}_i$ ($v_{e\|} = \hbar q_z/m$ is the electron unperturbed classical velocity component parallel to $\boldsymbol{b}$) is the classical relative velocity of the guiding center of the electron helical motion with respect to the ion.

Now we consider the classical limit of (5.105). The classical regime can be realized by the two limits $\lambda_B^2 = \hbar/m\Omega_e \to 0$ ($\zeta \to 0$) and $n = E_\perp/\hbar\Omega_e \to \infty$ (large quantum numbers), where $E_\perp = mv_{e\perp}^2/2$ is the electron classical energy perpendicular to the magnetic field with $v_{e\perp} = \hbar q_y/m$. We also note that in this limit $n\zeta \to k_\perp^2 a^2/4$, where $a = v_{e\perp}/\Omega_e$ is the electron cyclotron radius. The limit of the function $F_{n;l+n}^2(\zeta)$ at $\zeta \to 0$, $n \to \infty$ and $n\zeta \to k_\perp^2 a^2/4$ as found in Appendix D is

$$F_{n;l+n}^2(\zeta)\Big|_{\lambda_B\to 0; n\to\infty} = J_l^2(k_\perp a) + \zeta Q_l(k_\perp a) + \mathrm{O}(\zeta^2). \tag{5.106}$$

The first term in (5.106) gives the full classical limit. The second term gives a quantum correction. Also we should expand the δ–function in (5.105) for small $\hbar k_z^2/2m$. We note that the first term of this expansion in the summand of (5.105) which is proportional to $J_l^2(k_\perp a)\,\delta(\zeta_l(\boldsymbol{k}))$ vanishes due to the antisymmetrical behavior of this expression with respect to the changes $l \to -l$ and $\boldsymbol{k} \to -\boldsymbol{k}$. The other nonvanishing terms contribute to the electron energy transfer which leads to

$$\begin{aligned} -\Delta E_\alpha^{\mathrm{CAR}} \to \frac{(2\pi)^4 e^2}{2m|v_{ix}|L^2} \int \mathrm{d}\boldsymbol{k}\, |\Phi_{\mathrm{ei}}(\boldsymbol{k})|^2 (\boldsymbol{k}\cdot\boldsymbol{v}_i) \sum_{l=-\infty}^{+\infty} \delta(\zeta_l(\boldsymbol{k})) \\ \times \left\{ \frac{(\boldsymbol{k}\cdot\boldsymbol{b})^2}{\zeta_l(\boldsymbol{k})} J_l^2(k_\perp a) + \frac{k_\perp^2}{2\Omega_e}\left[J_{l+1}^2(k_\perp a) - J_{l-1}^2(k_\perp a)\right]\right\}. \end{aligned} \tag{5.107}$$

The last expression is the classical limit of the quantum mechanical CAR energy transfer. In this case the classical energy transfer depends on $v_{e\perp}$, $v_{e\|}$ and $\boldsymbol{v}_i$. Note, however, that in a quantum treatment an electron forms a wave–packet with the amplitude $\propto F_{n;l+n}^2(\zeta)$ in $\boldsymbol{k}$–space (see, e.g., expression (5.105)) which decays exponentially on a scale λ_B. And this guarantees the convergence of the $\boldsymbol{k}$–integral in (5.105) at large $\boldsymbol{k}$ or at short distances (quantum diffraction). In a classical limit the quantum amplitude $F_{n;l+n}^2(\zeta)$ is transformed into $J_l^2(k_\perp a)$ with an infinite range $\propto k^{-1}$ according to (5.106). Obviously this aggravates the convergence of the $\boldsymbol{k}$–integral in (5.107) and one can expect that the classical limit (5.107) exhibits singularities at large $\boldsymbol{k}$ which can be avoided by introducing an upper cut–off parameter in (5.107) as discussed in Chaps. 2 and 3.

As shown in (5.105) the quantum CAR energy transfer behaves as $\sim 1/|v_{ix}|$ and diverges in a limit $v_{ix} \to 0$. This is a consequence of the perturbative treatment in a

CAR geometry. In fact for an ion moving in the yz–plane with $v_{ix} = 0$ the contact time with an electron 1D wave–packet moving parallel to the same plane becomes infinitely large. In the next section we show the similar behavior for the CYL energy transfer which also diverges in a limit $v_{i\perp} \to 0$.

As an illustration consider now the expression (5.105) in the presence of an infinitely strong magnetic field when only the term with $l = 0$ (i.e. $n' = n$) contributes to the CAR energy transfer. Since $\lambda_B \to 0$ in this limit we obtain from (5.105)

$$\Delta E_\alpha^{\mathrm{CAR}} = \frac{(2\pi)^4 e^2}{\hbar |v_{ix}| L^2} \int \mathrm{d}\boldsymbol{k}\, |\Phi_{\mathrm{ei}}(\boldsymbol{k})|^2 \, (\boldsymbol{k} \cdot \boldsymbol{v}_i) \delta\left(\boldsymbol{k} \cdot \bar{\boldsymbol{v}}_r + \frac{\hbar k_z^2}{2m}\right), \tag{5.108}$$

where the electron state α is fixed by only one parameter $q_z = m v_{e\|}/\hbar$. In the classical limit ($\hbar \to 0$) we must expand (5.108) with respect to $\hbar$. The first term of the expansion of δ–function vanishes for symmetry reasons. Then the second term for the quantity $-(|v_{ix}|/\bar{v}_r)\Delta E_\alpha^{\mathrm{CAR}}$ yields a result which is precisely the classical second order energy transfer $\left\langle \Delta E_i^{(2)} \right\rangle_{\mathrm{I0}}$, expression (3.138) with (3.54), of an ion in an infinitely strong magnetic field integrated with respect to the impact parameters $\bar{s}\mathrm{d}\bar{s}$ in a full space and multiplied by a factor $(2\pi/L^2)$. In Sect. 5.8 we derive the similar relation for any finite magnetic field.

Consider some particular examples of the energy transfer (5.108). For an ion motion parallel to the magnetic field with $v_{i\perp} = 0$ and for the spherical–symmetric interaction potential from (5.108) we obtain

$$-\frac{|v_{ix}|}{|\bar{v}_{r\|}|} \Delta E_\alpha^{\mathrm{CAR}} \to \frac{2\,(2\pi)^5\, m e^2 v_{i\|}}{\hbar^2 L^2 \bar{v}_{r\|}} \int_{k_m}^{\infty} \Phi_{\mathrm{ei}}^2(k) k \mathrm{d}k\,, \tag{5.109}$$

where $k_m = 2m|\bar{v}_{r\|}|/\hbar$ and $v_{i\|} = v_{iz}$ is the ion velocity parallel to $\boldsymbol{b}$. In particular, for the screened interaction (3.2) (we remind the reader that $U(\boldsymbol{r}) = -e\Phi_{\mathrm{ei}}(\boldsymbol{r})$)

$$-\frac{|v_{ix}|}{|\bar{v}_{r\|}|} \Delta E_\alpha^{\mathrm{CAR}} = \frac{2\pi Z^2 e^4}{m L^2} \frac{v_{i\|}}{\bar{v}_{r\|}\left(\bar{v}_{r\|}^2 + v_\lambda^2\right)}\,. \tag{5.110}$$

Here $v_\lambda = \hbar/2m\lambda$ and λ is the screening length. In a classical limit or for the unscreened Coulomb interaction with $\lambda \to \infty$ equation (5.110) yields the same result which reads

$$-\frac{|v_{ix}|}{|\bar{v}_{r\|}|} \Delta E_\alpha^{\mathrm{CAR}} = \frac{2\pi Z^2 e^4}{m L^2} \frac{v_{i\|}}{\bar{v}_{r\|}^3}\,. \tag{5.111}$$

One can make some interesting remarks about expressions (5.109)-(5.111). Although the classical energy transfers (3.143) and (3.138) for the Coulomb and screened interactions, respectively, vanish at $v_{i\perp} = 0$ their quantum counterparts (5.111) and (5.110) yield finite results. Apparently, in this case the vanishing classical energy transfers are compensated by an infinite range of integration in $\bar{s}$–space, see Sect. 5.8. From (5.109) it is straightforward to see that for any interaction potential which decays for $k \to \infty$ faster[2] than the Coulomb (3.1) and screened (3.2)

---

[2] In a real space this corresponds to the potential which is less singular at the origin $r \to 0$ than the potentials (3.1) and (3.2)

potentials the quantum energy transfer (5.109) vanishes in a classical limit. For instance, this can be easily checked using the regularized potential (3.3).

Let us assume now that the electron wave packet propagates in y–direction ($v_{e\|} = 0$). In this case for an ion transverse motion with $v_{iy} = v_{i\|} = 0$ and for the screened potential (3.2) expression (5.108) yields

$$\Delta E_{\alpha}^{\mathrm{CAR}} = \frac{4\sqrt{2}Z^2 e^4 \lambda}{\hbar v_i L^2} \int_0^{\infty} \frac{x\mathrm{d}x}{\sqrt{x^2+1}\left(\sqrt{x^2+1}+1\right)^{3/2}} \times \frac{1}{(1+\ell x)^{3/2}} \left[(2\ell x+1)\ln\left(\sqrt{\ell x+1}+\sqrt{\ell x}\right) - \sqrt{\ell x(\ell x+1)}\right], \tag{5.112}$$

where $v_i = |v_{ix}|$, $\ell = (v_i/2v_\lambda)^2$. In the large ($v_i \gg v_\lambda$) and small ($v_i \ll v_\lambda$) velocity limits this energy transfer becomes

$$\Delta E_{\alpha}^{\mathrm{CAR}} = \frac{16\sqrt{2}Z^2 e^4}{3mv_i^2 L^2} \left[\ln\left(\frac{v_i}{2v_\lambda}\right) + \frac{3\pi-2}{12}\right], \tag{5.113}$$

$$\Delta E_{\alpha}^{\mathrm{CAR}} = \frac{2\sqrt{2}Z^2 e^4}{mv_\lambda^2 L^2}, \tag{5.114}$$

respectively. As seen from (5.114) the energy transfer of an ion in a low–velocity limit is a nonvanishing constant. Thus the second order CAR quantum perturbation theory predicts in a limit $v_i \to 0$ an unphysical result for the energy transfer of an ion moving transverse to the magnetic field, the first relation cf. (5.83).

5.7.2 Cylindrical Basis

We now switch to another quantum mechanical formulation for the electronic states in the external magnetic field. Because of symmetry the Schrödinger equation for the free particle and in the presence of external magnetic field may also be solved in cylindrical coordinates. We consider an electron with given angular momentum ν ($\nu = 0, \pm 1, \pm 2 ...$) and momentum q_z along the magnetic field. If we choose the axial symmetric vector potential in the form $\boldsymbol{A} = (1/2)\,[\boldsymbol{B} \times \boldsymbol{r}] = (B/2)\,(x\hat{\boldsymbol{y}} - y\hat{\boldsymbol{x}})$ or $A_\varphi = B\rho/2$, $A_z = A_\rho = 0$ then the unperturbed Schrödinger equation has the following solutions [70, 73, 123]

$$\psi_{\alpha}^{(0)}(\boldsymbol{r}) = \frac{1}{(2\pi L)^{1/2}\lambda_B} \left[\frac{n!}{2^{|\nu|}\,(|\nu|+n)!}\right]^{1/2} u_\sigma \mathrm{e}^{\mathrm{i}\nu\varphi} \mathrm{e}^{\mathrm{i}q_z z}\,(2\xi)^{|\nu|/2}\,\mathrm{e}^{-\xi/2} L_n^{|\nu|}(\xi), \tag{5.115}$$

where $L_n^{|\nu|}$ are the generalized Laguerre polynomials. Here $\alpha = \{\nu, n, \sigma, q_z\}$, $\xi = \rho^2/2\lambda_B^2$, $n = 0, 1, 2...$, ρ, φ and z are the cylindrical coordinates. The energy of the free electrons in the αth CYL state is given by

$$E_\alpha = E_{n+\eta_\nu;\sigma}(q_z) = \frac{\hbar^2 q_z^2}{2m} + \hbar\Omega_e\left(n + \sigma + \eta_\nu + \frac{1}{2}\right) \tag{5.116}$$

with $\eta_\nu = (\nu + |\nu|)/2$, $\mu_\nu = (|\nu| - \nu)/2$. For $\nu \le 0$ the electron energy is the same as $E_{n\sigma}(q_z)$ in the CAR basis (5.13).

Using the wave functions (5.115) we calculate the matrix elements [52]

$$S_{\alpha\beta}(\boldsymbol{k}) = \frac{2\pi}{L} \mathrm{i}^{\nu'-\nu} \delta_{\sigma\sigma'} \delta(q_z' - q_z + k_z) \mathrm{e}^{\mathrm{i}(\nu'-\nu)\theta} \Im_{nn';\nu\nu'}(k_\perp \lambda_B) , \tag{5.117}$$

where $\beta = \{\nu', n', \sigma', q_z'\}$. When the quantum numbers ν and ν' have different signs ($\nu\nu' \le 0$) the function $\Im_{nn';\nu\nu'}(y)$ is given by

$$\Im_{nn';\nu\nu'}(y) = (-1)^{n+n'+\eta_\nu+\mu_{\nu'}} F_{n;n'+|\nu'|}\left(\frac{y^2}{2}\right) F_{n';n+|\nu|}\left(\frac{y^2}{2}\right) , \tag{5.118}$$

where $F_{nn'}$ is defined by (5.15). In the opposite case when $\nu\nu' \ge 0$

$$\Im_{nn';\nu\nu'}(y) = (-1)^{\mu_{\nu'}+\mu_\nu} F_{nn'}\left(\frac{y^2}{2}\right) F_{n+|\nu|;n'+|\nu'|}\left(\frac{y^2}{2}\right) . \tag{5.119}$$

The quantum mechanical expression for the electron energy transfer has been derived above (see (5.103)). Substituting (5.117) into (5.103) and using the relations $\Im_{n'n;\nu'\nu}(y) = (-1)^{\nu'-\nu} \Im_{nn';\nu\nu'}(y)$ one finds

$$\begin{aligned} \Delta E_\alpha^{\mathrm{CYL}} &= -\frac{(2\pi)^3 e^2}{\hbar L} \int \mathrm{d}\boldsymbol{k}\mathrm{d}\boldsymbol{k}' \Phi_{\mathrm{ei}}(\boldsymbol{k}')\Phi_{\mathrm{ei}}(\boldsymbol{k}) (\boldsymbol{k}\cdot\boldsymbol{v}_i)\, \delta(k_z + k_z') \\ &\times \delta\left((\boldsymbol{k}+\boldsymbol{k}')\cdot\bar{\boldsymbol{v}}_r\right) \sum_{n'=0}^{\infty} \sum_{\nu'=-\infty}^{\infty} (-1)^{\nu'-\nu} \mathrm{e}^{\mathrm{i}(\nu'-\nu)(\theta-\theta')} \\ &\times \delta\left(\zeta_{\eta_\nu-\eta_{\nu'}+n-n'}(\boldsymbol{k}) - \frac{\hbar k_z^2}{2m}\right) \Im_{nn';\nu\nu'}(k_\perp\lambda_B)\, \Im_{nn';\nu\nu'}(k_\perp'\lambda_B) . \end{aligned} \tag{5.120}$$

The last expression together with (5.118) and (5.119) gives the final result for the quantum mechanical electron energy transfer within the CYL treatment. In (5.120) the Dirac function $\delta(k_z + k_z')$ enforces a dependence on $\boldsymbol{v}_{i\perp}$ only. This feature shows that the CYL energy transfer diverges as $\sim 1/v_{i\perp}$ for parallel ion motion with $\boldsymbol{v}_{i\perp} \to 0$. Again, this is the peculiarity of the perturbative treatment which is not capable to account the infinitely large interaction time between an ion and electron wave–packet moving parallel each other.

The classical regime can be realized by the following limits: $\lambda_B^2 \to 0$ and large quantum numbers $\nu \to \infty$ and $n \to \infty$. Here two different regimes with positive or negative angular momentum can be distinguished. In the first regime, $\nu \ge 0$, $\eta_\nu = \nu$, the energy of an electron in the αth Landau state is $E_{n+\nu;\sigma}(q_z)$. The transition to classical mechanics therefore requires that $\nu + n \to m v_{e\perp}^2/2\hbar\Omega_e = a^2/2\lambda_B^2 \to \infty$. As will be shown below the quantum number n (which is always positive) is associated with the classical impact parameter $\bar{s}$ according to the relation $n \to \bar{s}^2/2\lambda_B^2 \to \infty$. It will be proven that the impact parameter within the quantum treatment must be quantized with the quantum numbers n. As the number ν is positive here it thus requires $\bar{s} \le a$ for $\nu \ge 0$. Using (5.118)-(5.120) we obtain for $\nu \ge 0$

$$
\begin{aligned}
\Delta E_\alpha^{\mathrm{CYL}} &= -\frac{(2\pi)^3 e^2}{\hbar L} \int \mathrm{d}\boldsymbol{k}\mathrm{d}\boldsymbol{k}' \Phi_{\mathrm{ei}}(\boldsymbol{k}')\Phi_{\mathrm{ei}}(\boldsymbol{k})\,(\boldsymbol{k}\cdot\boldsymbol{v}_i)\,\delta(k_z+k_z')\delta\left((\boldsymbol{k}+\boldsymbol{k}')\cdot\bar{\boldsymbol{v}}_r\right) \\
&\times \left\{ \sum_{n'=-n}^{\infty} \Psi_{n'+n;n}(\zeta,\zeta') \sum_{\nu'=-\nu}^{\infty} (-1)^{\nu'} \mathrm{e}^{\mathrm{i}\nu'(\theta-\theta')} \right. \\
&\times \delta\left(\zeta_{\nu'+n'}(\boldsymbol{k}') + \frac{\hbar k_z^2}{2m}\right) \Psi_{n+\nu;n'+\nu'+n+\nu}(\zeta,\zeta') \\
&+ \sum_{n'=-(n+\nu)}^{\infty} (-1)^{n'} \mathrm{e}^{\mathrm{i}n'(\theta-\theta')} \delta\left(\zeta_{n'}(\boldsymbol{k}') + \frac{\hbar k_z^2}{2m}\right) \Psi_{n'+n+\nu;n+\nu}(\zeta,\zeta') \\
&\times \left. \sum_{\nu'=1+\nu+n'}^{\infty} (-1)^{\nu'} \mathrm{e}^{-\mathrm{i}\nu'(\theta-\theta')} \Psi_{n;\nu'+n}(\zeta,\zeta') \right\},
\end{aligned} \tag{5.121}
$$

where $\Psi_{pq}(\zeta,\zeta') = F_{pq}(\zeta)\,F_{pq}(\zeta')$, $\zeta' = k_\perp'^2\lambda_B^2/2$. In the classical limit this corresponds to the energy transfer with small impact parameters, $\bar{s} \le a$.

In the second regime with $\nu \le 0$, $\eta_\nu = 0$. The energy of an electron in αth state is $E_{n\sigma}(q_z)$ and the transition to classical mechanics requires $n \to a^2/2\lambda_B^2 \to \infty$. Now the classical impact parameter $\bar{s}$ is associated with the quantum number $n-\nu = n+|\nu|$ according to the relation $n-\nu \to \bar{s}^2/2\lambda_B^2 \to \infty$. The case of negative ν thus requires $\bar{s} \ge a$.

It can be proved that in the classical limit both expressions for the energy transfer with $\nu \le 0$ and $\nu \ge 0$ (expression (5.121)) are equivalent and therefore we consider here the classical limit only for (5.121) with positive ν. The resulting classical expression will be valid for $\bar{s} \le a$ as well as for $\bar{s} \ge a$. The limit of the function $F_{\nu;l+\nu}(z)$ at $\nu \to \infty$, $z \to 0$ (but $\nu z \to$ const) is given in Appendix D. Using also the Taylor expansion of the Dirac δ-function with respect to the small parameter λ_B^2 we obtain from (5.121)

$$
\begin{aligned}
-\Delta E_\alpha^{\mathrm{CYL}} &\to \frac{(2\pi)^3 e^2}{2mL} \int \mathrm{d}\boldsymbol{k}\mathrm{d}\boldsymbol{k}' \Phi_{\mathrm{ei}}(\boldsymbol{k}')\Phi_{\mathrm{ei}}(\boldsymbol{k})\,(\boldsymbol{k}\cdot\boldsymbol{v}_i)\,\delta(k_z+k_z') \\
&\times\ \delta\left((\boldsymbol{k}+\boldsymbol{k}')\cdot\bar{\boldsymbol{v}}_r\right) J_0(q) \sum_{l=-\infty}^{\infty} (-1)^l \mathrm{e}^{\mathrm{i}l(\theta-\theta')} \delta\left(\zeta_l(\boldsymbol{k}')\right) H_l(\boldsymbol{k},\boldsymbol{k}'),
\end{aligned} \tag{5.122}
$$

where $q = |\boldsymbol{k}_\perp + \boldsymbol{k}_\perp'|\bar{s}$ and

$$
\begin{aligned}
H_l(\boldsymbol{k},\boldsymbol{k}') &= J_l(x)J_l(y)\frac{(\boldsymbol{k}\cdot\boldsymbol{b})(\boldsymbol{k}'\cdot\boldsymbol{b})}{\zeta_l(\boldsymbol{k}')} + \frac{k_\perp k_\perp'}{\Omega_e} \\
&\times \left[\frac{l}{x} J_l(x)\,J_l'(y) + \frac{l}{y} J_l(y)\,J_l'(x) + \mathrm{i}\sin(\theta-\theta')\,J_l(x)J_l(y)\right].
\end{aligned} \tag{5.123}
$$

Here $x = k_\perp a$ and $y = k_\perp' a$. The zero–order term changes its sign under the exchange $\boldsymbol{k} \leftrightarrows \boldsymbol{k}'$ and gives no contribution. For the same reason some further terms also vanish in (5.123) after integration over $\boldsymbol{k}$ and $\boldsymbol{k}'$.

The classical CYL expression (5.122) depends on $v_{e\perp}$, $v_{e\parallel}$ and $\boldsymbol{v}_i$, and in addition on the impact parameter $\bar{s}$. In this sense the conformity between quantum and classical descriptions is more complete here than within CAR. Since the purely classical problem does not depend on the specific coordinate basis one can expect that two different limits, (5.107) and (5.122), correspond to the classical energy transfer averaged over different sets of parameters. In the next Section, we consider the corresponding classical problem and look for the connection with the classical limits of quantum mechanical energy transfers.

5.8 Averaged Classical Second–Order Energy Transfer

As a basis for further consideration, we recall that the ion classical energy transfer $\Delta E_i(\bar{\boldsymbol{R}}_0, \varphi)$ during the collision with a magnetized electron should depend on the initial position $\bar{\boldsymbol{R}}_0$ and the initial phase φ of the electron helical motion (see Sect. 3.5). Since the initial phase of electrons and the azimuthal angle $\vartheta_{\bar{s}}$ of $\bar{s}$ are not observable in the quantum mechanical treatment we therefore consider the φ, $\vartheta_{\bar{s}}$-averaged energy transfer. As has been shown in Sect. 3.5 the first-order energy transfer, which is proportional to the ion charge vanishes due to symmetry after averaging with respect to φ and $\vartheta_{\bar{s}}$. Hence the ion energy change receives a contribution only from higher orders. The second-order energy transfer $\Delta E_i^{(2)}$ and its φ-averaged value, which are proportional to Z^2, have been found in Sect. 3.5.

We consider now some averaged quantities which can be obtained from expression (3.95) for the energy transfer. First, the φ-averaged ion energy change, (3.95), is integrated over the 2D impact parameter $\bar{s}$ in the full space. Then one obtains that the classical limit (5.107) corresponds to

$$-\frac{|v_{ix}|}{\bar{v}_r}\Delta E_\alpha^{\mathrm{CAR}} \to \frac{1}{L^2}\mathfrak{E}_1 \equiv \frac{2\pi}{L^2}\int_0^\infty \left\langle \Delta E_i^{(2)} \right\rangle_{\varphi,\hat{\bar{s}}} \bar{s}\mathrm{d}\bar{s}\,. \tag{5.124}$$

We now average the φ-averaged ion energy change over the variable $z_0 = \boldsymbol{b}\cdot\bar{\boldsymbol{R}}_0$, which is the component of the vector $\bar{\boldsymbol{R}}_0$ along the magnetic field, and with respect to the angle $\vartheta_{\bar{s}}$. This gives the classical limit of CYL energy transfer

$$-\Delta E_\alpha^{\mathrm{CYL}} \to \frac{1}{L}\mathfrak{E}_2(\bar{s}) \equiv \frac{1}{L}\int_{-\infty}^{\infty} \left\langle \Delta E_i^{(2)} \right\rangle_{\varphi,\hat{\bar{s}}} \mathrm{d}z_0\,. \tag{5.125}$$

Equations (5.124) and (5.125) for the averaged energy changes $\mathfrak{E}_1$ and $\mathfrak{E}_2(\bar{s})$ have a different physical meaning. The quantity $\mathfrak{E}_1$ is identical to the effective transport cross section for the classical electron–ion scattering in the presence of an external magnetic field introduced in Sect. 4.7, see (4.107). The second quantity $\mathfrak{E}_2(\bar{s})$ is the classical energy transfer averaged over the all initial positions z_0 of the electron guiding center. Therefore from the quantum mechanical treatment only the $\varphi, \vartheta_{\bar{s}}, \bar{s}$ (in CAR) or $\varphi, \vartheta_{\bar{s}}, z_0$ (in CYL) averaged classical energy transfers can be recovered because in quantum mechanics only these quantities are observable and the electron angular orientation and initial position of the guiding center cannot be fixed.

It should be noted that although the full quantum mechanical expressions (5.105) and (5.120) may strongly differ from each other one can expect that for many–electron systems the statistical averaged physical quantities are the same for both treatments. In particular, we briefly show below the equivalence of the dielectric response functions obtained within the CAR and CYL formulations. For this purpose we substitute (5.116)- (5.119) into the kernel of the integral equation (5.10) and use the expansion of the 2D Dirac function with respect to the orthogonal functions $F_{\nu\nu'}$,

$$\sum_{\nu;\nu'=0}^{\infty} \mathrm{e}^{\mathrm{i}(\nu-\nu')(\theta-\theta')} F_{\nu\nu'}(\zeta)\, F_{\nu\nu'}(\zeta') = \frac{2\pi}{\lambda_B^2} \delta(\boldsymbol{k}_\perp - \boldsymbol{k}'_\perp) \,. \tag{5.126}$$

After lengthy calculations one obtains

$$\sum_{\alpha,\beta} S^*_{\alpha\beta}(\boldsymbol{k})\, S_{\alpha\beta}(\boldsymbol{k}') \frac{f(E_\beta) - f(E_\alpha)}{E_\alpha - E_\beta - \hbar\omega - \mathrm{i}0} = (2\pi)^3\, \delta(\boldsymbol{k}' - \boldsymbol{k}) \tag{5.127}$$

$$\times \sum_{\sigma} \sum_{n;n'=0}^{\infty} F^2_{nn'}(\zeta) \int_{-\infty}^{\infty} \frac{\mathrm{d}q_z}{(2\pi\lambda_B)^2} \frac{f(E_{n'\sigma}(q_z + k_z)) - f(E_{n\sigma}(q_z))}{E_{n\sigma}(q_z) - E_{n'\sigma}(q_z + k_z) - \hbar\omega - \mathrm{i}0} \,.$$

Here the arguments of the Fermi–Dirac function $f(E)$ in the second line of (5.127) are given by the eigenvalues of the free particles, (5.13) in CAR. We now apply (5.127) for calculation of the dielectric function in CYL. Substituting (5.127) into the integral equation (5.10) we arrive at the expression for the dielectric function of the magnetized quantum plasma obtained within CAR treatment (5.18). This shows the complete conformity between the CAR and CYL approaches.

6 Applications and Illustrating Examples

Examples of applications where the energy loss of ions in a magnetized plasma plays a prominent role are the stopping of the alpha particles created by fusion events in the hot plasma of magnetic confinement fusion devices and the cooling of ion beams or bunches of ions by magnetized electrons in electron cooler sections of storage rings or in traps. In a strong magnetic field the cyclotron motion of the target electrons sets the smallest relevant length and time scale. In that sense the fields usually applied in magnetic confinement fusion are still moderate, while stronger fields are employed to guide the electrons in the cooling sections of storage rings and even more so for the cooling in traps. We will therefore focus on these latter cases.

Electron cooling in storage rings is a well-established and powerful technique to improve the phase space quality of ion beams by the exchange of energy between the ions and a superimposed electron beam. In the rest frame of the beams this may be viewed as the stopping of ions in an electron plasma. We will present the essentials of electron cooling in storage rings and compare the theoretical description with experimental results in the next section, before turning then to electron cooling in Penning traps. This application attracts a lot of attention in connection with the production of antihydrogen by first cooling antiprotons and then merging them with positrons in such traps. And more recently, experiments are performed or in preparation where heavy ions or antiprotons are decelerated by electrons in traps for precision measurements on QED and fundamental symmetries. During the cooling in a trap the particles move under the influence of the external electric and magnetic fields, the space charge fields produced by the particles themselves and the stopping force $\mathcal{F}$ resulting from the interaction of the trapped ions with the enclosed electrons. The stopping of the heavy particles causes a strong heating of the electrons which are themselves cooled by synchrotron radiation in the applied strong magnetic field.

6.1 Electron Cooling in Storage Rings

In most experiments with charged particle beams it is desired that the beams are well concentrated and have a low temperature i.e. that they occupy only a small phase space volume. One method to achieve this is electron cooling in storage rings as proposed by Budker [29]. The ion beam is superimposed by a comoving electron beam in a cooling section of about one to two meter length. The electron beam is guided through this section by a magnetic field of about $B = 0.02 - 0.2$ T and separated from the ion beam thereafter. In the rest frame of the beams the cooling process may be regarded as a thermal equilibration between the ion beam and a cold electron plasma, or in a more detailed and microscopic view as the stopping of ions in an electron plasma with a density of about $n_e = 10^6 - 10^7$ cm^{-3}. Due to the acceleration in the electron gun the velocity distribution of the electrons in the rest frame is highly anisotropic. The very small velocity spread of the electrons parallel to the beam and its magnetic guiding field corresponds to a temperature $T_{\|}$ of a few K, while the transverse temperature $T_{\perp}$ is by two to three orders of magnitude larger and about 100 – 1000 K. As the transverse motion of the electrons is quenched by the magnetic field, the rather small longitudinal thermal velocity typically sets the relevant velocity scale for stopping. Reviews on electron cooling can be found e.g. in [85, 107, 120].

6.1.1 Energy Loss and Drag Force

The basic ingredient for a description and theoretical understanding of electron cooling is the stopping of ions in magnetized electrons in the typical regime of parameters given above. Due to the large transverse temperature $T_{\perp}$ of the electrons, the coupling parameter $\mathcal{Z}$ (2.42) is even for highly charged ions still small enough to allow the use of the perturbative BC (Chap. 3) and LR (Chap. 4) treatments. But closed evaluations of the general expressions for the stopping power in LR (2.60) and in BC based on the energy transfer (3.99) are only possible for the limiting cases treated in Chaps. 3 and 4. Here we have neither parallel ion motion nor an infinitely strong magnetic field or can restrict us to the low and high velocity regimes. In the present coupling regime, we may, however, assume that the dynamic polarization of the electrons can be accounted for by a velocity dependent screening length (cf. Sect. 4.6.2). We are thus looking for an appropriate approximate BC expression. In the BC picture different types and regimes of trajectories can be identified by comparing the impact parameter with the cyclotron radius and the pitch of the electron helix, which have been introduced in Sect. 2.2 and Fig. 2.5. There we also gave explicit expressions for the energy transfer in close unmagnetized (2.8) (Coulomb scattering), fast weakly magnetized (2.32) (stretched helices) and distant strongly magnetized (2.36) (tight helices) collisions. The energy loss which follows from the energy transfer by an integration with respect to the impact parameter $\bar{s}$ can now be composed out of these distinct cases by piecewise using (2.8), (2.32) and (2.36) in the corresponding $\bar{s}$–interval, as it has been advocated in [86, 124]. In a similar manner the approximations given e.g. in [35, 85] have been obtained by combining expression for the energy transfer in weakly and strongly magnetized collisions.

For a comparison with experiments at storage rings we propose the following approximation made up of the energy transfer for nonmagnetized Rutherford scattering (2.8) (which is the limit $\lambda \to \infty$ of (3.59)) for close collisions with $\bar{s} < \delta$ and of the energy transfer for the strongly magnetized, tight helix case with Coulomb interaction (2.36) (which has been re-derived from (3.99) in Sect. 3.8 (3.143)) for distant collisions with $\delta < \bar{s} < \lambda$. This yields after impact parameter integration according to (2.9) the energy loss

$$\frac{\mathrm{d}E_i}{\mathrm{d}l} = \frac{2\pi n_e Z^2 e^4}{m} \left\{ \frac{\boldsymbol{v}_r \cdot \hat{\boldsymbol{v}}_i}{v_r^3} \ln\left(1 + \frac{\delta^2}{s_{90}^2}\right) + \frac{v_{i\perp}^2}{\bar{v}_r^5} \left[\bar{\boldsymbol{v}}_{r\parallel} + \frac{\bar{\boldsymbol{v}}_{r\perp}}{2} \left(1 - \frac{\bar{v}_{r\parallel}^2}{\bar{v}_{r\perp}^2}\right)\right] \cdot \hat{\boldsymbol{v}}_i \ln\left(1 + \frac{\lambda^2}{\delta^2}\right) \right\}, \tag{6.1}$$

where $s_{90} = Ze^2/mv_r^2$ (see (2.6)), $\delta = \bar{v}_r/\Omega_e$ is the pitch of the helix and λ the velocity dependent screening length $\lambda(v_i) = (\bar{v}_{\mathrm{th}}/\omega_p)(1 + v_i^2/\bar{v}_{\mathrm{th}}^2)^{1/2}$. To avoid any negative values we replaced the conventional Coulomb logarithm (2.23) originally appearing in the second term with the modified Coulomb logarithm (2.11). Expression (6.1) is the energy loss of an ion in a monochromatic beam of electrons. In electron cooling of ion beams the velocity distribution of the electrons is anisotropic. It is usually modeled by a two-temperature Maxwellian (2.28)

$$f_0(\boldsymbol{v}_e) = \frac{1}{2\pi v_{\mathrm{th}\perp}^2} \frac{1}{(2\pi v_{\mathrm{th}\parallel}^2)^{1/2}} \exp\left(-\frac{v_{e\perp}^2}{2v_{\mathrm{th}\perp}^2} - \frac{v_{e\parallel}^2}{2v_{\mathrm{th}\parallel}^2}\right). \tag{6.2}$$

The stopping power then follows from the energy loss (6.1) by a velocity average with respect to (6.2), that is,

$$\frac{\mathrm{d}\mathcal{E}_i}{\mathrm{d}l} = \int \mathrm{d}\boldsymbol{v}_e\, f_0(\boldsymbol{v}_e)\, \frac{\mathrm{d}E_i}{\mathrm{d}l}. \tag{6.3}$$

The stopping force $\mathcal{F}$ on the ion, defined by $\mathrm{d}\mathcal{E}_i/\mathrm{d}l = \mathcal{F} \cdot \hat{\boldsymbol{v}}_i$ (2.1), can now easily be extracted from (6.3) and (6.1) as

$$\mathcal{F} = \frac{2\pi n_e Z^2 e^4}{m} \int \mathrm{d}\boldsymbol{v}_e f_0(\boldsymbol{v}_e) \left\{ \frac{\boldsymbol{v}_r}{v_r^3} \ln\left(1 + \frac{\delta^2}{s_{90}^2}\right) + \frac{v_{i\perp}^2}{\bar{v}_r^5} \left[\bar{\boldsymbol{v}}_{r\parallel} + \frac{\bar{\boldsymbol{v}}_{r\perp}}{2} \left(1 - \frac{\bar{v}_{r\parallel}^2}{\bar{v}_{r\perp}^2}\right)\right] \ln\left(1 + \frac{\lambda^2}{\delta^2}\right) \right\}. \tag{6.4}$$

For explicit results the velocity average in expression (6.4) will be provided by numerical integration. A comparison of this approximation of the stopping force with measurements in the ESR storage ring at GSI will be made in the next section.

6.1.2 Cooling Forces

For using electron cooling routinely as a tool to improve the quality of ion beams global quantities like cooling time and final emittances are of primary interest. For an insight into the cooling process itself a more detailed study of basic quantities like the stopping forces is needed. Such kind of measurements have been performed at several storage rings during the last ten years, like e.g. at the ESR at GSI [130–132], the TSR at Heidelberg [19, 20], CRYRING at Stockholm [32, 33] and CELSIUS at Uppsala [45]. In these experiments a so-called cooling force is extracted, which can be view as an averaged stopping force where the averaging runs over the ion distribution in the beam. As an example we focus on the measurements of longitudinal cooling forces for different fully stripped heavy ions as conducted at the electron cooler of the ESR storage ring. Two different methods have been used here to determine the cooling force. At low ion velocities the cooling force is extracted from the equilibrium between cooling and longitudinal heating with rf noise. At high relative velocities between the rest frames of the beams the cooling force is deduced from the momentum drift of the ion beam after a rapid change of the electron energy. Details of these methods are given in [130, 131]. The measured cooling forces are shown in Fig. 6.1 (filled circles). The electron beam in these experiments has a typical density of $n_e \approx 10^6$ cm^{-3} and can be described by an anisotropic velocity distribution (6.2) with $T_\perp = mv_{\text{th}\perp}^2 \approx 0.11$ eV and $T_\| = mv_{\text{th}\|}^2 \approx 0.1$ meV. As the ion beam had a radial position some mm off the electron beam axis in these experiments, an additional transverse cyclotron velocity v_c of the electrons has to be taken into account, see [132] for details. The velocity distribution relevant for the averaging in (6.3) and (6.4) is thus given by

$$f_0(\mathbf{v}_e) = \frac{\mathcal{N}}{2\pi v_{\text{th}\perp}^2} \frac{1}{(2\pi v_{\text{th}\|}^2)^{1/2}} \exp\left(-\frac{v_{e\perp}^2}{2v_{\text{th}\perp}^2} - \frac{(v_{e\|} - v_c)^2}{2v_{\text{th}\|}^2} \right) . \tag{6.5}$$

where $\mathcal{N}$ is a normalization factor and $v_c \approx v_{\text{th}\perp}$ in the present case. The strength of the magnetic guiding field was $B = 0.1$ T. The measured longitudinal cooling force represents an average over the stopping forces on individual ions. For a comparison with the theoretical model (6.4) the cooling force is thus interpreted as the average $\langle \mathcal{F}_\| \rangle$ of the component $\mathcal{F}_\|$ of the stopping force $\mathcal{F}$ (6.4) parallel to the beam axis (and the magnetic field) over the ion distribution $f_i(\mathbf{v}_i)$ in the beam (see also e.g. [40]), that is,

$$\langle \mathcal{F}_\| \rangle = \int \mathrm{d}\mathbf{v}_i \, f_i(\mathbf{v}_i) \, \mathcal{F}_\|(\mathbf{v}_i) , \tag{6.6}$$

where

$$\mathcal{F}_\|(\mathbf{v}_i) = \frac{2\pi n_e Z^2 e^4}{m} \int \mathrm{d}\mathbf{v}_e f_0(\mathbf{v}_e) \left\{ \frac{v_{r\|}}{v_r^3} \ln\left(1 + \frac{\delta^2}{s_{90}^2}\right) + \frac{v_{i\perp}^2 \bar{v}_{r\|}}{\bar{v}_r^5} \ln\left(1 + \frac{\lambda^2}{\delta^2}\right) \right\} . \tag{6.7}$$

For low ion velocities this average is taken with respect to the transverse ion velocity only and the cooling force depends on the parallel ion velocity, i.e. $\langle \mathcal{F}_\| \rangle = \langle \mathcal{F}_\| \rangle (v_{i\|})$. In the experimental procedure used for high ion velocities the cooling force is an average over the complete ion distribution. This average $\langle \mathcal{F}_\| \rangle = \langle \mathcal{F}_\| \rangle (\langle v_{i\|} \rangle)$ depends now on the velocity of the CM of the ion beam relative to the rest frame of the electron beam $\langle v_{i\|} \rangle$. Both velocities are denoted as relative ion velocity in Fig. 6.1. To perform the average (6.6) the distribution $f_i(\mathbf{v}_i)$ must be known. However, this distribution was not determined in detail, but there exists an estimate of the beam divergence $\langle \theta_i \rangle \lesssim 0.5$ mrad [131]. This yields after transformation to the rest frame of the ion beam for the transverse ion velocities $\langle v_{i\perp} \rangle \approx \gamma\beta c \langle \theta_i \rangle$, where γ, β are the relativistic factors related to the beam velocity in the lab frame and c is the speed of light. For the measurements at hand with an ion energy of 250 keV/u this results in $\langle v_{i\perp} \rangle \lesssim 10^5$ m/s. Due to the lack of a more detailed knowledge about the ion distribution, the average (6.6) is replaced by $\langle \mathcal{F}_\| \rangle = \mathcal{F}_\|(v_{i\perp} = \langle v_{i\perp} \rangle, v_{i\|})$ where

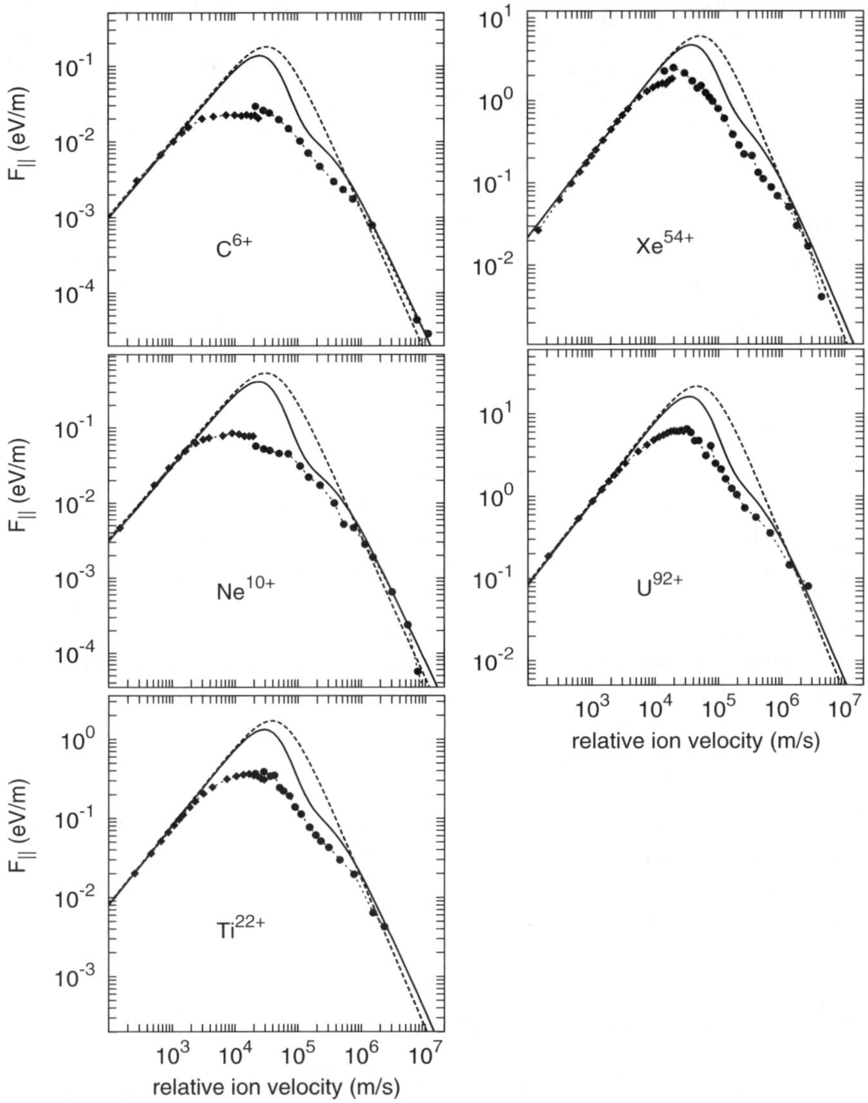

Fig. 6.1. Longitudinal cooling force for various fully stripped ions as function of the relative ion velocity with respect to the rest frame of the electron beam. Filled circles: experimental data from measurements at the electron cooler of the ESR storage ring [130–132]. Solid curve: binary collision approximation (6.7). Dashed curve: empiric formula (6.8) for the cooling force as proposed in [102]. Both theoretical descriptions of the cooling force are calculated for an electron beam with $n_e = 10^6$ cm^{-3}, $T_\perp = 0.11$ eV and $T_\parallel = 0.1$ meV in a magnetic field of $B = 0.1$ T, and are fitted to the experimental results at low relative velocities by treating the transverse ion velocity $v_{i\perp}$ as a free parameter (see the text for details).

$v_{i\perp} = \langle v_{i\perp} \rangle$ is treated as a free parameter to fit the BC stopping force (6.7) to the experimental data. As $\mathcal{F}_{\|}(v_{i\perp}, v_{i\|})$ (6.7) is rather sensitive to a variation of $v_{i\perp}$ at low parallel velocities $v_{i\|}$ this fit is done for the linear increase of the cooling force at low relative velocities. The related values for $v_{i\perp} = \langle v_{i\perp} \rangle$ are in the range $v_{i\perp}/v_{\mathrm{th}\|} = 10-17$ which corresponds to $\langle \theta_i \rangle \approx 0.2-0.3$ mrad in good agreement with the estimated beam divergence. The resulting theoretical predictions are given by the solid curves in Fig. 6.1. They agree well with the experimental data at low and high velocities but overestimate the cooling force at medium velocities $v_{\mathrm{th}\|} = 4.2 \times 10^3\mathrm{m/s} \lesssim v_{i\|} \lesssim v_{\mathrm{th}\perp} = 5 \times 10^3\mathrm{m/s}$. We mainly ascribe this difference to the rather rough approximation made here for the averaging over the ion distribution in the beam. The experimental data are also compared to an empirical formula for the cooling force

$$\mathcal{F}_{\|}(\mathbf{v}_i) = -\frac{4\pi n_e Z^2 e^4}{m} \frac{v_{i\|}}{(v_{i\|}^2 + v_{i\perp}^2 + v_{\mathrm{eff}}^2)^{3/2}} \ln\left(1 + \frac{\langle s_{\max} \rangle}{\langle s_{\min} \rangle + a_e}\right) \tag{6.8}$$

as proposed by Parkhomchuk [102]. Here $\langle s_{\min} \rangle = Ze^2/m(v_i^2 + \bar{v}_{\mathrm{th}}^2)$ and $\langle s_{\max} \rangle = (v_i^2 + \bar{v}_{\mathrm{th}}^2)^{1/2}/\omega_p$ are the minimal and maximal impact parameters, (2.59), $a_e = v_{\mathrm{th}\perp}/\Omega_e$ is the cyclotron radius of the electrons, and v_{eff} is an effective electron velocity related to the transverse magnetic and electric fields in the electron cooler, see [102], which can be viewed as a fitting parameter. The force (6.8), which is the stopping force on a single ion, must also be averaged over the ion distribution like in (6.6). As above the force (6.8) with an average $\langle v_{i\perp} \rangle$ has been used instead. If the additional parameter is chosen as $v_{\mathrm{eff}} = v_{\mathrm{th}\|}$ (that is, rather small) the experimental data at low velocities are fitted by nearly the same values for the beam divergence $\langle \theta_{i\perp} \rangle$ as before. The resulting cooling force is shown by the dashed curves in Fig. 6.1. The agreement with the experimentally deduced cooling force is as good as for the BC force (6.7) at low and high velocities. At medium velocities the deviation from the experimental results is even larger, in particular for $v_{i\|} \approx v_{\mathrm{th}\perp}$, compared to the more detailed BC treatment. For a more accurate comparison with the measurements and a critical evaluation of the theoretical approaches a detailed knowledge of the ion distribution is indispensable.

6.1.3 Emittance and momentum spread

The aim of electron cooling is the improvement of the quality of the ion beam which is measured by the longitudinal momentum spread $\Delta p/p$ and the horizontal rms emittance

$$\varepsilon = \left(\langle x^2 \rangle \langle x'^2 \rangle - \langle x x' \rangle^2\right)^{1/2} . \tag{6.9}$$

Here x is the horizontal distance from the reference orbit and $x' = \mathrm{d}x/\mathrm{d}s$ its derivative with respect to the pathlength s along this orbit. In Fig. 6.2 we compare measurements of these observables at the ESR storage ring [51] with theoretical results [125] which where obtained by inserting different cooling force models into the beam dynamics simulation code BETACOOL by I. Meshkov and coworkers [16, 117]. Similar to equations (6.1)-(6.4) the theoretical cooling force is composed

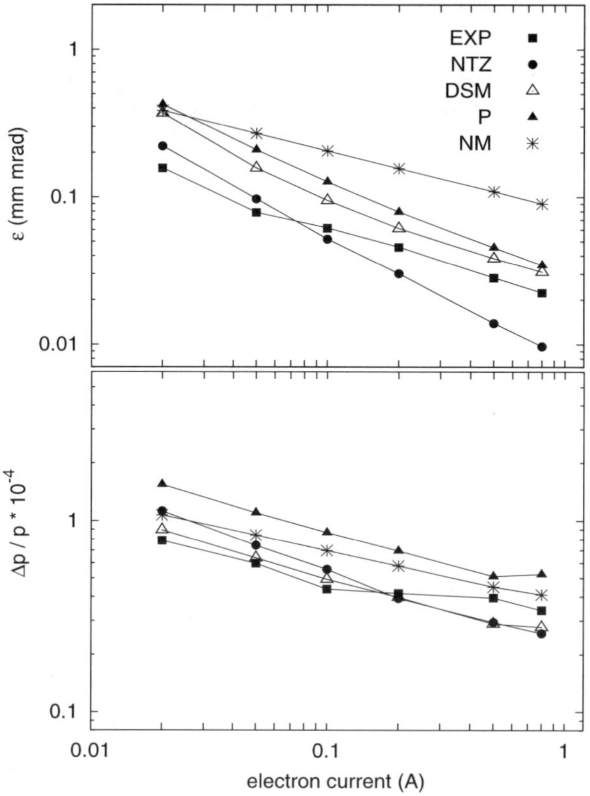

Fig. 6.2. Horizontal rms emittance ε (6.9) (upper part) and longitudinal momentum spread $\Delta p/p$ (lower part) as function of the electron current. Experimental values from measurements performed at the ESR (EXP) [51] compared to the perturbative binary collision treatment as outlined in Sect. 2.2 (NTZ) [86], the Derbenev, Skrinsky and Meshkov model (DSM) [35,85], the empirical model of Parkhomchuk (P) [102] and the non-magnetic cooling force (NM).

from the energy transfers for Coulomb scattering (2.8), stretched helices (2.32) and tight helices (2.36) according to the relative size of the impact parameter, the pitch and the radius of the cyclotron motion as it is described in detailed in [86]. This perturbative BC treatment (denoted here as NTZ model) yields a good agreement with the experiment except for the the emittance at large electron currents. In Sect. 2.2 we mentioned the similar DSM model of Derbenev, Skrinsky and Meshkov [35,85] which overestimates the emittance somewhat. Neglecting the magnetic field altogether (NM) results in too large values for both the emittance ε and the momentum spread $\Delta p/p$ as does the empirical model P (6.8) of Parkhomchuk [102]. For further details on the actual implementation of the employed cooling force models and the BETACOOL code see [17].

6.2 Electron Cooling in Penning Traps

Ions and antiprotons are also cooled by electrons or positrons in traps, as e.g. in the Penning traps employed in the recent and planned experiments for the production of antihydrogen at CERN [6, 44] or the generation of slow highly charged ions at GSI [12, 109]. Although electron cooling is routinely used in connection with the antihydrogen production a lot of observations are not yet satisfactorily explained and understood, like e.g. the spatial correlations observed in the antiproton energy spectra by the ATHENA collaboration [7]. More comprehensive theoretical studies are needed as well for the envisaged cooling of highly charged ions up to bare Uranium in the HITRAP facility [12, 109] at GSI. Here the specific conditions in Penning traps have to be taken into account. These are the rather strong magnetic field of a few Tesla and, in contrast to electron cooling in storage rings, an intially isotropic velocity distribution of the electrons together with a strong and fast heating of the electrons by the ions as both are confined together during the whole cooling process. As we will see later, the electron temperature T_e typically rises by about four orders of magnitude from $T_e \sim 1$ meV in the initial state to a few eV. The energy transfered thereby from the ions to the electrons finally leaves the trap by the synchrotron radiation which is emitted by the electrons due to their gyration in the strong magnetic field. An important subject are also losses by radiative or three-body ion-electron recombination or charge exchange with residual gas atoms/molecules. In cases like the HITRAP cooler trap which aims at providing cold highly charged ions for subsequent experiments such losses are clearly unwanted and the cooling rates must be large compared to e.g. radiative recombination rates. In this respect, the strong heating of the electrons is highly welcome. It can considerably reduce the recombination of ions with free electrons, at least if the electrons can be removed from the trap or spatially separated from the ions as soon as the cooling of the ions is completed and before the cooling of the electrons by synchrotron radiation increases the recombination rates again. Sufficiently large cooling rates are, however, not only mandatory for avoiding too large recombination losses, but also to enable the envisaged repetition rate for HITRAP. During one operation cycle of about 10 s the trap is loaded with electrons, the ions are caught, cooled by electrons and after that by resistive cooling. Finally the ions have to be extracted out of the trap in a sufficiently smooth and slow manner to avoid heating them up again. That is, the electron cooling itself may only take a small fraction of the whole time of one cycle. To check whether such cooling times are feasible and considering possible optimizations requires numerical studies based on an appropriate theoretical description. The ultimate aim of such simulations of the cooling process is the determination of the time evolution of the phase-space distribution $f(\boldsymbol{r}_i, \boldsymbol{v}_i, t)$ of the ions. In the following we will discuss a few results on the time evolution of the ion energy, the energy distribution and the electron temperature during the cooling process in a trap, starting with the explanation of the model we employed for this purpose.

6.2.1 Modeling of the Cooling Process in a Trap

In our theoretical treatment [140, 141] of the cooling process we follow in large parts the basic ideas developed in the earlier approaches [14,59,110]. There are two essential effects to be considered: the energy loss of the ions and the change of the temperature T_e of the trapped electrons which are heated by the ions and cooled by synchrotron radiation. Assuming an instantaneous conversion of the energy lost by the trapped ions into thermal electron energy, the increase of T_e follows from energy conservation as

$$\frac{3}{2}N_e k_B \frac{dT_e}{dt} = \frac{dE_e}{dt} = -\frac{dE_i}{dt} = -\sum_{\mu}^{N_i} \frac{d\mathcal{E}_{i\mu}}{dt} = -\sum_{\mu}^{N_i} M\frac{d\boldsymbol{v}_{i\mu}}{dt} \cdot \boldsymbol{v}_{i\mu} . \tag{6.10}$$

Here dE_e/dt is the change per unit time of the total kinetic energy of all electrons and dE_i/dt the change of the kinetic energy of all ions. This is the sum of the energy changes $d\mathcal{E}_{i\mu}/dt$ of the individual ions due to the force $M d\boldsymbol{v}_{i\mu}/dt$ on the μ-th ion which includes the stopping force when passing through the cloud of trapped electrons. With the heating given by (6.10) and the cooling of the electrons by synchrotron radiation the evolution of the electron temperature is determined by the rate equation

$$\frac{dT_e(t)}{dt} = -\frac{2}{3k_B N_e} \sum_{\mu}^{N_i} \frac{dE_\mu(t)}{dt} - \frac{1}{\tau_s}(T_e - T_0) . \tag{6.11}$$

Here T_0 is the ambient temperature (typically about 4 K) supplied by the cryostat, and τ_s the time constant for synchrotron radiation [28]

$$\frac{1}{\tau_s} = \frac{1}{3\pi\varepsilon_0}\frac{e^4 B^2}{m^3 c^3} , \tag{6.12}$$

where $\tau_s \approx 0.1$ s for electrons gyrating in a homogeneous magnetic field of $B = 6$ T. The use of a single electron temperature T_e in Eqs. (6.10) and (6.11) implies that an isotropic electron distribution, i.e. a Maxwellian with temperature T_e, is assumed.

The energy change $d\mathcal{E}_{i\mu}/dt$ of an ion is determined by the equations of motion

$$M\frac{d\boldsymbol{v}_{i\mu}}{dt} = \mathcal{F}[n_e, T_e(t), \boldsymbol{B}, \boldsymbol{v}_{i\mu}] + Ze\Big(-\nabla\phi(\boldsymbol{r}_{i\mu}) + \boldsymbol{v}_{i\mu} \times \boldsymbol{B}\Big) , \quad \frac{d\boldsymbol{r}_{i\mu}}{dt} = \boldsymbol{v}_{i\mu} , \tag{6.13}$$

where $\mathcal{F}[n_e, T_e, \boldsymbol{B}, \boldsymbol{v}_i]$ is the stopping force (2.1) exerted on the ion by collisions with the magnetized trapped electrons which have a Maxwellian velocity distribution of the actual temperature $T_e(t)$. This stopping force acts on the ions together with the Lorentz force due to the homogeneous magnetic field $\boldsymbol{B}$ in the trap and the (static) electric field $-\nabla\phi(\boldsymbol{r})$ composed of the electric potential of the trap and the space charge field of the electrons. The interaction between the ions is neglected here in view of the low ion density. Since the stopping force $\mathcal{F}$ depends on T_e, Equations (6.10)-(6.13) are coupled and have to be solved numerically. The trajectories $\{\boldsymbol{r}_{i\mu}(t), \boldsymbol{v}_{i\mu}(t)\}$ of the ions and thus the evolution of the ionic distribution $f(\boldsymbol{r}_i, \boldsymbol{v}_i, t)$ are given by the solutions of (6.13). To that end the electric field $-\nabla\phi(\boldsymbol{r})$ in the trap must be known and implemented. For the HITRAP design such work is in progress [87]. For our purposes the typical nested trap with harmonic trap potentials for the electrons and spheroidal electron clouds is replaced by the simpler trap configuration depicted in Fig. 6.3. Here, the field of the trap with length L is given by a square well potential, and the electrons are spatially distributed as a cylindrical cloud of constant density n_e, length Λ and radius ρ. The radius ρ is chosen somewhat larger than the largest radial initial position plus the initial cyclotron radius of the trapped ions. This provides in transverse direction a spatial overlap of all the ions with the electron cloud during the whole cooling process. The space charge field of this electron cloud is neglected.

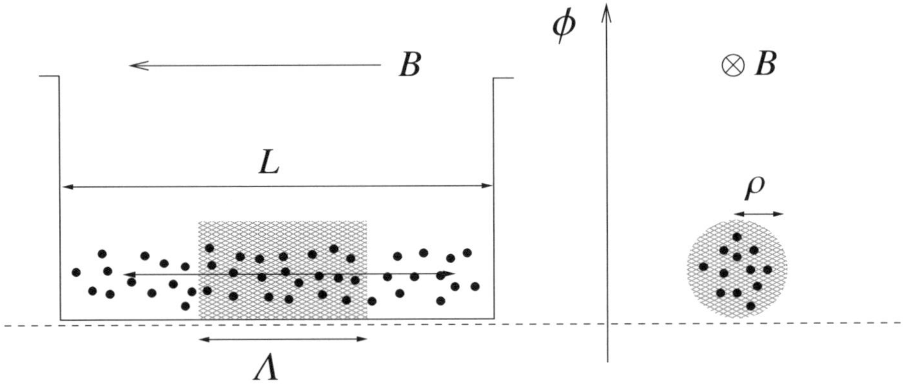

Fig. 6.3. Simplified trap configuration as used for the present calculations of electron cooling. After [141].

The basic feature of the present treatment is its detailed description of the cooling by equation (6.13) and the individual stopping force $\mathcal{F}[T_e, \boldsymbol{v}_{i\mu}]$ on each ion. In earlier treatments, like [14, 59, 110], only the total (or average) ionic energy E_i expressed by an ion temperature $T_i = 2E_i/3k_\mathrm{B}N_i$ has been considered, and the equation of motion (6.13) was replaced by a rate equation for T_i. But such kind of averaged description cannot take into account effects which are, for instance, related to the anisotropy of the velocity distribution of the ions after their injection into the trap. The present approach enables an advanced treatment of electron cooling based on the time evolution of the ion distribution, for which the stopping force $\mathcal{F}$ is a key quantity. At strong coupling, i.e. high ion charges and low relative velocities, and at large magnetic fields, the method of choice would be the numerically expensive solution of the Vlasov-Poisson equation by PIC simulations. It accounts for both the nonlinearity and the collectivity of the response. In cases where the collectivity is not so important the CTMC implementation of binary collisions would come next. In the present application the electron temperature will become rather large, as we will see below. We are thus mostly in a weak coupling regime, even for the cooling of highly charged ions like bare Uranium, which allows a perturbative treatment. In view of including the dynamic collective response of the electrons we are using here, as a reasonable approximation for the stopping force $\mathcal{F}$ needed in equation (6.13), the LR expression for an infinitely strong magnetic field (4.32) given by

$$\mathcal{F}_{\text{inf}} = -\frac{Z^2 k_B^2}{2\pi^2 \lambda_{D\|}^2} \int_0^1 d\mu \int_0^{\pi} d\phi \left(\cos\phi (1-\mu^2)^{1/2}, \sin\phi (1-\mu^2)^{1/2}, \mu \right) Q_0 \left(x \frac{\cos\Theta}{\mu}, \xi \right) . \tag{6.14}$$

Here $\mu = \hat{\boldsymbol{k}} \cdot \boldsymbol{b}$, $x = v_i / v_{\text{th}\|}$, $\xi = k_{\max} \lambda_{D\|}$ and the function Q_0 is given by equation (4.6). Expression (6.14) is evaluated by numerical integration which can be done with relatively low computational effort. As discussed in Sect. 4.6, such a LR treatment predicts, however, unrealisticly large stopping forces on ions moving almost parallel to $\boldsymbol{b}$. But this will affect only a small number of ions out of the considered ensemble of ions. We estimate the related error in the global observables (cooling times, electron temperatures) of at most 20%. Improved stopping forces based on a fit to PIC and CTMC results are in progress. Then also the role of the ion gyration on the stopping force and the assumption of a straight ion trajectory during the interaction with the electrons made so far needs some further attention, as the screening length and the ionic cyclotron radius are often of the same size at the typical conditions for electron cooling in traps.

Besides the cooling process itself, also losses of highly charged ions by recombination with free electrons or charge exchange with residual gas particles might be important. Charge exchange is expected to be sufficiently suppressed by the envisaged ultra high vacuum. Three body recombination which scales like $T_e^{-9/2}$ [53, 82] will be strongly reduced by the large heating of the electrons (see below). There remains radiative ion-electron recombination (RR) which depends on the electron temperature like $T_e^{-1/2}$ [65]. Thus RR is is expected to give the dominant contribution to the losses of highly charged ions into lower charge states during the cooling. The present scheme, (6.10)-(6.13), can be supplemented by calculating instantaneous radiative recombination rates $\nu_{RR,i\mu}(t)$ or each ion from the actual ion velocity $\boldsymbol{v}_{i\mu}(t)$ and electron temperature $T_e(t)$ which then provides a surviving probability of the initial charge state. The losses, e.g. for bare Uranium, turn out to be about 10% to 15% for the typical conditions and parameters in HITRAP, see [141] for details and explicit results.

6.2.2 Cooling of Protons and Highly Charged Ions

The coupled differential equations (6.10)-(6.13), involving the stopping force $\mathcal{F}$ due to the interaction with magnetized electrons (6.14) have been applied to the cooling of protons and bare Uranium. In the case of electron cooling of protons experimental results for the time evolution of the average proton energy and the electron temperature are available [59] and can be used to evaluate the theoretical treatment and its numerical implementation. In these experiments $N_p \sim 5 \times 10^5$ protons of a few keV have been cooled in a multi-ring Penning trap by $N_e \sim 10^8$ electrons with a density of about $n_e \sim 3.5 \times 10^7$ cm^{-3} at $B = 1$ T and $T_0 \lesssim 10$

K. The time evolution of the mean energy of the protons was determined by time-of-flight measurements after opening the trap at certain times. The electron energy (temperature) was inferred from the measured frequency of plasma oscillation of the confined electron cloud. For details on these techniques and the experimental parameter see [59] and the reference therein. A comparison of our calculations with these experimental results for protons with an initial energy of $E_{p0} = 2$ keV is shown in Fig. 6.4. The agreement with the time evolution of the proton energy is very

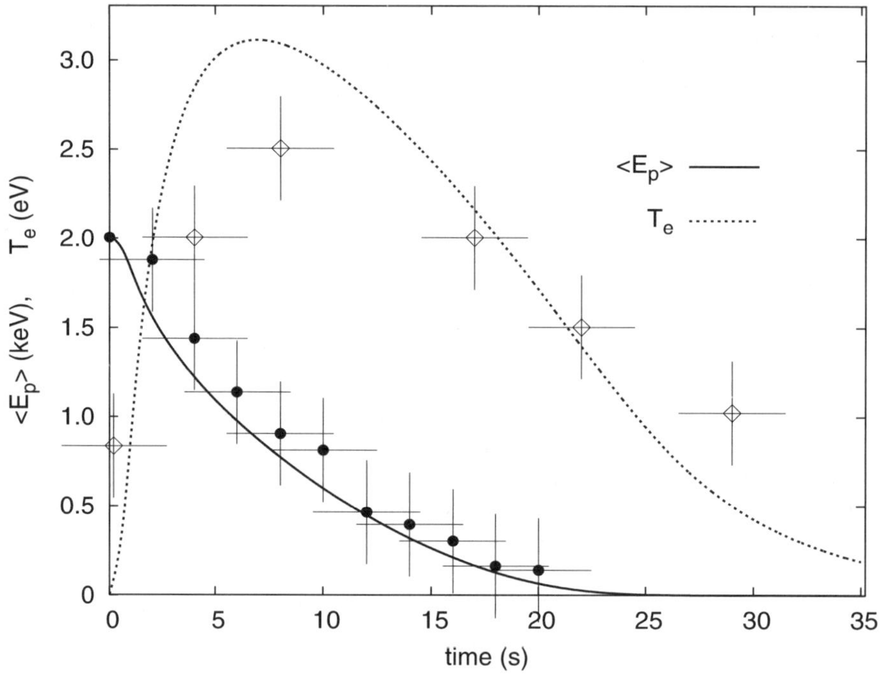

Fig. 6.4. Time evolution of the average energy $\langle E_p \rangle(t)$ and the electron temperature $T_e(t)$ during cooling of an ensemble of protons by magnetized electrons with $n_e = 3.5 \times 10^7$ cm^{-3}, $N_p/N_e = 5 \times 10^{-3}$, $T_0 = 10$ K and $B = 1$ T. Curves: theoretical results obtained from a numerical solution of (6.10)-(6.14) assuming the simplified trap of Fig. 6.3 and $\Lambda/L = 0.2$. Symbols with errorbars: experimental results for the proton energy (filled circles) and the electron temperature (open diamonds) from [59].

good in particular when taking the experimental errors into account. Compared to the calculations the experiments show a slightly slower cooling, i.e. a larger cooling time. Concerning the electron temperature the overall trend, with a strong increase at the very beginning and a slow decay due to synchrotron radiation, is well reproduced by the theoretical predictions. In detail the agreement is, however, less satisfactory than for the proton energy. Here some further investigations are needed. A possible explanation might be the development of an anisotropic velocity distribution during the cooling process.

For bare Uranium, calculations have been performed with an ion distribution of 500 U^{92+} ions representing the ion ensemble of roughly 10^5 ions which are expected to be trapped during a typical cooling cycle at HITRAP [12]. To get the correct heating rate corresponding to the actual number of ions in the trap the total energy loss $\sum_{\mu}^{N_i} \mathrm{d}E_{\mu}/\mathrm{d}t$ of the $N_i = 500$ simulated ions is multiplied with a factor $10^5/500$ in (6.10) and (6.11). The initial distribution, i.e. at time $t = 0$, of the 500 ions of bare Uranium was obtained from a preceding simulation of the injection of an ion bunch into the cooler trap for the designed HITRAP setup [57]. The same ion ensemble was, after proper rescaling, also used for the cooling of protons shown in Fig. 6.4.

As a first example we consider the case of $N_e = 10^9$ trapped electrons with a density $n_e = 10^7$ cm^{-3} and $N_i/N_e = 10^{-4}$ when assuming that the trap is filled with a bunch of 10^5 ions. For all following calculations for the HITRAP setup an effective size of the electron cloud Λ with $\Lambda/L = 0.4$ is assumed. The resulting time evolution of the average energy per ion $\langle E_i \rangle(t)$ and the electron temperature $T_e(t)$ are shown in Fig. 6.5 (top) together with the variation of the width $\langle E_i^2 - \langle E_i \rangle^2 \rangle^{1/2}$ of the energy distribution of the ions (bottom). Some snapshots of the energy distribution $(\mathrm{d}N/\mathrm{d}E)$ itself are plotted in Fig. 6.6. While the ions cool down in this example within roughly one second from their initial energy of about 1.7 keV/u, a strong and rapid heating of the electrons from $T_0 \approx 0.34$ meV to about 2.2 eV takes place already within the first 0.1 s (see Fig 6.5 top). After this initial strong heating due to the large energy loss of the highly charged U^{92+} ions, the energy and the energy loss of the ions continuously decrease and with it the heating of the electrons. Now the cooling by synchrotron radiation dominates and the electrons start to cool down. This behavior is qualitative exactly the same as we have already seen in Fig. 6.4 for the cooling of protons. At the final stage of the cooling (after about 1.5 s), the energy of the ions reaches the equilibrium value given by T_e (not resolved in this figure), the heating of the electrons stops and the electrons temperature decays like $\exp(-t/\tau_s)$ towards T_0 due to synchrotron radiation. During cooling the energy distribution of the ions (Fig. 6.6) shifts, in accordance with the evolution of $\langle E_i \rangle$, towards lower energy. It thereby develops an increasingly pronounced peak, that is, a strongly shrinking width. This evolution of the width is also shown in the lower part of Fig. 6.5 as a continuous function of time.

To get some idea on the optimal conditions for the cooling process the number of trapped electrons N_e (i.e. the ratio N_i/N_e) and their density n_e can be varied. In the real nested trap such density changes can be achieved within some limits by varying the applied electrostatic potential. Some examples for the observed dependencies are given in Fig. 6.7. As expected, a decreasing number of electrons at a given number of ions, i.e. an increasing N_i/N_e, at fixed electron density ($n_e = 10^7$ cm^{-3}) results in a stronger heating of the electrons (going from bottom to top in the left panel of Fig. 6.7). This increases the cooling time because the stopping force gets smaller at larger electron temperatures. A closer look reveals a more complicated feedback between electron temperature and energy loss than naively expected.

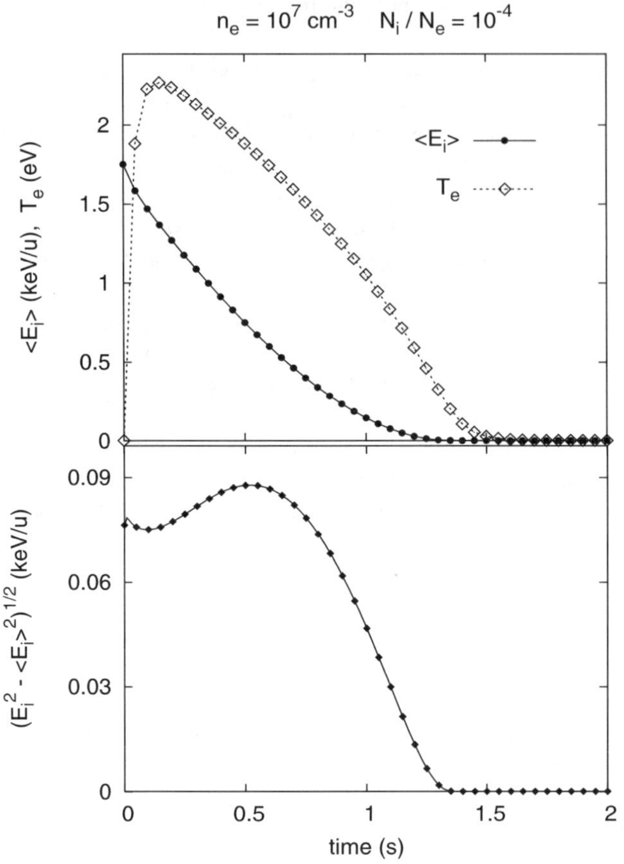

Fig. 6.5. Time evolution of an ensemble of bare Uranium ions cooled by magnetized electrons with $n_e = 10^7$ cm^{-3}, $N_i/N_e = 10^{-4}$, $T_0 = 4$ K and $B = 6$ T in the simplified trap sketched in Fig. 6.3 as obtained from a numerical solution of the coupled equations (6.10)-(6.14). Top: Average energy per ion $\langle E_i \rangle(t)$ and electron temperature $T_e(t)$. Bottom: Width of the energy distribution of the ions as function of time. After [141].

Changing the electron number by a factor 10 should result in a 10 time larger/lower T_e. But the stopping force depends strongly and nonlinearly on T_e and is larger for lower temperatures. This then implies a stronger heating of the electrons. A variation of the electron density n_e acts in a similar manner. An increase of n_e results in an (almost linear) increase of the stopping force and thus enhances the heating of the electrons. The higher T_e in turn reduces the energy loss of the ions in a nonlinear manner. The corresponding variation of cooling times and electron temperatures are shown in the right part of Fig. 6.7. Instead of a scaling of the cooling time like $1/n_e$, i.e. here a change by a factor 10, it only changes by factors 2 . . . 4. The cooling time reduces from about 1.2 s only to 0.5 s when increasing the density n_e from 10^7 cm^{-3}

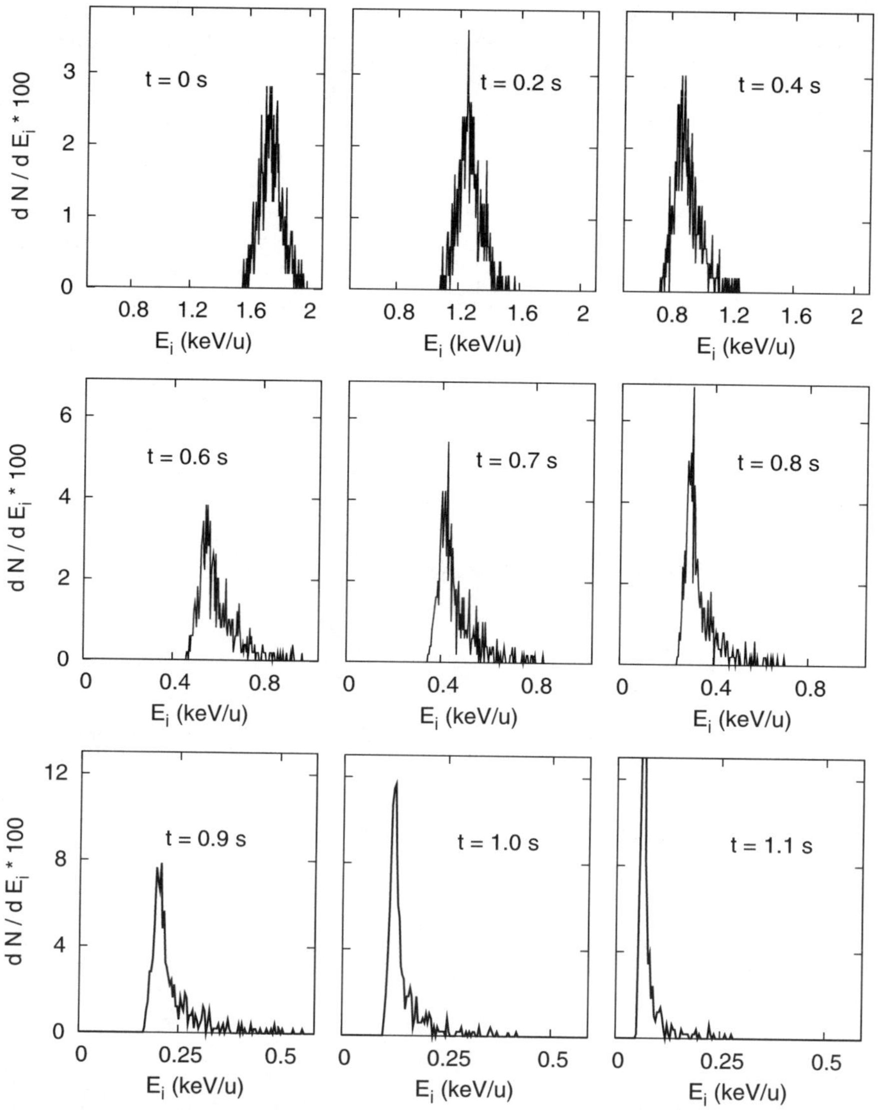

Fig. 6.6. Energy distribution of the ensemble of bare Uranium ions considered in Fig. 6.5 during cooling by magnetized electrons ($n_e = 10^7$ cm^{-3}, $N_i/N_e = 10^{-4}$, $T_0 = 4$ K, $B = 6$ T) at different times. The variation of the scales between the top, center and bottom panels should be noted. After [141].

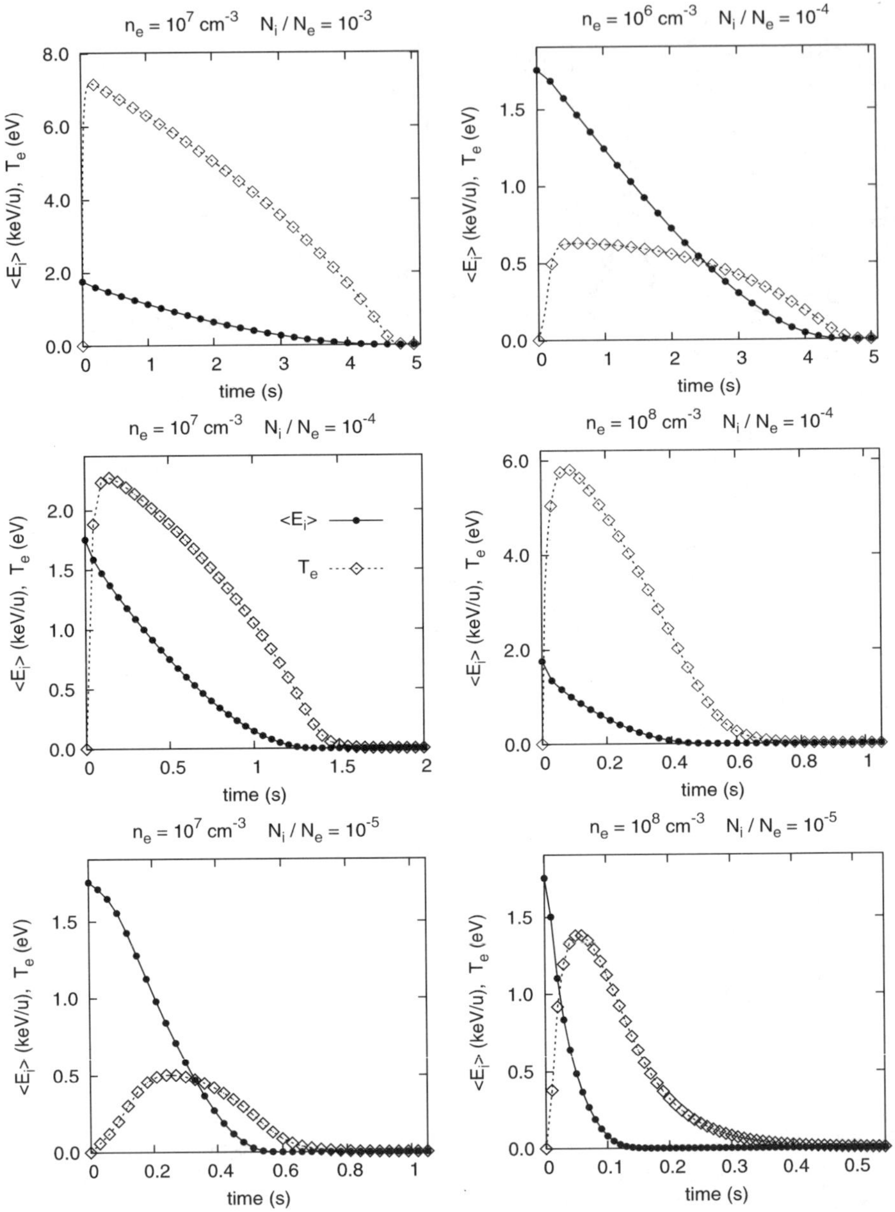

Fig. 6.7. Electron temperature and average ion energy as function of time for the ensemble of bare Uranium ions of Fig. 6.5 cooled in the trap configuration of Fig. 6.3 by magnetized electrons with $T_0 = 4$ K and $B = 6$ T, now for different electron densities n_e and different numbers of electrons N_e at given N_i, i.e. different ratios N_i/N_e. The variation of the scales should be noted. After [141].

to 10^8 cm^{-3} at fixed $N_i/N_e = 10^{-4}$ (from left center to right center in Fig. 6.7), and increases from about 1.2 s to 4.3 s for decreasing $n_e = 10^7$ cm^{-3} to $n_e = 10^6$ cm^{-3} (from left center to right top).

Such calculations of electron cooling in Penning traps show an expected strong dependence of the observed cooling times on the number of trapped electrons N_e, or more precisely on the ratio N_i/N_e, and on the electron density n_e, but in a more complex manner than one would naively estimate. The whole cooling process is subject to a nonlinear feedback between electron heating and stopping force. This prevents simple predictions of the variation of the cooling time. An optimization for fast cooling thus requires a fully numerical treatment and careful modeling. The situation becomes even more involved if also recombination losses have to be minimized, as the optimal parameters for a high cooling rate may run into a conflict with unacceptably high recombination rates; see [141] for the related variation of the recombination losses. Bearing in mind the simplifications and assumptions employed here these theoretical studies demonstrate that cooling times < 1 s (and RR-losses $< 10\%$) will be feasible for U^{92+} in HITRAP.

6.2.3 Cooling of Antiprotons and Protons by Electrons and Positrons

In the experiments for antihydrogen production by the ATRAP and ATHENA collaborations at CERN [6, 44] the antiprotons are cooled in Penning traps both by electrons and, in the final stage, by positrons. In this context we would like to consider in some detail the difference between the energy loss of antiprotons in strongly magnetized electrons compared to positrons or likewise the difference between electron cooling of antiprotons and protons, or more general, between repulsive and attractive electron(positron)-ion interaction. To our knowledge there are, however, no dedicated quantitative measurements like those presented in Sect. 6.2.2 available in the repulsive case. We will hence discuss this question on the basis of CTMC simulations (Sect. 2.3), and as an illustrating example for the difficulties and problems arising for a perturbation treatment in cases of a strong magnetic field and an ion motion under small angles α to the magnetic field lines. Under these conditions a perturbative treatment, for example in third order [10, 81], is not sufficient. We immediately expect a significant difference between the attractive and repulsive interactions in a binary collisions picture just by simple symmetry arguments. For large magnetic fields the electrons tend to move like beads on a wire along the magnetic field lines, as depicted in Fig. 6.8. For attractive interaction and for ions also moving along the magnetic field lines the acceleration of the electron when entering the potential is fully canceled by the subsequent deceleration when leaving it. In this case the velocity- and energy transfer between ion and electron vanishes. For a repulsive interaction, however, large velocity transfers occur when the particles are reflected from each other which results in a nonzero velocity- and energy transfer between ion and electrons. For ions moving with a finite angle to the magnetic field this symmetry argument does not apply any longer, but as we will see from the results of CTMC simulations some effect of this completely different scattering behavior persists and leads to a considerable difference between attractive and

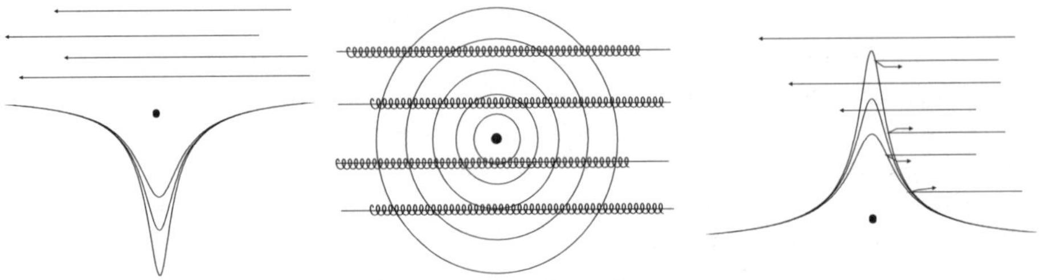

Fig. 6.8. In a strong magnetic field electrons are constrained to move along the field lines (center). For an attractive potential there is no scattering (left), for a repulsive potential low energy electrons are backscattered (right).

repulsive interaction also for nonzero angles α. This is shown in Fig. 6.9 for the stopping of protons versus antiprotons under the typical conditions in a Penning trap (see the previous section). The very different behavior of the stopping power S (top) and the stopping force parallel to the magnetic field $\mathcal{F}_{\|}$ (bottom) for protons and antiprotons is clearly visible. While S and $\mathcal{F}_{\|}$ for a proton steeply fall off towards zero for $v_{i\perp} \to 0$, i.e, when the proton is moving along the magnetic field, S and $\mathcal{F}_{\|}$ for an antiproton tend to a maximum. The transverse stopping force which anyway becomes zero for parallel motion shows only a slight sensitivity on the sign of the projectile charge. For a more quantitative comparison the stopping power is plotted in Fig. 6.10 as function of the velocity v_i for some selected angles α ($v_{i\|} = v_i \cos\alpha, v_{i\perp} = v_i \sin\alpha$) (left column) and also for a somewhat larger electron temperature, that is, weaker coupling (right column). In terms of the angle α we find, as expected, the largest difference between protons and antiprotons at small α. This difference diminishes with increasing angle, but remains significant up to angles of about $\alpha = 30^o$ and larger (not shown). For decreasing coupling strength (2.42), e.g. for an increasing electron temperature T_e, that is, when going towards or deeper into the weak coupling regime these differences are strongly reduced, as shown in Fig. 6.10 for a temperature increase by a factor of five (going from the left column to the right one). For $\alpha = 0$, of course, a large difference between attractive and repulsive interaction, i.e. between $S = 0$ and some finite value, always remains. But the range of angles around $\alpha = 0$ where the stopping power is markedly distinct form each other strongly shrinks down in the very weak coupling regime. This behavior found in the framework of a binary collision picture remains qualitatively unchanged when adding the full nonlinear dynamic polarization of the electrons (as in the numerical PIC treatment discussed in Sect. 2.5 and Sect. 4.6.2, Fig. 4.12). Although there is now a finite stopping force on a proton also for $\alpha = 0$ (the above symmetry argument does not apply any more), it is still small compared to that on an antiproton and this difference diminishes again for increasing angle.

The strong sensitivity of the stopping power on the sign of the projectile charge at strong magnetic field and small angles between $\boldsymbol{v}_i$ and $\boldsymbol{B}$ suggests that any per-

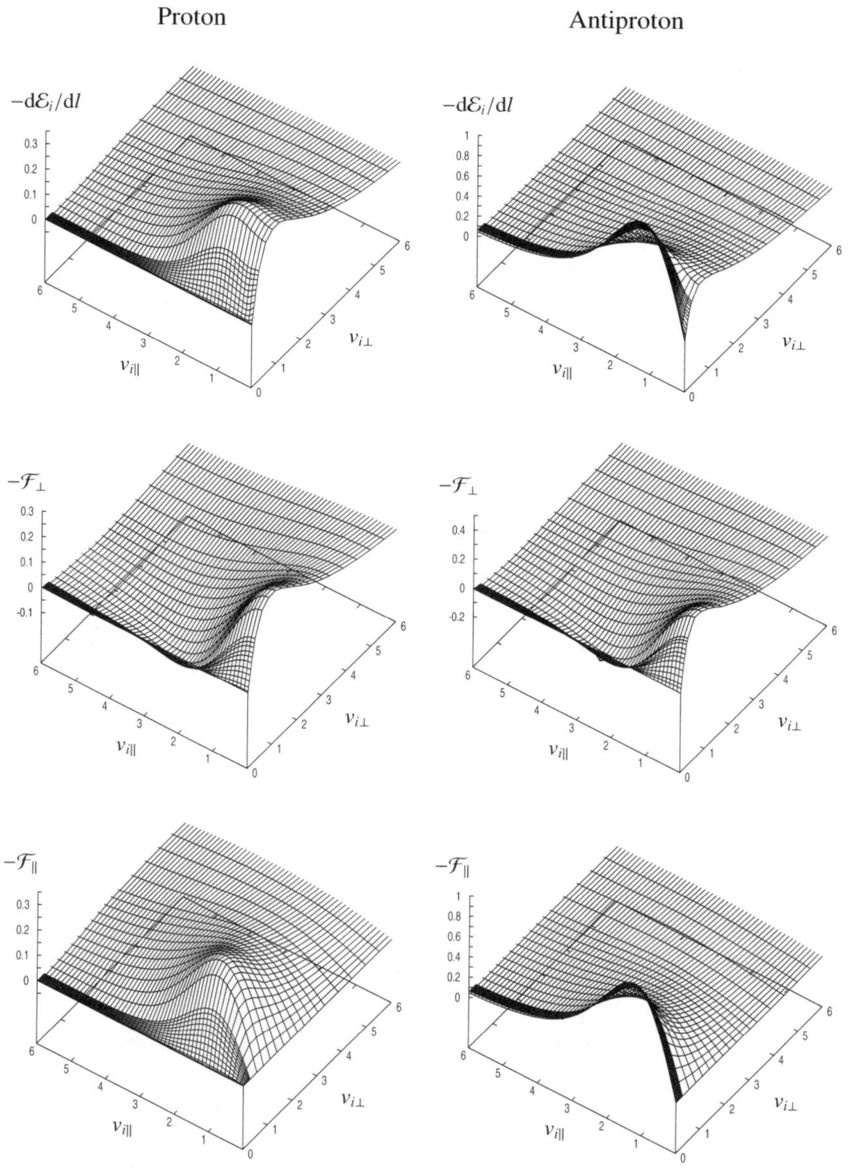

Fig. 6.9. Stopping power $S = -\mathrm{d}\mathcal{E}_i/\mathrm{d}l = -\mathcal{F} \cdot \hat{\boldsymbol{v}}_i$ (top) and stopping force transverse $\mathcal{F}_\perp$ (center) and along $\mathcal{F}_\parallel$ the magnetic field $\boldsymbol{B}$ as functions of the transverse $v_{i\perp}$ and parallel $v_{i\parallel}$ components of the projectile velocity in units of the thermal velocity v_{th} for the stopping of protons (left column) and antiprotons (right column) in magnetized electrons with $n_e = 7\times10^6$ cm^{-3}, $T_e = 4.2$ K and at $B = 6$ T ($S, \mathcal{F}_\perp, \mathcal{F}_\parallel$ in arbitrary units).

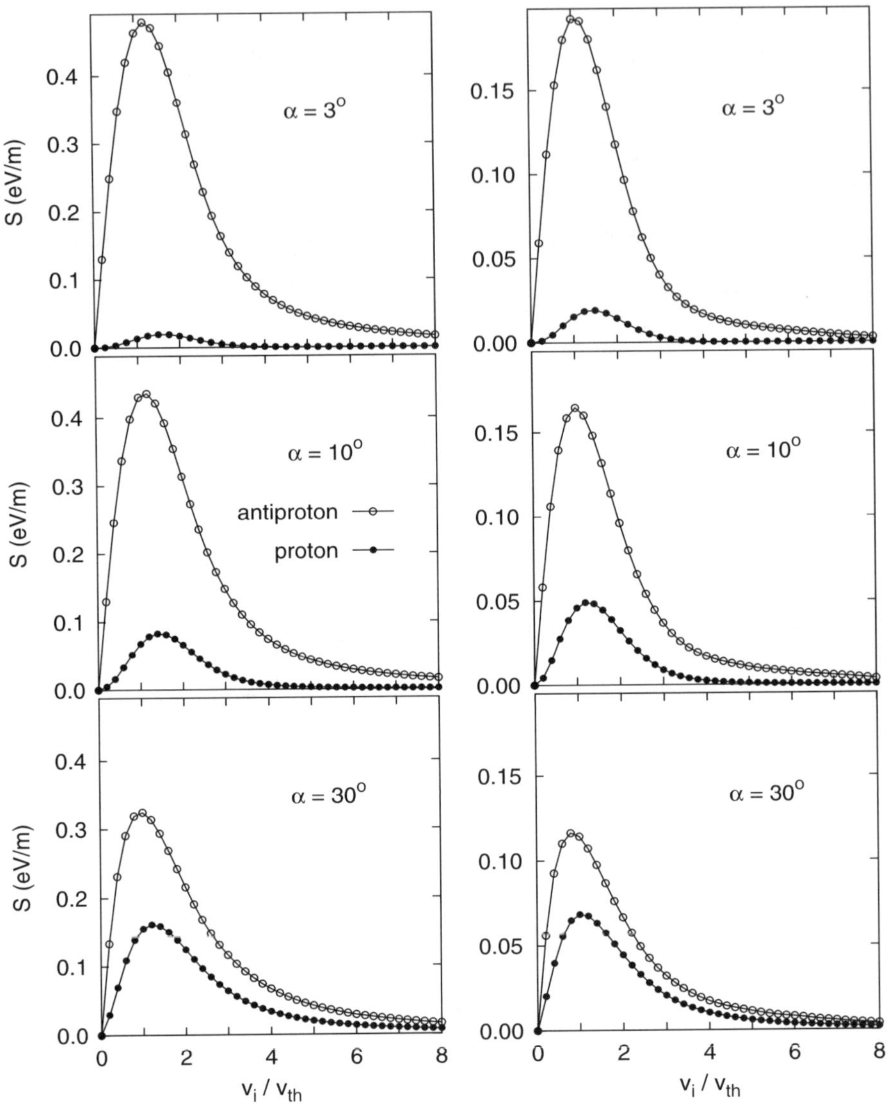

Fig. 6.10. Stopping power S of antiprotons (open circles) and protons (filled circles) in a magnetized electron plasma with $n_e = 7 \times 10^6$ cm^{-3}, $T_e = 4.2$ K (left panel) and $T_e = 21$ K (right panel), and $B = 6$ T for different directions α of the ion motion calculated by CTMC simulations. After [140].

turbative treatment as the linear response (LR) or binary collision to O(Z^2) (BC) approach which yields results independent of the sign of the interaction runs into difficulties and becomes doubtful under these conditions. There exists apparently no suitable parameter of smallness in case of this quasi-one-dimensional electron motion. We have already been confronted with this question when investigating the perturbative approaches and the difference which shows up between LR and BC in the limiting case of $\alpha \to 0$ and a strong magnetic field (see in particular Sects. 3.10 and 4.6.2). But the region where perturbative treatments fail considerably, i.e. here the region of small α, turns out to be rather small in the weak coupling regime. There the deficiencies of the perturbative approaches ought to be of minor significance in applications where an ensemble of ions with non–vanishing transverse motion is involved, like the electron cooling in storage rings and traps discussed in the previous Sections of this Chapter.

7 Summary and Conclusion

In the wide field of the interaction of charged particle beams with matter this monograph is primarily concerned with a special topic, namely the influence of external magnetic fields on the passage of heavy ions or antiprotons through an electron plasma. Observables are the stopping power which determines the range of the particles and the time until they come to rest as well as their straggling. The independent variables are the strength of the magnetic field $\boldsymbol{B}$, the velocity $\boldsymbol{v}_i$ of the projectile (both its magnitude and the angle α with $\boldsymbol{B}$), the charge Ze of the projectile and properties of the target like its density n_e and the velocity distribution $f(\boldsymbol{v}_e)$ of the target electrons. Theoretical estimates of the stopping power and the straggling in terms of these variables are desired for applications in nuclear fusion, electron cooling of charged particle beams in storage rings and the deceleration of particles in traps.

We studied in detail the two complementary pictures in which the passage of charged particles through a plasma can be treated. In the binary collision approximation (BC, Chap. 3) the focus is on the pairwise interaction of the particles. The desired observables can be obtained numerically from classical trajectory Monte Carlo (CTMC, Sect. 2.3) simulations. Of course closed analytical treatments can offer more physical insight provided the underlying assumptions are valid. Within the BC we expanded the velocity transfer with respect to the projectile–electron interaction. As straggling is described by the second moment of the velocity transfer, an expansion up to first order is sufficient, but the second order is needed for the stopping power S. For this observable the energy transfer ΔE_i to the test particle is first averaged with respect to unobserved quantities like the phase φ of the cyclotron motion and integrated with respect to the impact parameter $\bar{\boldsymbol{s}}$ (or $\boldsymbol{s}$ in the unmagnetized case). This yields the energy loss $\mathrm{d}E_i/\mathrm{d}l$ of the test particle with monochromatic electrons. For the stopping power S and the stopping (drag) force $\mathcal{F}$ the energy loss must then be averaged with respect to the electron velocity distribution.

With increasing magnetic field the unperturbed motion of the electrons becomes quasi one–dimensional, they move along the field lines like beads on a wire. For comoving projectiles and attractive interactions the velocity transfer vanishes, while it becomes maximal in the repulsive case. This indicates a breakdown of the perturbation treatment. For a strong field the stopping power behaves in a logarithmic manner at small projectile velocities. While this divergence can be lifted by using velocity–dependent cut–off parameters as suggested by quantum diffraction, the validity of the perturbation treatment at small velocities must be checked by comparing with CTMC results (Sect. 3.10). These must be averaged in the chaotic regimes of phase space, see Sect. 3.7.

In the dielectric theory (DT, Chap. 4) one calculates the response of the target, here the magnetized plasma to the perturbation caused by the projectile. In this continuum theory the underlying nonlinear Vlasov–Poisson equation can be solved numerically by a particle-in-cell (PIC, Sect. 2.5) simulation. More physical insight can be obtained by linearization of the response (LR). One calculates first the dynamic dielectric response function $\varepsilon(\boldsymbol{k}, \omega)$ of the target, here a spatially homogeneous, magnetized, anisotropic Maxwellian electron plasma, see Sect. 2.4. In a second step the stopping power is calculated from the force between the test particle and the polarization cloud it has created in its wake. For that purpose the inverse of the dielectric function must be integrated in $\boldsymbol{k}$–space. Generally this cannot be done in closed form (Chap. 4). More manageable expressions are obtained by disregarding the dynamic collectivity (reduced linear response, RLR, Sect. 4.6), which corresponds to the second order perturbation treatment of the BC (Sect. 4.7). Some account of the dynamics can be restored in an RLR with a dynamic screening of the Coulomb potential. But as in BC both the LR and RLR stopping powers behave in an anomalous manner at low ion velocities where the response is too strong for linearization. The linear response results must be validated by comparing with PIC simulations, the role of collectivity can be studied in comparison to CTMC simulations, see Fig. 4.12.

The electrons must be treated quantum mechanically whenever their magnetic (Landau) energy becomes comparable or is larger than their thermal energy. In Chap. 5 the quantum theory of the stopping power is developed both within the linearized DT and in the BC picture. As in the classical case both approaches agree up to the collective dynamic response. The quantum energy transfers agree with the classical results in that limit, provided that the appropriate averages with respect to unobserved quantum numbers and variables, respectively, are taken, see Sect. 5.8.

In view of the strengths and weaknesses of these approaches we recommend a pragmatic approach for the calculation of the stopping of charged particles in a magnetized electron plasma: For large projectile velocities (compared to the thermal velocity of the electrons) the dielectric linear response treatment is preferred as it includes collectivity. The linearization becomes increasingly doubtful for strong magnetic fields and low projectile velocities. In the attractive case this is cured to some extent within the BC by self–cutting Coulomb logarithms as motivated by quantum diffraction. In the repulsive case, e.g. for the cooling of antiprotons by electrons the CTMC is indicated to account for the hard collisions at low velocities. The method of choice is the PIC simulation which also accounts for the collective response. On the other hand, the simulation methods offer less physical insight, are numerically expensive and suffer from statistical fluctuations.

In applications (Chap. 6) like electron cooling in heavy ion storage rings or the deceleration of charged, heavy particles in traps the region of phase space where the perturbation expansion or the linearization are doubtful is small. Observables like cooling rates which require some averaging with respect to the initial data turn out to be quite robust. So with sufficient care predictions can be done with good confidence. There exists, for example, a wide range of parameters for which ions like U^{92+} survive recombination while brought to rest in a trap.

A Dielectric Function of the Magnetized Electron-Ion Plasma

Here we describe the evaluation of the dielectric function with the anisotropic Maxwell distribution (2.43). We introduce the Fourier transformations of $f_{1\nu}(\boldsymbol{r}, \boldsymbol{v}, t)$ with respect to variables $\boldsymbol{r}$ and t, $f_{1\nu}(\boldsymbol{k}, \omega, \boldsymbol{v})$. Because of the cylindrical symmetry (around the magnetic field direction $\boldsymbol{b} = \boldsymbol{B}/B = \hat{z}$) of the problem, we choose

$$\boldsymbol{v} = \hat{\boldsymbol{x}} v_\perp \cos\sigma + \hat{\boldsymbol{y}} v_\perp \sin\sigma + v_\parallel \hat{z}\,. \tag{A.1}$$

Then the Vlasov equation (2.44) for the distribution function becomes

$$\frac{\partial}{\partial\sigma} f_{1\nu}(\boldsymbol{k}, \omega, \boldsymbol{v}) - \frac{\mathrm{i} c_\nu}{\Omega_\nu} (\boldsymbol{k} \cdot \boldsymbol{v} - \omega - \mathrm{i}0) f_{1\nu}(\boldsymbol{k}, \omega, \boldsymbol{v}) = -\frac{\mathrm{i}\,|q_\nu|}{m_\nu \Omega_\nu} \phi(\boldsymbol{k}, \omega) \left(\boldsymbol{k} \cdot \frac{\partial f_{0\nu}}{\partial \boldsymbol{v}}\right), \tag{A.2}$$

where $\phi(\boldsymbol{k}, \omega)$ is the Fourier transformation of $\phi(\boldsymbol{r}, t)$. The positive infinitesimal $+\mathrm{i}0$ in equation (A.2) serves to assure the adiabatic turning on of the disturbance and guarantees thereby the causality of the response. The solution of the equation (A.2) has the form

$$\begin{aligned} f_{1\nu}(\boldsymbol{k}, \omega, \boldsymbol{v}) = &-\frac{\mathrm{i}\,|q_\nu|}{m_\nu \Omega_\nu} \phi(\boldsymbol{k}, \omega) \int_\infty^\sigma \mathrm{d}\sigma_2 \left(\boldsymbol{k} \cdot \frac{\partial f_{0\nu}}{\partial \boldsymbol{v}}\right)_{\sigma=\sigma_2} \\ &\times \exp\left[\frac{\mathrm{i} c_\nu}{\Omega_\nu} \int_\sigma^{\sigma_2} \mathrm{d}\sigma_1 \left[\omega + \mathrm{i}\,0 - (\boldsymbol{k} \cdot \boldsymbol{v})_{\sigma=\sigma_1}\right]\right]. \end{aligned} \tag{A.3}$$

Combining the expression (A.3) with the Poisson equation (2.45) we find for the dielectric function

$$\begin{aligned} \varepsilon(\boldsymbol{k}, \omega) = 1 &+ \frac{\mathrm{i}}{\varepsilon_0 k^2} \sum_\nu \frac{c_\nu n_\nu q_\nu^2}{m_\nu \Omega_\nu} \int_0^\infty v_\perp \mathrm{d}v_\perp \int_0^{2\pi} \mathrm{d}\sigma \int_{-\infty}^{+\infty} \mathrm{d}v_\parallel \int_\infty^\sigma \mathrm{d}\sigma_2 \\ &\times \left[k_\parallel \frac{\partial f_{0\nu}}{\partial v_\parallel} + k_\perp \cos(\theta - \sigma_2) \frac{\partial f_{0\nu}}{\partial v_\perp}\right] \\ &\times \exp\left\{\frac{\mathrm{i} c_\nu}{\Omega_\nu} \int_\sigma^{\sigma_2} \mathrm{d}\sigma_1 \left[\omega + \mathrm{i}0 - k_\parallel v_\parallel - k_\perp v_\perp \cos(\theta - \sigma_1)\right]\right\}, \end{aligned} \tag{A.4}$$

where $k_x = k_\perp \cos\theta$, $k_y = k_\perp \sin\theta$. After integration with respect to the variables σ_1, σ_2 and σ, and using the expression [52]

$$\mathrm{e}^{-iz\sin\theta} = \sum_{n=-\infty}^{+\infty} J_n(z)\,\mathrm{e}^{-in\theta}, \tag{A.5}$$

where J_n is the Bessel function of the nth order, we obtain [5, 67, 77]

$$\varepsilon(\boldsymbol{k}, \omega) = 1 - \sum_{\nu} \frac{2\pi\omega_{p\nu}^2}{k^2} \sum_{n=-\infty}^{+\infty} \int_0^{\infty} v_{\perp} \mathrm{d}v_{\perp} \int_{-\infty}^{+\infty} \mathrm{d}v_{\parallel} \times \left(\frac{n\Omega_{\nu}}{v_{\perp}} \frac{\partial f_{0\nu}}{\partial v_{\perp}} + k_{\parallel} \frac{\partial f_{0\nu}}{\partial v_{\parallel}} \right) \frac{J_n^2(k_{\perp} v_{\perp}/\Omega_{\nu})}{n\Omega_{\nu} + k_{\parallel} v_{\parallel} - \omega - \mathrm{i}0} . \tag{A.6}$$

Substituting expression (2.43) for the unperturbed distribution function $f_{0\nu}$ into (A.6) we finally obtain [67]

$$\varepsilon(\boldsymbol{k}, \omega) = 1 + \sum_{\nu} \frac{1}{k^2 \lambda_{\mathrm{D}\nu\parallel}^2} \left\{ 1 + \sum_{n=-\infty}^{+\infty} \left(1 + \frac{1}{\tau_{\nu}} \frac{n\Omega_{\nu}}{\omega - n\Omega_{\nu}} \right) \left[W\left(\frac{\omega - n\Omega_{\nu}}{|k_{\parallel}| v_{\mathrm{th}\nu\parallel}} \right) - 1 \right] \Lambda_n(z_{\nu}) \right\} , \tag{A.7}$$

where $z_{\nu} = k_{\perp}^2 v_{\mathrm{th}\nu\perp}^2 / \Omega_{\nu}^2 = k_{\perp}^2 a_{\nu}^2$, $\Lambda_n(z) = \mathrm{e}^{-z} I_n(z)$, $I_n(z)$ is the modified Bessel function of the nth order, and $W(z)$ is the plasma dispersion function, equations (4.2)-(4.4), see also [43]. For the derivation of equation (A.7) we have used the formula [52, 67]

$$\sum_{n=-\infty}^{\infty} \Lambda_n(z) = 1 . \tag{A.8}$$

To show the identity of the two forms, expressions (2.48) and (A.7) of the dielectric function we use the expansion in modified Bessel functions [52]

$$\mathrm{e}^{-z(1-\cos\theta)} = \sum_{n=-\infty}^{\infty} \Lambda_n(z) \, \mathrm{e}^{\mathrm{i}n\theta} . \tag{A.9}$$

This allows to rewrite $\exp[-X_{\nu}(t)]$ with $X_{\nu}(t)$ from (2.49) as

$$\exp[-X_{\nu}(t)] = \exp\left(-t^2 \cos^2\beta\right) \sum_{n=-\infty}^{+\infty} \Lambda_n(z_{\nu}) \exp\left(\frac{\mathrm{i}n\Omega_{\nu} t\sqrt{2}}{k v_{\mathrm{th}\nu\parallel}} \right) . \tag{A.10}$$

Substituting (A.10) into expression (2.48) and integration over the variable t leads to equation (A.7).

B Anomalous Term

Consider the more detailed derivation of the anomalous contribution which arises from the first term of equation (4.42). We start with the expression

$$S_{\mathrm{I}} = \frac{4Z^2 e^2}{(2\pi)^{3/2}\lambda_{\mathrm{D}\|}^2}\sum_\nu \varrho_\nu \frac{v_i}{v_{\mathrm{th}\nu\|}}\int_0^\xi k^3 \mathrm{d}k \int_0^\pi \mathrm{d}\phi\, Q_\nu(k,\phi,v_i)\,, \tag{B.1}$$

where the variable k is scaled in units of $\lambda_{\mathrm{D}\|}^{-1}$ and $\xi = k_{\max}\lambda_{\mathrm{D}\|}$. Here

$$Q_\nu(k,\phi,v_i) = \int_0^1 \frac{\mathrm{d}\mu}{\mu}\Phi_\nu(\mu,k,\phi,\omega)\exp\left(-\frac{v_i^2\cos^2\Theta}{2v_{\mathrm{th}\nu\|}^2\mu^2}\right), \tag{B.2}$$

where

$$\Phi_\nu(\mu,k,\phi,\omega) = \frac{\Lambda_0(z_\nu)\cos^2\Theta}{\left[k^2+\mathcal{G}(\omega)\right]^2+\mathcal{F}^2(\omega)} \tag{B.3}$$

with $z_\nu = (k^2\tau_\nu/\eta_\nu^2\varrho_\nu)(1-\mu^2)$ and $\eta_\nu = \Omega_\nu/\omega_{p\nu}$. For $v_i \to 0$ a leading–term approximation of (B.2) leads to

$$Q_\nu(k,\phi,v_i) \simeq \frac{1}{2}\Phi_\nu^{(0)}(0,k,\phi)\ln\left(\frac{2v_{\mathrm{th}\nu\|}^2}{\gamma_0 v_i^2\sin^2\alpha\cos^2\phi}\right) + \int_0^1\frac{\mathrm{d}\mu}{\mu}\left[\Phi_\nu^{(0)}(\mu,k,\phi)-\Phi_\nu^{(0)}(0,k,\phi)\right]+\mathrm{O}(v_i)\,, \tag{B.4}$$

where $\gamma_0 = \mathrm{e}^\gamma$ and γ is the Euler's constant, $\Phi_\nu^{(0)}(\mu,k,\phi)$ is the value of the function $\Phi_\nu(\mu,k,\phi,\omega)$ at $v_i \to 0$ $(\omega \to 0)$. Note that in this limit $\mathcal{F}(\omega) \to 0$ and $\mathcal{G}(\omega)|_{\omega\to 0} = \mathcal{H}(k,\mu)$ with

$$\mathcal{H}(k,\mu) = 1 + \frac{2\sqrt{2}}{|\mu|\,k}\sum_\nu \varrho_\nu^{3/2}\eta_\nu\left(\frac{1}{\tau_\nu}-1\right)\sum_{n=1}^\infty n\Lambda_n(z_\nu)\,\mathrm{Di}\left(\frac{n\eta_\nu\sqrt{\varrho_\nu}}{\sqrt{2}k\,|\mu|}\right). \tag{B.5}$$

Thus

$$\Phi_\nu^{(0)}(\mu,k,\phi) = \frac{\Lambda_0(z_\nu)\cos^2\Theta}{\left[k^2+\mathcal{H}(k,\mu)\right]^2} \tag{B.6}$$

and

$$\Phi_\nu^{(0)}(0,k,\phi) = \frac{\Lambda_0(k^2\varsigma_\nu)\sin^2\alpha\cos^2\phi}{\left[k^2 + \mathcal{H}_0(k)\right]^2} \tag{B.7}$$

with $\varsigma_\nu = \tau_\nu/\eta_\nu^2\varrho_\nu$ and $\mathcal{H}_0(k) = \mathcal{H}(k,0)$,

$$\mathcal{H}_0(k) = 1 + 2\sum_\nu \varrho_\nu\left(\frac{1}{\tau_\nu} - 1\right)\sum_{n=1}^{\infty}\Lambda_n(k^2\varsigma_\nu)\,. \tag{B.8}$$

Using the relation (A.8) (see also [52]) the function $\mathcal{H}_0(k)$ finally takes the form

$$\mathcal{H}_0(k) = \sum_\nu \varrho_\nu\left[\frac{1}{\tau_\nu} + \left(1 - \frac{1}{\tau_\nu}\right)\Lambda_0(k^2\varsigma_\nu)\right] \tag{B.9}$$

with $\Lambda_0(z) = \mathrm{e}^{-z}I_0(z)$.

In small velocity limit, $v_i \to 0$, substituting equation (B.4) into (B.1) and after μ and ϕ-integrations we finally obtain $S_{\mathrm{I}} = S_{\mathrm{an}} + S'_{\mathrm{I}}$, where S_{an} and S'_{I} are determined by (4.43) and (4.45), respectively.

C Dielectric Function of the Magnetized Quantum Plasma

In this Appendix we derive an alternative integral representation of the dielectric function of the quantum plasma discussed in Sect. 5.1 (see equation (5.18)). This representation can be treated as the quantum-mechanical analogue of classical dielectric function, (2.48).

Consider the imaginary part of the dielectric function (5.18)

$$\mathrm{Im}\varepsilon(\boldsymbol{k},\omega) = \frac{4\pi^2 e^2}{k^2(2\pi\lambda_{\mathrm{B}})^2} \sum_{\sigma=\pm 1/2} \sum_{n;n'=0}^{\infty} F_{nn'}^2(\zeta) \int_{-\infty}^{+\infty} dq_z \tag{C.1}$$
$$\times \left[f\left(E_{n'\sigma}(q_z+k_z)\right) - f\left(E_{n\sigma}(q_z)\right)\right] \delta\left[E_{n\sigma}(q_z) - E_{n'\sigma}(q_z+k_z) - \hbar\omega\right] .$$

Here the eigenvalues of the free particle $E_{n\sigma}(q_z)$ are defined by (5.13). Inserting equation (5.13) into (C.1) and employing the integral representation of the δ-function,

$$\delta(E - E') = \frac{1}{2\pi\mathrm{i}} \int_{\delta-\mathrm{i}\infty}^{\delta+\mathrm{i}\infty} d\lambda\, \mathrm{e}^{\lambda(E-E')} , \tag{C.2}$$

equation (C.1) can be written as

$$\mathrm{Im}\varepsilon(\boldsymbol{k},\omega) = \frac{2\pi e^2}{k^2(2\pi\lambda_{\mathrm{B}})^2} \int_0^{\infty} dE f(E) \frac{1}{2\pi\mathrm{i}} \int_{\delta-\mathrm{i}\infty}^{\delta+\mathrm{i}\infty} d\lambda\, \mathrm{e}^{\lambda E}\left(1+\mathrm{e}^{-\lambda\hbar\Omega_e}\right)$$
$$\times \int_{-\infty}^{+\infty} d\varsigma \mathrm{e}^{-\mathrm{i}\varsigma\hbar\omega} \int_{-\infty}^{+\infty} dq_z \exp\left(-\mathrm{i}\hbar^2\varsigma\frac{k_z^2+2q_zk_z}{2m}\right) \tag{C.3}$$
$$\times \sum_{n'=-\infty}^{\infty}\left[\mathrm{e}^{\mathrm{i}\varsigma\hbar\Omega_e n'} \exp\left(-\lambda\frac{\hbar^2(q_z+k_z)^2}{2m}\right) - \mathrm{e}^{-\mathrm{i}\varsigma\hbar\Omega_e n'} \exp\left(-\lambda\frac{\hbar^2 q_z^2}{2m}\right)\right]$$
$$\times \sum_{n=0}^{\infty} \mathrm{e}^{-\lambda\hbar\Omega_e n} F_{n;n'+n}^2(\zeta) ,$$

where δ is an arbitrary real parameter. In (C.3) the summation over n is performed using (5.53) and (5.55). The result reads as

$$
\begin{aligned}
\mathrm{Im}\varepsilon(\boldsymbol{k},\omega) = {} & \frac{2\pi e^2}{k^2(2\pi\lambda_{\mathrm{B}})^2}\int_0^{\infty}\mathrm{d}E f(E)\frac{1}{2\pi\mathrm{i}}\int_{\delta-\mathrm{i}\infty}^{\delta+\mathrm{i}\infty}\mathrm{d}\lambda\, \mathrm{e}^{\lambda E}\frac{\mathrm{e}^{-\zeta\coth(\lambda\hbar\Omega_e/2)}}{\tanh(\lambda\hbar\Omega_e/2)} \\
& \times\int_{-\infty}^{+\infty}\mathrm{d}\varsigma \mathrm{e}^{-\mathrm{i}\varsigma\hbar\omega}\int_{-\infty}^{+\infty}\mathrm{d}q_z \exp\left(-\mathrm{i}\hbar^2\varsigma\frac{k_z^2+2q_zk_z}{2m}\right) \\
& \times\left\{\exp\left(-\lambda\frac{\hbar^2(q_z+k_z)^2}{2m}\right)\sum_{n'=-\infty}^{\infty}\mathrm{e}^{\mathrm{i}n'\hbar\Omega_e(\varsigma-\mathrm{i}\lambda/2)}I_{n'}\left(\frac{\zeta}{\sinh(\lambda\hbar\Omega_e/2)}\right)\right. \\
& \left.-\exp\left(-\lambda\frac{\hbar^2q_z^2}{2m}\right)\sum_{n'=-\infty}^{\infty}\mathrm{e}^{-\mathrm{i}n'\hbar\Omega_e(\varsigma+\mathrm{i}\lambda/2)}I_{n'}\left(\frac{\zeta}{\sinh(\lambda\hbar\Omega_e/2)}\right)\right\}.
\end{aligned}
\tag{C.4}
$$

The sum over n' can be evaluated using the summation formula (A.9). After some transformations in the ς-integral this yields

$$
\begin{aligned}
\mathrm{Im}\varepsilon(\boldsymbol{k},\omega) = {} & \frac{4\pi e^2}{k^2(2\pi\lambda_{\mathrm{B}})^2}\int_0^{\infty}\mathrm{d}E f(E)\frac{1}{2\pi\mathrm{i}}\int_{\delta-\mathrm{i}\infty}^{\delta+\mathrm{i}\infty}\mathrm{d}\lambda\, \mathrm{e}^{\lambda E}\frac{\sinh(\lambda\hbar\omega/2)}{\tanh(\lambda\hbar\Omega_e/2)}\mathrm{e}^{-\zeta\coth(\lambda\hbar\Omega_e/2)} \\
& \times\int_{-\infty}^{+\infty}\mathrm{d}\varsigma \exp\left[-\mathrm{i}\varsigma\left(\frac{\hbar^2k_z^2}{2m}+\hbar\omega\right)\right]\exp\left[\frac{\zeta}{\sinh(\lambda\hbar\Omega_e/2)}\cos(\hbar\Omega_e\varsigma)\right] \\
& \times\int_{-\infty}^{+\infty}\mathrm{d}q_z \exp\left[-\lambda\frac{\hbar^2}{2m}\frac{(q_z+k_z)^2+q_z^2}{2}\right]\exp\left(-\mathrm{i}\varsigma\hbar^2\frac{q_zk_z}{m}\right).
\end{aligned}
\tag{C.5}
$$

In equation (C.5) there remains the q_z-integral. Straightforward calculation of this integral finally yields

$$
\begin{aligned}
\mathrm{Im}\varepsilon(\boldsymbol{k},\omega) = {} & \frac{4\pi e^2}{k^2}\frac{\sqrt{2\pi m}}{2\hbar(2\pi\lambda_{\mathrm{B}})^2}\int_0^{\infty}\mathrm{d}E f(E)\frac{1}{2\pi\mathrm{i}}\int_{\delta-\mathrm{i}\infty}^{\delta+\mathrm{i}\infty}\frac{\mathrm{d}\lambda}{\sqrt{\lambda}}\mathrm{e}^{\lambda E}\frac{\sinh(\lambda\hbar\omega/2)}{\tanh(\lambda\hbar\Omega_e/2)} \\
& \times\int_{-\infty}^{+\infty}\mathrm{d}\varsigma \mathrm{e}^{-\mathrm{i}\varsigma\hbar\omega/2}\exp\left[-\frac{k_\perp^2\lambda_{\mathrm{B}}^2}{2}\frac{\cosh(\lambda\hbar\Omega_e/2)-\cos(\hbar\Omega_e\varsigma/2)}{\sinh(\lambda\hbar\Omega_e/2)}\right] \\
& \times\exp\left[-\frac{\hbar^2k_z^2}{8m\lambda}\left(\lambda^2+\varsigma^2\right)\right].
\end{aligned}
\tag{C.6}
$$

For deriving equation (C.6) we have made the change $\varsigma\to\varsigma/2$. The integral form of the real part of the dielectric function can be evaluated from Kramers-Kronig relation, equation (5.21).

D Some Properties of the Function $F_{nn'}(\zeta)$

We consider some properties of the function $F_{nn'}(\zeta)$. First we give a detailed derivation of the classical limit of the function $F_{n;l+n}(\zeta)$ with fixed l and $n \to \infty$, $\zeta \to 0$ but $n\zeta = c^2/4 = \text{const}$. We express the Laguerre polynomials through the confluent hypergeometric sum [52]. Using (5.15) this results in

$$F_{n;l+n}(\zeta) = \left[\frac{\Gamma(n+l+1)}{\Gamma(n+1)}\right]^{1/2} \zeta^{l/2} \mathrm{e}^{-\zeta/2} \sum_{k=0}^{n} \frac{\Gamma(-n+k)}{\Gamma(-n)} \frac{1}{\Gamma(k+l+1)} \frac{\zeta^k}{k!}, \tag{D.1}$$

where $\Gamma(x)$ is the Euler function. For forthcoming considerations the following limit will be useful

$$\left.\frac{\Gamma(q+k)}{q^k \Gamma(q)}\right|_{q\to\infty} = 1 + \frac{k^2-k}{2q} + \mathrm{O}\left(\frac{1}{q^2}\right). \tag{D.2}$$

Here k is an arbitrary but fixed number. Using the last relation and the standard representation of the Bessel functions [52] in the limits $n \to \infty$, $\zeta \to 0$ but $n\zeta = c^2/4 = \text{const}$ (D.1) becomes

$$\left.F_{n;l+n}(\zeta)\right|_{n\to\infty;\,\zeta\to 0} = J_l(c) + \zeta P_l(c) + \mathrm{O}\left(\zeta^2\right), \tag{D.3}$$

where $P_l(c) = ((l+1)/c)\, J_l'(c)$. The function Q_l in (5.106) is given by $Q_l(c) = 2J_l(c)\, P_l(c)$.

Consider two sums involving the functions $F_{nn'}(\zeta)$. The first one is the summation of $F_{nn'}^2(\zeta)$ with respect to n',

$$\mathfrak{B}_n(\zeta) = \sum_{n'=0}^{\infty} F_{nn'}^2(\zeta). \tag{D.4}$$

For an evaluation of (D.4) we use the integral representation of the function $F_{nn'}(\zeta)$

$$F_{nn'}(\zeta) = \int_{-\infty}^{\infty} \varphi_n(t)\, \varphi_{n'}\left(t + \sqrt{2\zeta}\right) \mathrm{d}t, \tag{D.5}$$

which follows from equations (5.4), (5.12) and (5.14). Here $\varphi_n(t) = A_n \mathrm{e}^{-t^2/2} H_n(t)$ with $A_n^{-2} = 2^n n! \sqrt{\pi}$, where $H_n(t)$ are the Hermite polynomials. Note that the functions $\varphi_n(t)$ are orthogonal with $\int_{-\infty}^{\infty} \varphi_n(t)\varphi_{n'}(t)\mathrm{d}t = \delta_{nn'}$. Moreover these functions obey the differential equation [1, 52]

$$\varphi_n''(\zeta) = (\zeta^2 - 1 - 2n)\varphi_n(\zeta)\,. \tag{D.6}$$

Thus substituting (D.5) into (D.4) we arrive at

$$\mathfrak{B}_n(\zeta) = \int_{-\infty}^{\infty} \varphi_n\left(t - \sqrt{2\zeta}\right) \mathrm{d}t \int_{-\infty}^{\infty} \varphi_n\left(\tau - \sqrt{2\zeta}\right) \mathrm{d}\tau \sum_{n'=0}^{\infty} \varphi_{n'}(t)\varphi_{n'}(\tau)\,. \tag{D.7}$$

Now we recall the completeness of the functions $\varphi_n(t)$

$$\sum_{n=0}^{\infty} \varphi_n(t)\varphi_n(\tau) = \delta(t - \tau)\,. \tag{D.8}$$

Using this relation in (D.7) we finally obtain

$$\mathfrak{B}_n(\zeta) = \int_{-\infty}^{\infty} \varphi_n^2(t)\,\mathrm{d}t = 1\,. \tag{D.9}$$

In the last step we have introduced the new variable of integration in (D.7).

Consider the second sum which is defined as

$$\mathfrak{D}_n(\zeta) = \sum_{n'=1}^{\infty} n' F_{nn'}^2(\zeta)\,. \tag{D.10}$$

Again we use the integral representation of the functions $F_{nn'}(\zeta)$ (D.5). With this representation equation (D.10) becomes

$$\mathfrak{D}_n(\zeta) = \int_{-\infty}^{\infty} \varphi_n\left(t - \sqrt{2\zeta}\right) \mathrm{d}t \int_{-\infty}^{\infty} \varphi_n\left(\tau - \sqrt{2\zeta}\right) \mathrm{d}\tau \sum_{n'=1}^{\infty} n'\varphi_{n'}(t)\varphi_{n'}(\tau)\,. \tag{D.11}$$

In the next step we express $n'\varphi_{n'}(t)$ through the functions $\varphi_{n'}(t)$ and its second derivative using the differential equation (D.6). After some calculations this yields

$$\mathfrak{D}_n(\zeta) = \frac{1}{2}\int_{-\infty}^{\infty} \left[(\tau^2 - 1)\varphi_n\left(\tau - \sqrt{2\zeta}\right) - \varphi_n''\left(\tau - \sqrt{2\zeta}\right)\right] \varphi_n\left(\tau - \sqrt{2\zeta}\right) \mathrm{d}\tau\,. \tag{D.12}$$

In this equation the second derivative φ_n'' can be replaced by φ_n using (D.6). Then changing the integration variable we obtain

$$\mathfrak{D}_n(\zeta) = \int_{-\infty}^{\infty} \left(n + \zeta + \tau\sqrt{2\zeta}\right) \varphi_n^2(\tau)\mathrm{d}\tau = n + \zeta\,. \tag{D.13}$$

Recalling the relation (D.9) this result can be alternatively written in the form

$$\sum_{n'=0}^{\infty} (n' - n) F_{nn'}^2(\zeta) = \zeta\,. \tag{D.14}$$

References

1. M. Abramowitz, I.A. Stegun: *Handbook of Mathematical Functions*, 10th edn (Dover, New York 1972)
2. R. Abrines, I.C. Percival: Proc. Phys. Soc. London **88**, 861 (1966)
3. I.A. Akhiezer: Sov. Phys. JETP **13**, 667 (1961)
4. A.I. Akhiezer, I.A. Akhiezer, R.V. Polovin, A.G. Sitenko, K.N. Stepanov: *Plasma Electrodynamics*, 1st edn (Pergamon, Oxford 1975) vol. 1
5. A.F. Alexandrov, L.S. Bogdankevich, A.A. Rukhadze: *Principles of Plasma Electrodynamics* (Springer-Verlag, Berlin New York 1984)
6. M. Amoretti et al: Nature **419**, 456 (2002)
7. M. Amoretti et al: Phys. Lett. B **590**, 133 (2004)
8. N.R. Arista, W. Brandt: Phys. Rev. A **23**, 1898 (1981)
9. J. D'Avanzo, M. Lontano, P.F. Bortignon: Phys. Rev. E **47**, 3574 (1993)
10. W.H. Barkas, J.N. Dyer, H.H. Heckman: Phys. Rev. Lett **11**, 26 (1963)
11. G. Basbas, R.H. Ritchie: Phys. Rev. A **25**, 1943 (1982)
12. Th. Beier et al: *HITRAP, Technical Design report*, GSI Darmstadt, 2003
13. G. Benford, N. Rostoker: Phys. Rev. **181**, 729 (1969)
14. J. Bernard et al: Nucl. Instrum. Meth. Phys. Res. A **532**, 224 (2004)
15. I.B. Bernstein: Phys. Rev. **109**, 10 (1958)
16. BETACOOL program for simulation of beam dynamics in storage rings: http://lepta.jinr.ru/betacool/betacool.htm
17. BETACOOL manual: http://lepta.jinr.ru/betacool/sources/sources.htm
18. H. Bethe: Ann. Physik **5**, 325 (1930)
19. M. Beutelspacher, M. Grieser, D. Schwalm, A. Wolf: Nucl. Instrum. Meth. Phys. Res. A **441**, 110 (2000)
20. M. Beutelspacher: Thesis, University Heidelberg, Heidelberg (2000)
21. K. Binder, D.W. Heermann: *Monte Carlo Simulation in Statistical Physics*, 3rd edn (Springer, Berlin 1997)
22. C.K. Birdsall, A.B. Langdon: *Plasma Physics via Computer Simulations* (McGraw-Hill, New York 1985)
23. F. Bloch: Ann. Physik **16**, 285 (1933); Ann. Physik **30**, 72 (1933)
24. N. Bohr: Phil. Mag. **25**, 10 (1913); Phil. Mag. **30**, 581 (1915)
25. O. Boine-Frankenheim, J. D'Avanzo: Phys. Plasmas **3**, 792 (1996)
26. J. Bosser: CERN 95-06, 673 (1995)
27. S.V. Bozhokin, É.A. Choban: Sov. J. Plasma Phys. **10**, 452 (1984)
28. L.S. Brown, G. Gabrielse: Rev. Mod. Phys. **58**, 233 (1986)
29. G.I. Budker: Atomnaya Energiya **22**, 346 (1967)
30. V. Celli, N.D. Mermin: Ann. Phys. (N.Y.) **30**, 249 (1964)
31. C. Cereceda, C. Deutsch, M. De Peretti, M. Sabatier, H.B. Nersisyan: Phys. Plasmas **7**, 2884 (2000)

32. H. Danared: Nucl. Instrum. Meth. Phys. Res. A **391**, 24 (1997)
33. H. Danared et al: Nucl. Instrum. Meth. Phys. Res. A **441**, 123 (2000)
34. A.K. Das: Physica A **85**, 575 (1976)
35. Ya.S. Derbenev, A.N. Skrinsky: Part. Accel. **8**, 235 (1978)
36. C. Deutsch: Phys. Lett. A **60**, 317 (1977)
37. C. Deutsch, P. Fromy: Phys. Rev. E **51**, 632 (1995)
38. L. De Ferrariis, N.R. Arista: Phys. Rev. A **29**, 2145 (1984)
39. H. Ehrenreich, M. Cohen: Phys. Rev. **115**, 786 (1959)
40. A.V. Fedotov et al: Experimental Benchmarking of the Magnetized Friction Force. In: *AIP Conference Proceedings*, vol 821, ed by S. Nagaitsev, R.J. Pasquinelli (AIP, New York 2006) pp 265–269
41. E. Fermi, E. Teller: Phys. Rev. **72**, 399 (1947)
42. G.S. Fishman: *Monte Carlo*, 3rd print (Springer, New York 1999)
43. D.B. Fried, S.D. Conte: *The Plasma Dispersion Function* (Academic Press, New York 1961)
44. G. Gabrielse et al: Phys. Rev. Lett. **89**, 213401 (2002)
45. B. Galnander et al: Cooling Force Measurements at CELSIUS. In: *AIP Conference Proceedings*, vol 821, ed by S. Nagaitsev, R.J. Pasquinelli (AIP, New York 2006) pp 259–264
46. D.K. Geller, C. Weisheit: Phys. Plasmas **4**, 4258 (1997)
47. M.E. Glinsky, T.M. O'Neil, M.N. Rosenbluth, K. Tsuruta, S. Ichimaru: Phys. Fluids B **4**, 1156 (1992)
48. H. Goldstein, C. Poole, J. Safko: *Classical Mechanics*, 3rd edn (Addison Wesley, San Francisco 2002)
49. J. Goldstone, K. Gottfried: Nuovo Cimento **13**, 849 (1959)
50. L.M. Gorbunov, H.H. Matevosyan, H.B. Nersisyan: Sov. Phys. JETP **75**, 460 (1992)
51. V. Gostishchev, C. Dimopoulou, K Beckert, P. Beller, A. Dolinskii, F. Nolden, M. Steck, A. Smirnov, I. Meshkov, A. Sidorin, G. Trubnikov: http://accelconf.web.cern.ch/accelconf/e06/PAPERS/MOPCH075.PDF
52. I.S. Gradshteyn, I.M. Ryzhik: *Tables of Integrals, Series and Products*, 2nd edn (Academic, New York 1980)
53. A.V. Gurevich, L.P. Pitaevskii: Sov. Phys. JETP **19**, 870 (1964)
54. M.C. Gutzwiller: The Diamagnetic Kepler Problem. In: *Chaos in Classical and Quantum Mechanics*, ed by F. John, L. Kadanoff, J.E. Marsden, L. Sirovich, S. Wiggins (Springer–Verlag, New York 1990) pp 323–339
55. T. Hamada: Austr. J. Phys. **31**, 291 (1978)
56. H. Hasegawa, M. Robnik, G. Wunner: Prog. Theor. Phys. Suppl. **98**, 198 (1989)
57. F. Herfurth: private communications
58. L. Hernquist: Astrophys. J. Suppl. Ser. **56**, 325 (1984)
59. H. Higaki et al: Phys. Rev. E **65**, 046410 (2002)
60. R.W. Hockney, J.W. Eastwood: *Computer Simulations Using Particles* (McGraw–Hill, New York 1981)
61. N. Honda, O. Aona, T. Kihara: J. Phys. Soc. Jpn. **18**, 256 (1963)
62. N.J. Horing: Ann. Phys. (N.Y.) **31**, 1 (1965)
63. N.J. Horing: Ann. Phys. (N.Y.) **54**, 405 (1969)
64. B. Hu, W. Horton, C. Chiu, T. Petrosky: Phys. Plasmas **9**, 1116 (2002)
65. J.D. Huba: *NRL Plasma Formulary*, Revised edn (Naval Research Laboratory, Washington 2006) p 55
66. S. Ichimaru, M.N. Rosenbluth: Phys. Fluids **13**, 2778 (1970)

67. S. Ichimaru: *Basic Principles of Plasma Physics*, 1st edn (Benjamin, MA 1973)
68. J.D. Jackson: *Classical Electrodynamics*, 3rd edn (John Wiley, New York 1998)
69. G. Kelbg: Ann. Physik **12**, 219 (1963)
70. V.I. Kogan, V.M. Galitskiy: *Problems in Quantum Mechanics* (Prentice-Hall, Englewood Cliffs, NJ 1963) p 22 problem 1
71. O.V. Konstantinov, V.I. Perel': Sov. Phys. JETP **26**, 1151 (1968)
72. N.A. Krall, A.W. Trivelpiece: *Principles of Plasma Physics* (McGraw-Hill, New York 1973)
73. L.D. Landau, E.M. Lifshitz: *Quantum Mechanics*, 3rd edn (Butterworth-Heinemann, Oxford 1996)
74. L.D. Landau, E.M. Lifshitz: *Electrodynamics of Continuous Media*, 2nd edn (Pergamon Press, Oxford 1984)
75. L.D. Landau, E.M. Lifshitz: *Statistical Physics*, 3rd edn (Butterworth-Heinemann, Oxford 1984) Part I
76. L.D. Landau, E.M. Lifshitz: *Statistical Physics* (Butterworth-Heinemann, Oxford 1980) Part II
77. E.M. Lifshitz, L.P. Pitaevskii: *Physical Kinetics*, 1st edn (Pergamon Press, Oxford 1981)
78. C. Lehmann, G. Leibfried: Z. Phys. **172**, 465 (1963)
79. J. Lindhard: Mat. Fys. Medd. K. Dan. Vidensk. Selsk. **28**, 8 (1954)
80. J. Lindhard, A. Winther: Mat. Fys. Medd. K. Dan. Vidensk. Selsk. **34**, 1 (1964)
81. J. Lindhard: Nucl. Instrum. Meth. **132**, 1 (1976)
82. P. Mansback, J.C. Keck: Phys. Rev. **181**, 275 (1969)
83. R.M. May, N.F. Cramer: Phys. Fluids **13**, 1766 (1970)
84. N.D. Mermin, E. Canel: Ann. Phys. (N.Y.) **26**, 247 (1964)
85. I.N. Meshkov: Phys. Part. Nuclei **25**, 631 (1994)
86. B. Möllers, M. Walter, G. Zwicknagel, C. Carli, C. Toepffer: Nucl. Instrum. Meth. Phys. Res. B **207**, 462 (2003)
87. B. Möllers: Elektronenkühlung hochgeladener Ionen in Penningfallen, Thesis, University Erlangen, Erlangen (2007)
88. H.B. Nersisyan: Phys. Rev. E **58**, 3686 (1998)
89. H.B. Nersisyan, C. Deutsch: Phys. Lett. A **246**, 325 (1998)
90. H.B. Nersisyan, M. Walter, G. Zwicknagel: Phys. Rev. E **61**, 7022 (2000)
91. H.B. Nersisyan, A.K. Das: Phys. Rev. E **62**, 5636 (2000)
92. H.B. Nersisyan, A.K. Das: Phys. Lett. A **296**, 131 (2002)
93. H.B. Nersisyan, G. Zwicknagel, C. Toepffer: Phys. Rev. E **67**, 026411 (2003)
94. H.B. Nersisyan: Nucl. Instrum. Meth. Phys. Res. B **205**, 276 (2003)
95. H.B. Nersisyan, G. Zwicknagel, C. Toepffer: Eur. Phys. J. D **28**, 235 (2004)
96. H.B. Nersisyan, G. Zwicknagel: J. Phys. A: Math. Gen. **37**, 11073 (2004)
97. H.B. Nersisyan: Contrib. Plasma Phys. **45**, 46 (2005)
98. H.B. Nersisyan, A.K. Das: Interaction of ion beams with plasmas: Energy loss and equipartition sum rules. In: *Advances in Plasma Physics Research*, vol 6, ed by F. Gerard (Nova, New York 2007), in press
99. J. Neufeld, R.H. Ritchie: Phys. Rev. **98**, 1632 (1955)
100. P. Neugebauer, H. Riffert, H. Herold, H. Ruder: Phys. Rev. A **54**, 467 (1996)
101. J. Ortner, V.M. Rylyuk, T.M. Tkachenko: Phys. Rev. E **50**, 4937 (1994)
102. V.V. Parkhomchuk: Nucl. Instrum. Meth. Phys. Res. A **441**, 9 (2000)
103. G.G. Pavlov, D.G. Yakovlev: Sov. Phys. JETP **43**, 389 (1976)
104. T. Peter: J. Plasma Phys. **44**, 269 (1990)
105. T. Peter, J. Meyer-ter-Vehn: Phys. Rev. A **43**, 1998 (1991)

106. H. Poth: CERN 87-03, 534 (1987)
107. H. Poth: Phys. Reports **196**, 135 (1990)
108. W.H. Press, B.P. Flannery, S.A. Teukolsky, W.T. Vetterling: *Numerical Recipes* (Cambridge University Press, Cambridge 1989)
109. W. Quint et al: Hyperfine Int. **132**, 457 (2001)
110. S.L. Rolston, G. Gabrielse: Hyperfine Int. **44**, 233 (1988)
111. A. Ron: Phys. Rev. **134**, A70 (1964)
112. M.N. Rosenbluth, W.M. MacDonald, D.L. Judd: Phys. Rev. **107**, 1 (1957)
113. N. Rostoker, M.N. Rosenbluth: Phys. Fluids **3**, 1 (1960)
114. G. Schmidt, E.E. Kunhardt, J.L. Godino: Phys. Rev. E **62**, 7512 (2000)
115. C. Seele, G. Zwicknagel, C. Toepffer, P.-G. Reinhard: Phys. Rev. E **57**, 3368 (1998)
116. J.G. Siambis: Phys. Rev. Lett. **37**, 1750 (1976)
117. A.O. Sidorin, I.N. Meshkov, I.A. Seleznev, A.V. Smirnov, E.M. Syresin, G.V. Trubnikov: Nucl. Instrum. Meth. Phys. Res. A **558**, 325 (2006)
118. P. Sigmund: *Stopping of Heavy Ions. A Theoretical Approach* (Springer, Berlin 2004)
119. P. Sigmund: *Particle Penetration and Radiation Effects. General Aspects and Stopping of Swift Point Charges* (Springer, Berlin 2006)
120. A.H. Sørensen, E. Bonderup: Nucl. Instrum. Meth. **215**, 27 (1983)
121. Q. Spreiter, M. Walter: J. Comput. Phys. **152**, 102 (1999)
122. M. Steinberg, J. Ortner: Phys. Rev. E **63**, 046401 (2001)
123. D. Ter Haar: *Selected Problems in Quantum Mechanics*, 2nd edn (Infosearch, London 1964) p 36 problem 8
124. C. Toepffer: Phys. Rev. A **66**, 022714 (2002)
125. G.V. Trubnikov: private communication
126. I. Villó–Pérez, N.R. Arista, R. Garcia–Molina: J. Phys. B: Mol. Opt. Phys. **35**, 1455 (2002)
127. M. Walter, C. Toepffer, G. Zwicknagel: Nucl. Instrum. Meth. Phys. Res. B **168**, 347 (2000)
128. M. Walter, G. Zwicknagel, C. Toepffer: Eur. Phys. J. D **35**, 527 (2005)
129. A.A. Ware, J.C. Wiley: Phys. Fluids B **5**, 2764 (1993)
130. Th. Winkler et al: Hyperfine Int. **99**, 277 (1996)
131. Th. Winkler: Untersuchungen zur Elektronenkühlung hochgeladener schwerer Ionen, Thesis, University Heidelberg, Heidelberg (1996)
132. Th. Winkler, K. Beckert, F. Bosch, H Eickhoff, B. Franzke, F. Nolden, H. Reich, B. Schlitt, M. Steck: Nucl. Instrum. Meth. Phys. Res. A **391**, 12 (1997)
133. P.A. Wolff: Phys. Rev. B **1**, 164 (1970)
134. G. Zwicknagel, C. Toepffer, P.-G. Reinhard: Phys. Reports **309**, 117 (1999)
135. G. Zwicknagel: Ion–Electron Collisions in a Homogeneous Magnetic Field. In: *AIP Conference Proceedings*, vol 498, ed by J.J. Bollinger, R.L. Spencer, R.C. Davidson (AIP, New York 1999) pp 469–474
136. G. Zwicknagel: Theory and Simulation of the Interaction of Ions with Plasmas: Nonlinear Stopping, Ion-Ion Correlation Effects and Collisions of Ions with Magnetized Electrons. Thesis, Erlangen University, Erlangen (2000)
137. G. Zwicknagel: Nucl. Instrum. Meth. Phys. Res. A **441**, 44 (2000)
138. G. Zwicknagel, C. Toepffer: Energy Loss of Ions by Collisions with Magnetized Electrons. In: *AIP Conference Proceedings*, vol 606, ed by F. Anderegg, L. Schweikhard, C.F. Driscoll (AIP, New York 2002) pp 499–508
139. G. Zwicknagel: Nucl. Instrum. Meth. Phys. Res. B **197**, 22 (2002)

140. G. Zwicknagel: Electron Cooling of Ions and Antiprotons in Traps. In: *AIP Conference Proceedings*, vol 821, ed by S. Nagaitsev, R.J. Pasquinelli (AIP, New York 2006) pp 513–522
141. G. Zwicknagel: Electron Cooling of Highly Charged Ions in Penning Traps. In: *AIP Conference Proceedings*, vol 862, ed by M. Drewsen, U. Uggerhoj, H. Knudsen (AIP, New York 2006) pp 281–291
142. P.S. Zyryanov: Sov. Phys. JETP **13**, 751 (1961)
143. P.S. Zyryanov, V.P. Kalashnikov: Sov. Phys. JETP **14**, 799 (1962)

List of Symbols and Abbreviations

a	cyclotron radius of the electrons in BC treatment.
a_ν	thermal cyclotron radius of the plasma species ν, $v_{\text{th}\nu\perp}/\Omega_\nu$.
a_c	cyclotron radius of the projectile ion.
a_B	Bohr radius $\hbar^2/m\not{e}^2$.
a_{WS}	Wigner–Seitz radius = $(4\pi n_e/3)^{-1/3}$.
$\boldsymbol{A}$	vector potential.
$\boldsymbol{C}_\perp^{(r)}$, $C_\parallel^{(r)}$	components of an arbitrary vector $\boldsymbol{C}$ transverse or parallel to $\boldsymbol{v}_r$, respectively.
$\boldsymbol{C}_\perp$, $C_\parallel$	components of an arbitrary vector $\boldsymbol{C}$ transverse or parallel to $\boldsymbol{b}$, respectively.
$\boldsymbol{b}$	unit vector along the magnetic field.
$\boldsymbol{B}$	magnetic field.
BC	binary collisions.
CAR	Cartesian coordinate basis.
CM	center of mass.
CTMC	Classical–Trajectory–Monte–Carlo.
CYL	cylindrical coordinate basis.
DT	dielectric theory.
$\text{Di}(\zeta)$	Dawson integral.
$-\mathrm{d}E/\mathrm{d}l$	energy loss of monochromatic particle beam.
$-\mathrm{d}\mathcal{E}/\mathrm{d}l$	stopping power (velocity averaged energy loss).
$\boldsymbol{E}_{\text{ext}}$	electrical field created by a moving ion $-\nabla\Phi_{\text{ei}}$.
E_α, $E_{n\sigma}(q_z)$	energy eigenvalues in the Landau state α (n, σ, q_z).
E_F	Fermi energy of unmagnetized electrons.
$-e$	charge of the electron q_e.
$\not{e}^2$	$e^2/4\pi\varepsilon_0$.
$f(\boldsymbol{r}, \boldsymbol{v}, t)$	nonequilibrium electron distribution function.
$f_0(\boldsymbol{v})$	equilibrium electron distribution function.
$F_\nu(z)$	Fermi function.
$\boldsymbol{F}_{\text{tot}}$	total averaged electrical force acting on the plasma.
$\boldsymbol{F}(\boldsymbol{r})$	interaction force.
$\boldsymbol{F}(\boldsymbol{k})$	Fourier transformed interaction force.
$g(\zeta)$, $f(\zeta)$	real and imaginary parts of the plasma dispersion function.
g	statical ion-plasma coupling parameter $\sqrt{3}\lvert Z\rvert\Gamma^{3/2}$.

$H_n(x)$	Hermite polynomials.
$\hat{H}$	Hamiltonian of the system.
$\hat{H}_0$, $\hat{H}_1(t)$	zero and first order Hamiltonians, respectively.
$\hbar$	Planck's constant divided by 2π.
ICF	inertial confinement fusion.
Im	imaginary part.
I_n	modified Bessel function of the first kind.
$\boldsymbol{j}_\alpha(\boldsymbol{r}, t)$	probability current density in the Landau state $\|\alpha\rangle$.
$\boldsymbol{j}_\alpha^{(0,1)}(\boldsymbol{r}, t)$	zero and first order probability current densities in the state $\|\alpha\rangle$, respectively.
$\boldsymbol{J}(\boldsymbol{r}, t)$	electrical current density.
$\boldsymbol{J}^{(1)}(\boldsymbol{r}, t)$	first-order electrical current density.
J_n	Bessel function of the first kind.
K	constant of motion for the magnetized electrons.
k_{B}	Boltzmann's constant.
$k_{\max}$	maximum cut–off parameter in k-space.
$k_{\min}$	minimum cut–off parameter in k-space.
K_n	modified Bessel function of the second kind.
$\boldsymbol{k}$, $\boldsymbol{k}'$	wave vectors.
L	Lagrangian of the system.
L	normalization length.
$L(s_{\max}, s_{\min})$	conventional Coulomb logarithm $\ln(s_{\max}/s_{\min})$.
L_{th}	velocity-averaged Coulomb logarithm $\ln(\langle s_{\max}\rangle/\langle s_{\min}\rangle)$.
$\mathrm{Li}_2(z)$	dilogarithm function.
$L_n^\mu(z)$	generalized Laguerre polynomials.
LR	linear response.
m	mass of the electron.
m_i	mass of the plasma ions.
M	mass of the projectile ion.
n_e	electron density.
n_i	ion density.
$\boldsymbol{n}_r$	unit vector along $\boldsymbol{v}_r$.
$\bar{\boldsymbol{n}}_r$	unit vector along $\bar{\boldsymbol{v}}_r$.
PIC	particle-in-cell.
$\hat{\boldsymbol{p}}$	momentum operator.
q_ν	charge of the plasma species ν.
Re	real part.
$\boldsymbol{R}_0$	initial position of the electron in a field-free case.
$\bar{\boldsymbol{R}}_0$	initial position of the guiding center for magnetized electrons.
$\boldsymbol{r}_0(t)$	initial relative radius–vector.
$\boldsymbol{r}$	position vector.
RLR	reduced linear response.
$\boldsymbol{s}$	impact parameter in field-free case.

$\bar{\boldsymbol{s}}$	impact parameter of the guiding center for magnetized electrons.
$s_{\max}$	maximum cut–off parameter.
$s_{\min}$	minimum cut–off parameter.
S_0	SP in the absence of magnetic field.
S_{inf}	SP in the presence of infinitely strong magnetic field.
S_{an}	anomalous SP.
S_{LR}	SP within LR treatment.
S_{RLR}	SP within RLR treatment.
S_{BC}	SP within BC treatment.
$S_{\alpha\beta}(\boldsymbol{k})$	matrix elements of $e^{i\boldsymbol{k}\cdot\boldsymbol{r}}$.
$S_{ei}(\boldsymbol{q})$	electron-ion static structure factor.
SP	stopping power.
t	time.
$\overline{T}$	average plasma temperature $\frac{2}{3}T_\perp + \frac{1}{3}T_\parallel$.
T_{eff}	effective temperature
$T_\perp$	plasma temperature transverse to the magnetic field.
$T_\parallel$	plasma temperature parallel to the magnetic field.
Tr	trace operation.
$U_{\text{p}}(\boldsymbol{r}, t)$	density of plasma potential energy
$\boldsymbol{u}$	unit vector transverse to the magnetic field $(\cos\varphi, \sin\varphi)$.
$\boldsymbol{u}_p$	unit vector transverse to the magnetic field $(\cos\varphi_p, \sin\varphi_p)$.
$\boldsymbol{v}_r$	relative velocity in the field-free case.
$\bar{\boldsymbol{v}}_r$	relative velocity of the guiding center.
$\boldsymbol{v}_i$	ion velocity.
$\boldsymbol{v}_e$	electron velocity.
$\boldsymbol{v}_0(t)$	initial relative velocity $\dot{\boldsymbol{r}}_0(t)$.
v_{th}	thermal velocity of plasma electrons.
$v_{\text{th}\perp}$	electron transverse thermal velocity $(k_{\text{B}}T_\perp/m)^{1/2}$.
$v_{\text{th}\parallel}$	electron longitudinal thermal velocity $(k_{\text{B}}T_\parallel/m)^{1/2}$.
$\overline{v}_{\text{th}}$	average thermal velocity of the electrons $(k_{\text{B}}\overline{T}/m)^{1/2}$.
w	$\Omega_\perp/\omega_p$.
$W(\zeta)$	plasma dispersion function.
x	scaled projectile ion velocity $v_i/v_{\text{th}\parallel}$.
$\hat{\boldsymbol{x}}, \hat{\boldsymbol{y}}, \hat{\boldsymbol{z}}$	unit vectors.
$Z_i e$	charge of the plasma ion q_i.
Ze	charge of the projectile ion.
$\mathfrak{E}_1$	is identical to $\sigma(\bar{\boldsymbol{v}}_r, \boldsymbol{v}_i)$.
$\mathfrak{E}_2(\bar{s})$	classical BC energy transfer averaged over the all initial positions z_0 of the electron guiding center.
$\mathcal{E}_F$	Fermi energy of magnetized electrons.
$\mathcal{F}$	macroscopic stopping force.
$\mathcal{R}$	friction coefficient.

$\mathcal{Z}$	dynamical ion-plasma coupling parameter.
$\mathcal{G}, \mathcal{F}$	real and imaginary parts of the dispersion function of a magnetized plasma, respectively.
$\mathcal{S}_0$	$(k_B T/\not{e})^2$.
α	angle between the magnetic field $\boldsymbol{B}$ and the ion velocity $\boldsymbol{v}_i$.
β	angle between the wave vector $\boldsymbol{k}$ and the magnetic field $\boldsymbol{B}$.
α, β	Landau states.
γ	scaled magnetic field ω_p/Ω_c.
γ	Euler's constant $\simeq 0.5772$.
γ_0	exponent of the Euler's constant $= e^{\gamma}$.
Γ	Coulomb coupling parameter.
$\Gamma(z)$	Euler's Γ-function.
$\Delta\tau_{\text{rel}}$	relaxation time in two-temperature plasma.
δ	pitch of the electron helices divided by 2π, $\bar{v}_r/\Omega_e$.
$\delta E_i(t),\ \delta E_e(t)$	ion and electron energy changes at the time t, respectively.
$\delta \boldsymbol{v}(t),\ \delta \boldsymbol{p}(t)$	relative velocity and momentum changes at the time t, respectively.
$\delta(x)$	Dirac function.
$\delta_{\alpha\beta}$	Kronecker symbol.
$\Delta\boldsymbol{v}, \Delta\boldsymbol{p}$	relative velocity and momentum transfers.
$\Delta E_i^{(1,2)}$	first and second order ion energy transfers.
$\Delta E_e^{(1,2)}$	first and second order electron energy transfers.
$\langle\Delta E_i^{(2)}\rangle_{\varphi}$	φ-averaged second order ion energy transfer.
$\langle\Delta E_i^{(2)}\rangle_{\varphi,\vartheta_s}$	φ, ϑ_s-averaged second order ion energy transfer.
$\Delta E_{\alpha}^{\text{CAR}}$	quantum mechanical energy transfer of an electron in the state $\|\alpha\rangle$ in a Cartesian basis.
$\Delta E_{\alpha}^{\text{CYL}}$	quantum mechanical energy transfer of an electron in the state $\|\alpha\rangle$ in a cylindrical basis.
ε_0	vacuum permittivity.
$\varepsilon(\boldsymbol{k}, \omega)$	longitudinal dielectric function of the magnetized plasma.
$\varepsilon_{ij}(\boldsymbol{k}, \omega)$	dielectric tensor of the magnetized plasma.
$\varepsilon_{ij}(\omega)$	dielectric tensor of the magnetized cold plasma.
$\varepsilon_{\perp,\parallel}(\omega)$	diagonal elements of the tensor $\varepsilon_{ij}(\omega)$.
ζ_ν	scaled frequency $\omega/kv_{\text{th}\nu\parallel}$.
$\zeta(z)$	Riemann function.
η	scaled magnetic field Ω_e/ω_p.
$\theta,\ \theta'$	azimuthal angles of $\boldsymbol{k}_\perp$ and $\boldsymbol{k}'_\perp$ respectively.
ϑ_b	angle between $\boldsymbol{b}$ and $\bar{\boldsymbol{v}}_r$.
$\vartheta_s,\ \vartheta_{\bar{s}}$	azimuthal angle of the impact parameters $\boldsymbol{s}$ and $\bar{\boldsymbol{s}}$.
Θ	angle between ion velocity $\boldsymbol{v}_i$ and wave vector $\boldsymbol{k}$.
$\Theta(z)$	unit step (Heaviside) function.
$\Lambda(s_{\max}, s_{\min})$	modified Coulomb logarithm.
Λ_{th}	velocity-averaged modified Coulomb logarithm.

Λ_Q	Coulomb logarithm in quantum–mechanical treatment.
$\Lambda_n(z)$	$e^{-z}I_n(z)$.
$\lambda_{D\nu\|}$	Debye screening length of the plasma species ν, $v_{th\nu\|}/\omega_{p\nu}$.
$\lambda_{D\|}$	Debye screening length for electrons $v_{th\|}/\omega_p$.
λ	screening length.
λ_B	magnetic length.
$\lambda\!\!\!^-$	thermal wavelength.
μ	reduced mass.
ν	plasma species index ($\nu = e, i$).
ν_e	effective collision frequency.
$\hat{\rho}$	density matrix operator.
$\rho_i(\boldsymbol{r}, t)$	charge density of projectile ion.
$\rho_\alpha(\boldsymbol{r}, t)$	probability density.
$\rho_\alpha^{(0)}(\boldsymbol{r}, t)$	zero-order probability density.
$\rho(\boldsymbol{r}, t)$	induced charge density.
$\sigma(\bar{\boldsymbol{v}}_r, \boldsymbol{v}_i)$	effective transport cross–section.
$\hat{\sigma}_z$	spin operator.
σ	spin variable.
τ_ν	temperature anisotropy of the species ν in temperature-anisotropic two-component plasma $T_{\nu\perp}/T_{\nu\|}$.
τ	temperature anisotropy in two-temperature electron plasma $T_\perp/T_\|$.
τ_s	time constant for synchrotron radiation.
$\phi(\boldsymbol{r}, t)$	scalar potential.
$\Phi_{ei}(\boldsymbol{r})$	electric potential created by ion on electron.
$\Phi_{ei}(\boldsymbol{k})$	Fourier transformed electric potential created by ion on electron.
φ, φ'	initial phases of magnetized electrons.
φ_p	initial phase of projectile ion.
$\chi_\nu(\boldsymbol{k}, \omega)$	susceptibility of the plasma species ν.
$\chi(\boldsymbol{k}, \omega)$	susceptibility of the plasma electrons $\chi_e(\boldsymbol{k}, \omega)$.
$\psi_\alpha^{(0,1)}$	unperturbed and first order electron wave functions in the Landau state α.
ω, ω'	frequencies.
ω, ω'	resonance frequencies $\boldsymbol{k} \cdot \boldsymbol{v}_i$ and $\boldsymbol{k}' \cdot \boldsymbol{v}_i$, respectively.
$\omega_{p\nu}$	plasma frequency of the species ν.
ω_p	plasma frequency of electrons = ω_{pe}.
ω_H	upper hybrid frequency.
Ω_ν	cyclotron frequency of the plasma species ν.
Ω_c	cyclotron frequency of the projectile ion.
Ω_α	energy eigenvalue in the state $\|\alpha\rangle$ in units of $\hbar$, $E_\alpha/\hbar$.
$\Omega_{\alpha\beta}$	$\Omega_\alpha - \Omega_\beta$.
$\Omega_{1,2,3}(\beta)$	plasma resonances.
$\Omega_{\|,\perp}$	strengths of the Coulomb correlations.

Printing: Krips bv, Meppel
Binding: Stürtz, Würzburg